Lecture Notes in Mathematics 1702

Editors:
J.-M. Morel, Cachan
F. Takens, Groningen
B. Teissier, Paris

T0155772

Jin Ma
Jiongmin Yong

Forward-Backward Stochastic Differential Equations and their Applications

 Springer

Author

Jin Ma
Department of Mathematics
Purdue University
150 N. University Street
West Lafayette
IN 47906-1395, USA

e-mail: majin@math.purdue.edu

Jiongmin Yong
Department of Mathematics
Fudan University
Shanghai 200433
People's Republic of China

e-mail: jyong@fudan.edu.cn

Cataloging-in-Publication Data available

Mathematics Subject Classification (2000):
PRIMARY: 60H10, 15, 20, 30; 93E03;
SECONDARY: 35K15, 20, 45, 65; 65M06, 12, 15, 25; 65U05;90A09, 10, 12, 16

ISSN print edition: 0075-8434
ISBN-10 3-540-65960-9 Springer Berlin Heidelberg New York
ISBN-13 978-3-540-65960-0 Springer Berlin Heidelberg New York

Corrected 3rd printing

DOI 10.1007/BFb0092524

Springer is a part of Springer Science + Business Media
springer.com
© Springer-Verlag Berlin Heidelberg 2007

Typesetting: Camera-ready LATEX output by the authors

Printed on acid-free paper   SPIN: 11367352   VA41/3100/SPi   5 4 3 2 1 0

To

Yun and Meifen

Preface

This book is intended to give an introduction to the theory of forward-backward stochastic differential equations (FBSDEs, for short) which has received strong attention in recent years because of its interesting structure and its usefulness in various applied fields.

The motivation for studying FBSDEs comes originally from stochastic optimal control theory, that is, the adjoint equation in the Pontryagin-type maximum principle. The earliest version of such an FBSDE was introduced by Bismut [1] in 1973, with a decoupled form, namely, a system of a usual (forward) stochastic differential equation and a (linear) backward stochastic differential equation (BSDE, for short). In 1983, Bensoussan [1] proved the well-posedness of general linear BSDEs by using martingale representation theorem. The first well-posedness result for nonlinear BSDEs was proved in 1990 by Pardoux–Peng [1], while studying the general Pontryagin-type maximum principle for stochastic optimal controls. A little later, Peng [4] discovered that the adapted solution of a BSDE could be used as a probabilistic interpretation of the solutions to some semilinear or quasilinear parabolic partial differential equations (PDE, for short), in the spirit of the well-known Feynman-Kac formula. After this, extensive study of BSDEs was initiated, and potential for its application was found in applied and theoretical areas such as stochastic control, mathematical finance, differential geometry, to mention a few.

The study of (strongly) coupled FBSDEs started in early 90s. In his Ph.D thesis, Antonelli [1] obtained the first result on the solvability of an FBSDE over a "small" time duration. He also constructed a counterexample showing that for coupled FBSDEs, large time duration might lead to non-solvability. In 1993, the present authors started a systematic investigation on the well-posedness of FBSDEs over arbitrary time durations, which has developed into the main body of this book. Today, several methods have been established for solving a (coupled) FBSDE. Among them two are considered effective: the *Four Step Scheme* by Ma Protter–Yong [1] and the *Method of Continuation* by Hu–Peng [2], and Yong [1]. The former provides the explicit relations among the forward and backward components of the adapted solution via a quasilinear partial differential equation, but requires the non-degeneracy of the forward diffusion and the non-randomness of the coefficients; while the latter relaxed these conditions, but requires essentially the "monotonicity" condition on the coefficients, which is restrictive in a different way.

The theory of FBSDEs have given rise to some other problems that are interesting in their own rights. For example, in order to extend the Four Step Scheme to general random coefficient case, it is not hard to see that one has to replace the quasilinear parabolic PDE there by a quasilinear *backward stochastic* partial differential equation (BSPDE for short), with a strong degeneracy in the sense of stochastic partial differential equations. Such BSPDEs can be used to generalize the Feynman-Kac formula and even the Black-Scholes option pricing formula to the case when the coefficients of the diffusion are allowed to be random. Other interesting subjects generated by FBSDEs but with independent flavors include FBSDEs with reflecting boundary conditions as well as the numerical methods for FBSDEs. It is worth pointing out that the FBSDEs have also been successfully applied to model and to resolve some interesting problems in mathematical finance, such as problems involving term structure of interest rates (consol rate problem) and hedging contingent claims for large investors, etc.

The book is organized as follows. As an introduction, we present several interesting examples in Chapter 1. After giving the definition of solvability, we study some special FBSDEs that are either non-solvable or easily solvable (e.g., those on small durations). Some comparison results for both BSDE and FBSDE are established at the end of this chapter. In Chapter 2 we content ourselves with the linear FBSDEs. The special structure of the linear equations enables us to treat the problem in a special way, and the solvability is studied thoroughly. The study of general FBSDEs over arbitrary duration starts from Chapter 3. We present virtually the first result regarding the solvability of FBSDE in this generality, by relating the solvability of an FBSDE to the solvability of an optimal stochastic control problem. The notion of *approximate solvability* is also introduced and developed. The idea of this chapter is carried on to the next one, in which the Four Step Scheme is established. Two other different methods leading to the existence and uniqueness of the adapted solution of general FBSDEs are presented in Chapters 6 and 7, while in the latter even reflections are allowed for both forward and backward equations. Chapter 5 deals with a class of linear backward SPDEs, which are closely related to the FBSDEs with random coefficients; Chapter 8 collects some applications of FBSDEs, mainly in mathematical finance, which in a sense is the inspiration for much of our theoretical research. Those readers needing stronger motivation to dig deeply into the subject might actually want to go to this chapter first and then decide which chapter would be the immediate goal to attack. Finally, Chapter 9 provides a numerical method for FBSDEs.

In this book all "headings" (theorem, lemma, definition, corollary, example, etc.) will follow a single sequence of numbers within one chapter (e.g., Theorem 2.1 means the first "heading" in Section 2, possibly followed immediately by Definition 2.2, etc.). When a heading is cited in a different chapter, the chapter number will be indicated. Likewise, the numbering for the equations in the book is of the form, say, (5.4), where 5 is the section number and 4 is the equation number. When an equation in different chapter is cited, the chapter number will precede the section number.

We would like to express our deepest gratitude to many people who have inspired us throughout the past few years during which the main body of this book was developed. Special thanks are due to R. Buckdahn, J. Cvitanic, J. Douglas Jr., D. Duffie, P. Protter, with whom we enjoyed wonderful collaboration on this subject; to N. El Karoui, J. Jacod, I. Karatzas, N. V. Krylov, S. M. Lenhart, E. Pardoux, S. Shreve, M. Soner, from whom we have received valuable advice and constant support. We particularly appreciate a special group of researchers with whom we were students, classmates and colleagues in Fudan University, Shanghai, China, among them: S. Chen, Y. Hu, X. Li, S. Peng, S. Tang, X. Y. Zhou. We also would like to thank our respective Ph.D. advisors Professors Naresh Jain (University of Minnesota) and Leonard D. Berkovitz (Purdue University) for their constant encouragement.

JM would like to acknowledge partial support from the United States National Science Fundation grant #DMS-9301516 and the United States Office of Naval Research grant #N00014-96-1-0262; and JY would like to acknowledge partial support from Natural Science Foundation of China, the Chinese Education Ministry Science Foundation, the National Outstanding Youth Foundation of China, and Li Foundation at San Francisco, USA.

Finally, of course, both authors would like to take this opportunity to thank their families for their support, understanding and love.

Jin Ma, West Lafayette
Jiongmin Yong, Shanghai

January, 1999

Contents

Chapter 1

Introduction

§1. Some Examples

To introduce the *forward-backward stochastic differential equations* (FBS-DEs, for short), let us begin with some examples. Unless otherwise specified, throughout the book, we let $(\Omega, \mathcal{F}, \{\mathcal{F}_t\}_{t \geq 0}, \mathbf{P})$ be a complete filtered probability space on which is defined a d-dimensional standard Brownian motion $W(t)$, such that $\{\mathcal{F}_t\}_{t \geq 0}$ is the natural filtration of $W(t)$, augmented by all the $\mathbf{P}$-null sets. In other words, we consider only the *Brownian* filtration throughout this book.

§1.1. A first glance

One of the main differences between a *stochastic differential equation* (SDE, for short) and a (deterministic) ordinary differential equation (ODE, for short) is that one cannot reverse the "time". The following is a simple but typical example. Suppose that $d = 1$ (i.e., the Brownian motion is one-dimensional), and consider the following (trivial) differential equation:

$$(1.1) \qquad dY(t) = 0, \qquad t \in [0, T],$$

where $T > 0$ is a given *terminal time*. For any $\xi \in \mathbb{R}$ we can require either $Y(0) = \xi$ or $Y(T) = \xi$ so that (1.1) has a unique solution $Y(t) \equiv \xi$. However, if we consider (1.1) as a stochastic differential equation (with null drift and diffusion coefficients) in Itô's sense, things will become a little more complicated. First note that a solution of an Itô SDE has to be $\{\mathcal{F}_t\}_{t \geq 0}$-adapted. Thus specifying $Y(0)$ and $Y(T)$ will have essential difference. Consider again (1.1), but as a terminal value problem:

$$(1.2) \qquad \begin{cases} dY(t) = 0, \qquad t \in [0, T], \\ Y(T) = \xi, \end{cases}$$

where $\xi \in L^2_{\mathcal{F}_T}(\Omega; \mathbb{R})$, the set of all $\mathcal{F}_T$-measurable square integrable random variables. Since the only solution to (1.2) is $Y(t) \equiv \xi$, $\forall t \in [0, T]$, which is not necessarily $\{\mathcal{F}_t\}_{t \geq 0}$-adapted unless ξ is a constant, the equation (1.2), viewed as an Itô SDE, does not have a solution in general!

Intuitively, there are two ways to get around with this difficulty: (1) modify (or even remove) the *adaptedness* of the solution in its definition; (2) reformulate the *terminal value problem* of an SDE so that it may allow a solution which is $\{\mathcal{F}_t\}_{t \geq 0}$-adapted. We note here that method (1) requires techniques such as new definitions of a *backward* Itô integral, or more generally, the so-called *anticipating stochastic calculus*. For more on the discussion in that direction, one is referred to the books of, say, Kunita

[1] and Nualart [1]. In this book, however, we will content ourselves with method (2), because of its usefulness in various applications as we shall see in the following sections.

To reformulate (1.2), we first note that a reasonable way of modifying the solution $Y(t) = \xi$ so that it is $\{\mathcal{F}_t\}_{t \geq 0}$-adapted and satisfies $Y(T) = \xi$ is to define

$$(1.3) \qquad Y(t) \overset{\Delta}{=} E\{\xi | \mathcal{F}_t\}, \qquad t \in [0, T].$$

Let us now try to derive, if possible, an (Itô) SDE that the process $Y(\cdot)$ might enjoy. An important ingredient in this derivation is the *Martingale Representation Theorem* (cf. e.g., Karatzas-Shreve [1]), which tells us that *if the filtration $\{\mathcal{F}_t\}_{t \geq 0}$ is Brownian, then every square integrable martingale M with zero expectation can be written as a stochastic integral with a unique integrand that is $\{\mathcal{F}_t\}_{t \geq 0}$-progressively measurable and square integrable.* Since the process $Y(\cdot)$ defined by (1.3) is clearly a square integrable $\{\mathcal{F}_t\}_{t \geq 0}$-martingale, an application of the Martingale Representation Theorem leads to the following representation:

$$(1.4) \qquad Y(t) = Y(0) + \int_0^t Z(s)dW(s), \qquad \forall t \in [0, T], \quad \text{a.s.},$$

where $Z(\cdot) \in L_{\mathcal{F}}^2(0, T; \mathbb{R})$, the set of all $\{\mathcal{F}_t\}_{t \geq 0}$-adapted square integrable processes. Writing (1.4) in a differential form and combining it with (1.3) (note that ξ is $\mathcal{F}_T$-measurable), we have

$$(1.5) \qquad \begin{cases} dY(t) = Z(t)dW(t), & t \in [0, T], \\ Y(T) = \xi. \end{cases}$$

In other words, if we reformulate (1.2) as (1.5); and more importantly, instead of looking for a single $\{\mathcal{F}_t\}_{t \geq 0}$-adapted process $Y(\cdot)$ as a solution to the SDE, we look for a pair $(Y(\cdot), Z(\cdot))$ (although it looks a little strange at this moment), then finding a solution which is $\{\mathcal{F}_t\}_{t \geq 0}$-adapted becomes possible! It turns out, as we shall develop in the rest of the book, that (1.5) is *the* appropriate reformulation of a *terminal value problem* (1.2) that possesses an *adapted solution* (Y, Z). Adding the extra component $Z(\cdot)$ to the solution is the key factor that makes finding an *adapted* solution possible.

As was traditionally done in the SDE literature, (1.5) can be written in an integral form, which can be deduced as follows. Note from (1.4) that

$$(1.6) \qquad Y(0) = Y(T) - \int_0^T Z(s)dW(s) = \xi - \int_0^T Z(s)dW(s).$$

Plugging (1.6) into (1.4) we obtain

$$(1.7) \quad Y(t) = Y(0) + \int_0^t Z(s)dW(s) = \xi - \int_t^T Z(s)dW(s), \quad \forall t \in [0, T].$$

In the sequel, we shall not distinguish (1.5) and (1.7); each of them is called a *backward stochastic differential equation* (BSDE, for short). We would like to emphasize that the stochastic integral in (1.7) is the usual (forward) Itô integral.

Finally, if we apply Itô's formula to $|Y(t)|^2$ (here $|\cdot|$ denotes the usual Euclidean norm, see §2), then

$$(1.8) \qquad E|\xi|^2 = E|Y(t)|^2 + \int_t^T E|Z(s)|^2 ds, \qquad \forall t \in [0, T].$$

Thus $\xi = 0$ implies that $Y \equiv 0$ and $Z \equiv 0$. Note that equation (1.7) is linear, relation (1.8) leads to the uniqueness of the $\{\mathcal{F}_t\}_{t \geq 0}$-adapted solution $(Y(\cdot), Z(\cdot))$ to (1.7). Consequently, if ξ is a non-random constant, then by uniqueness we see that $Y(t) \equiv \xi$ and $Z(t) \equiv 0$ is the only solution of (1.7), as we expect. In the following subsections we give some examples in stochastic control theory and mathematical finance that have motivated the study of the backward and forward-backward SDEs.

§1.2. A stochastic optimal control problem

Consider the following controlled stochastic differential equation:

$$(1.9) \qquad \begin{cases} dX(t) = \big[aX(t) + bu(t)\big]dt + dW(t), & t \in [0, T], \\ X(0) = x, \end{cases}$$

where $X(\cdot)$ is called the *state process*, $u(\cdot)$ is called the *control process*. Both of them are required to be $\{\mathcal{F}_t\}_{t \geq 0}$-adapted and square integrable. For simplicity, we assume X, u and W are all one-dimensional, and a and b are constants. We introduce the so-called *cost functional* as follows:

$$(1.10) \qquad J(u) = \frac{1}{2} E \left\{ \int_0^T \big[|X(t)|^2 + |u(t)|^2 \big] dt + |X(T)|^2 \right\}.$$

An *optimal control problem* is then to minimize the cost functional (1.10) subject to the *state equation* (1.9). In the present case, it can be shown that there exists a unique solution to this optimal control problem (in fact, the mapping $u \mapsto J(u)$ is convex and coercive). Our goal is to determine this optimal control.

Suppose $u(\cdot)$ is an optimal control and $X(\cdot)$ is the corresponding (optimal) state process. Then, for any admissible control $v(\cdot)$ (i.e., an $\{\mathcal{F}_t\}_{t \geq 0}$-adapted square integrable process), we have

$$(1.11) \qquad \begin{aligned} 0 &\leq \frac{J(u + \varepsilon v) - J(u)}{\varepsilon} \\ &\to E \left\{ \int_0^T [X(t)\xi(t) + u(t)v(t)] dt + X(T)\xi(T) \right\}, \qquad \varepsilon \to 0, \end{aligned}$$

where $\xi(\cdot)$ satisfies the following *variational system*:

$$(1.12) \qquad \begin{cases} d\xi(t) = \big[a\xi(t) + bv(t)\big]dt, & t \in [0, T], \\ \xi(0) = 0. \end{cases}$$

In order to get more information from (1.11), we introduce the following
adjoint equation:

$$(1.13) \qquad \begin{cases} dY(t) = -[aY(t) + X(t)]dt + Z(t)dW(t), & t \in [0, T], \\ Y(T) = X(T). \end{cases}$$

and we require that the processes $Y(\cdot)$ and $Z(\cdot)$ both be $\{\mathcal{F}_t\}_{t \geq 0}$-adapted.
It is clear that (1.13) is a BSDE with a more general form than the one we
saw in §1.1, since $Y(\cdot)$ is specified at $t = T$, and $X(T)$ is $\mathcal{F}_T$-measurable in
general.

Now let us assume that (1.13) admits an adapted solution $(Y(\cdot), Z(\cdot))$.
Then, applying Itô's formula to $Y(t)\xi(t)$, one has

$$E[X(T)\xi(T)] = E[Y(T)\xi(T)]$$

$$(1.14) \qquad = E \int_0^T \left\{ [-aY(t) - X(t)]\xi(t) + Y(t)[a\xi(t) + bv(t)] \right\} dt$$

$$= E \int_0^T [-X(t)\xi(t) + bY(t)v(t)] dt.$$

Hence, (1.11) becomes

$$(1.15) \qquad 0 \leq E \int_0^T [bY(t) + u(t)]v(t)dt.$$

Since $v(\cdot)$ is arbitrary, we obtain that

$$(1.16) \qquad u(t) = -bY(t), \qquad \text{a.e.} \, t \in [0, T], \text{ a.s.}$$

We note that since $Y(\cdot)$ is required to be $\{\mathcal{F}_t\}_{t \geq 0}$-adapted, the process
$u(\cdot)$ is an *admissible* control (this is why we need the adapted solution for
(1.13)!). Substituting (1.16) into the state equation (1.9), we finally obtain
the following *optimality system*:

$$(1.17) \qquad \begin{cases} dX(t) = [aX(t) - b^2Y(t)]dt + dW(t), \\ dY(t) = -[aY(t) + X(t)]dt + Z(t)dW(t), & t \in [0, T], \\ X(0) = x, \qquad Y(T) = X(T). \end{cases}$$

We see that the equation for $X(\cdot)$ is *forward* (since it is given the initial
datum) and the equation for $Y(\cdot)$ is *backward* (since it is given the final
datum). Thus, (1.17) is a *coupled forward-backward stochastic differential
equation* (FBSDE, for short). It is clear that if we can prove that (1.17)
admits an adapted solution $(X(\cdot), Y(\cdot), Z(\cdot))$, then (1.16) gives an optimal
control, solving the original stochastic optimal control problem. Further, if
the adapted solution $(X(\cdot), Y(\cdot), Z(\cdot))$ of (1.17) is unique, so is the optimal
control $u(\cdot)$.

§1.3. Stochastic differential utility

Two of the most remarkable applications of the theory of BSDEs (a special case of FBSDEs) in finance theory have been the stochastic differential

utility and the contingent claim valuation. In this and the following subsections, we describe these problems from the perspective of FBSDEs.

Stochastic differential utility is an extension of the notion of *recursive
utility* to a continuous-time, stochastic setting. In the simplest discrete, deterministic model (see, e.g., Koopmans [1]), the problem of recursive utility
is to find certain utility functions that satisfy a recursive relation. For example, assume that the consumption plans are denoted by $c = \{c_0, c_1, \cdots\}$,
where c_t represents the consumption in period t, and the *current utility* is
denoted by V_t, then we say that $V = \{V_t : t = 0, 1, \cdots\}$ defines a recursive
utility if the sequence $V_0, V_1, \cdots$ satisfies the recursive relation:

$$(1.18) \qquad V_t = W(c_t, V_{t+1}), \qquad t = 0, 1, \cdots,$$

where the function W is called the *aggregator*. We should note that in
(1.18), the recursive relation is *backwards*. The problem can also be stated
as finding a utility function U defined on the space of consumption plans
such that, for any $t = 0, 1, \cdots$, it holds that $V_t = U(\{c_t, c_{t+1}, \cdots\})$, where V
satisfies (1.18). In particular, the utility function U can be simply defined
by $U(\{c_0, c_1, \cdots\}) = V_0$, once (1.18) is solved.

In the continuous-time model one often describes the consumption plan
by its *rate* $c = \{c(t) : t \geq 0\}$, where $c(t) \geq 0$, $\forall t \geq 0$ (hence the accumulate
consumption up to time t is $\int_0^t c(s)ds$). The *current utility* is denoted by
$Y(t) \overset{\Delta}{=} U(\{c(s) : s \geq t\})$, and the recursive relation (1.18) is replaced by a
differential equation:

$$(1.19) \qquad \frac{dY(t)}{dt} = -f(c(t), Y(t)),$$

where the function f is the *aggregator*. We note that the negative sign in
front of f reflects the time-reverse feature seen in (1.18). Again, once a
solution of (1.19) can be determined, then $U(c) = Y(0)$ defines a unitiliy
function.

An interesting variation of (1.18) and (1.19) is their *finite horizon* version, that is, there is a *terminal* time $T > 0$, such that the problem is restricted to $0 \leq t \leq T$. Suppose that the utility of the terminal consumption
is given by $u(c(T))$ for some prescribed utility function u, then the (backward) difference equation (1.18) with terminal condition $V_T = u(c(T))$ can
be solved uniquely. Likewise, we may pose (1.19), the continuous counter-
part of (1.18), as a terminal value problem with given $Y(T) = u(c(T))$, or
equivalently,

$$(1.20) \qquad Y(t) = u(c(T)) + \int_t^T f(c(s), Y(s))ds, \qquad t \in [0, T].$$

In a stochastic model (model with uncertainty) one assumes that both
consumption c and utility Y are stochastic processes, defined on some (fil-
tered) probability space $(\Omega, \mathcal{F}, \{\mathcal{F}_t\}_{t \geq 0}, \mathbf{P})$. A standard setting is that at
any time $t \geq 0$ the consumption rate $c(t)$ and the current utility $Y(t)$ can
only be determined by the information up to time t. Mathematically, this

axiomatic assumption amounts to saying that the processes c and Y are both adapted to the filtration $\{\mathcal{F}_t\}_{t \geq 0}$. Let us now consider (1.20) again, but bearing in mind that c and Y are $\{\mathcal{F}_t\}_{t \geq 0}$-adapted processes. Taking conditional expectation on both sides of (1.20), we obtain

$$(1.21) \qquad Y(t) = E\{Y(t)|\mathcal{F}_t\} = E\Big\{u(c(T)) + \int_t^T f(c(s), Y(s))ds \big| \mathcal{F}_t\Big\},$$

for all $t \in [0, T]$. In the special case when the filtration is generated by a given Brownian motion W, just as we have assumed in this book, we can apply the Martingale Representation Theorem as before to derive that

$$(1.22) \quad Y(t) = u(c(T)) + \int_t^T f(c(s), Y(s))ds - \int_t^T Z(s)dW_s, \quad t \in [0, T].$$

That is, (Y, Z) satisfies the BSDE (1.22). A more general BSDE that models the recursive utility is one in which the aggregator f depends also on Z. The following situation more or less justifies this point. Let $\overline{U}$ be another utility function such that $\overline{U} = \varphi \circ U$ for some C^2 function φ with $\varphi'(x) > 0$, $\forall x$ (in this case we say that U and $\overline{U}$ are *ordinally equivalent*). Let us define $\bar{u} = \varphi \circ u$, $\overline{Y}(t) = \varphi(Y(t))$, $\overline{Z}(t) = \varphi'(Y(t))Z(t)$, and

$$\bar{f}(c, y, z) = \varphi'(\varphi^{-1}(y))f(c, \varphi^{-1}(y)) - \frac{\varphi''(\varphi^{-1}(y))}{\varphi'(\varphi^{-1}(y))}z.$$

Then an application of Itô's formula shows that $(\overline{Y}, \overline{Z})$ satisfies the BSDE (1.22) with a new terminal condition $\bar{u}(c(T))$ and a new aggregator $\bar{f}$, which now depends on z.

The BSDE (1.22) can be turned into an FBSDE, if the consumption plan depends on other random sources which can be described by some other (stochastic) differential equations. The following scenario, studied by Duffie-Ceoffard-Skiadas [1], should be illustrative. Consider m agents sharing a total endowment in an economy. Assume that the total endowment, denoted by e, is a continuous, non-negative, $\{\mathcal{F}_t\}_{t \geq 0}$-adapted process; and that each agent has his own consumption process c^i and utility process Y^i satisfying

$$(1.23) \qquad Y^i(t) = u_i(c^i(T)) + \int_t^T f^i(c^i(s), Y^i(s))ds + \int_t^T Z^i(s)dW(s),$$

for $t \in [0, T]$. For a given *weight vector* $\alpha \in \mathbb{R}_+^m$, we say that an *allocation* $c_\alpha = (c_\alpha^1, \cdots, c_\alpha^m)$ is α-*efficient* if

$$(1.24) \quad \sum_{i=1}^m \alpha_i U_i(c_\alpha^i) = \sup\Big\{\sum_{i=1}^m \alpha_i U_i(c^i) \,\big|\, \sum_{i=1}^m c^i(t) \leq e(t), \ t \in [0, T], \text{a.s.}\Big\},$$

where $U_i(c^i) = Y^i(0)$.

It is conceivable that the α-efficient allocation c_α is no longer an independent process. In fact, using techniques of non-linear programming

it can be shown that, under certain technical conditions on the aggrega-
tors f^i's and the terminal utility functions u_i's, the process c_α takes the
form: $c_\alpha(t) = K(\lambda(t), e(t), Y(t))$, for some $\mathbb{R}^m$-valued function K, and
$\lambda = (\lambda^1, \cdots, \lambda^m)$, derived from a first-order necessary condition of the optimization problem (1.24), satisfies the differential equation:

$$(1.25) \qquad d\lambda^i(t) = \lambda^i(t)b^i(t, \lambda(t), Y(t))dt; \quad t \in [0, T],$$

with $b^i(t, \lambda, y, \omega) = \frac{\partial f^i(c, y^i)}{\partial y^i}\big|_{c=(K_i(\lambda, e(t,\omega), y))}$. Thus (1.23) and (1.25) form
an FBSDE.

§1.4. Option pricing and contingent claim valuation

In this subsection we discuss option pricing problems in finance and their
relationship with FBSDEs. Consider a security market that contains, say,
one bond and one stock. Suppose that their prices are subject to the
following system of stochastic differential equations:

$$(1.26) \qquad \begin{cases} dP_0(t) = r(t)P_0(t)dt, & \text{(bond)}; \\ dP(t) = P(t)b(t)dt + P(t)\sigma(t)dW(t), & \text{(stock)}, \end{cases}$$

where $r(\cdot)$ is the *interest rate* of the bond, $b(\cdot)$ and $\sigma(\cdot)$ are the *appreciation
rate* and *volatility* of the stock, respectively.

An *option* is by definition a contract which gives its holder the right to
sell or buy the stock. The contract should contain the following elements:

1) a specified price q (called the *exercise price*, or *striking price*);
2) a terminal time T (called the *maturity date* or *expiration date*);
3) an exercise time.

In this book we are particularly interested *European options*, which
specify the exercise time to be exactly equal to T, the maturity date. Let
us take the European *call* option (which gives its holder the right to buy)
as an example. The decision of the holder will depend, conceivably, on
$P(T)$, the stock price at time T. For instance, if $P(T) < q$, then the
holder would simply discard the option, and buy the stock directly from
the market; whereas if $P(T) > q$, then the holder should opt to exercise the
option to make profit. Therefore the total payoff of the writer (or seller) of
the option at time $t = T$ will be $(P(T) - q)^+$, an $\mathcal{F}_T$-measurable random
variable. The (option pricing) problem to the seller (and buyer alike) is
then how to determine a premium for this contract at present time $t = 0$.
In general, we call such a contract an *option* if the payoff at time $t = T$ can
be written explicitly as a function of $P(T)$ (e.g., $(P(T) - q)^+$). In all the
other cases where the payoff at time $t = T$ is just an $\mathcal{F}_T$-measurable random
variable, such a contract is called a *contingent claim*, and the corresponding
pricing problem is then called *contingent claim valuation problem*.

Now suppose that the agent sells the option at price y and then invests
it in the market, and we denote his total wealth at each time t by $Y(t)$.
Obviously, $Y(0) = y$. Assume that at each time t the agent invests a
portion of his wealth, say $\pi(t)$, called *portfolio*, into the stock, and puts

the rest $(Y(t) - \pi(t))$ into the bond. Also we assume that the agent can choose to consume so that the cumulative *consumption* up to time t is $C(t)$, an $\{\mathcal{F}_t\}_{t \geq 0}$-adapted, nondecreasing process. It can be shown that the dynamics of $Y(\cdot)$ and the *portfolio/consumption* process pair $(\pi(\cdot), C(\cdot))$ should follow an SDE as well:

(1.27)
$$\begin{cases} dY(t) = \{r(t)Y(t) + Z(t)\theta(t)\}dt + Z(t)dW(t) - dC(t), \\ Y(0) = y, \end{cases}$$

where $Z(t) = \pi(t)\sigma(t)$, and $\theta(t) \stackrel{\Delta}{=} \sigma^{-1}(t)[b(t) - r(t)]$ (called *risk premium* process). For any contingent claim $H \in L^2_{\mathcal{F}_T}(\Omega, \mathbb{R})$, the purpose of the agent is to choose such a pair (π, C) as to come up with enough money to "hedge" the payoff H at time $t = T$, that is, $Y(T) \geq H$. Such a consumption/investment pair, if exist, is called a *hedging strategy against* H. The *fair price* of the contingent claim is the smallest initial endowment for which the hedging strategy exists. In other words, it is defined by

(1.28) $y^* = \inf\{y = Y(0);\ \exists(\pi, C),\ \text{such that}\ Y^{\pi, C}(T) \geq H\}.$

Now suppose $H = g(P(T))$, and consider an agent who is so prudent that he does not consume at all (i.e., $C \equiv 0$), and is able to choose π so that $Y(T) = H = g(P(T))$. Namely, he chooses Z (whence π) by solving the following combination of (1.26) and (1.27):

(1.29)
$$\begin{cases} dP(t) = P(t)b(t)dt + P(t)\sigma(t)dW(t), \\ dY(t) = \{r(t)Y(t) + Z(t)\theta(t)\}dt + Z(t)dW(t), \\ P(0) = p, \qquad Y(T) = g(P(T)), \end{cases}$$

which is again an FBSDE (an *decouped* FBSDE, to be more precise). An interesting result is that if (1.29) has an adapted solution (Y, Z), then the pair $(\pi, 0)$, where $\pi = Z\sigma^{-1}$, is the optimal hedging strategy and $y = Y(0)$ is the fair price! A more complicated case in which we allow the interaction between the agent's wealth/strategy and the stock price will be studied in details in Chapter 8. In that case (1.29) will become a truly coupled FBSDE.

§2. Definitions and Notations

In this section we list all the notations that will be frequently used throughout the book, and give some definitions related to FBSDEs.

Let $\mathbb{R}^n$ be the n-dimensional Euclidean space with the usual Euclidean norm $|\cdot|$ and the usual Euclidean inner product $\langle \cdot, \cdot \rangle$. Let $\mathbb{R}^{m \times d}$ be the Hilbert space consisting of all $(m \times d)$-matrices with the inner product

(2.1) $\langle A, B \rangle \stackrel{\Delta}{=} \mathrm{tr}\{AB^T\}, \qquad \forall A, B \in \mathbb{R}^{m \times d}.$

Thus, the norm $|A|$ of A induced by inner product (2.1) is given by $|A| = \sqrt{\mathrm{tr}\{AA^T\}}$. Another natural norm for $A \in \mathbb{R}^{m \times d}$ could be taken as

$\|A\| \stackrel{\Delta}{=} \sqrt{\max \sigma(AA^T)}$ if we regard A as a linear operator from $\mathbb{R}^m$ to $\mathbb{R}^d$, where $\sigma(AA^T)$ is the set of all eigenvalues of AA^T. It is clear that the norms $|\cdot|$ and $\|\cdot\|$ are equivalent since $\mathbb{R}^{m\times d}$ is a finite dimensional space. In fact, the following relations hold:

$$(2.2) \qquad \|A\| \leq \sqrt{\mathrm{tr}\{AA^T\}} = |A| \leq \sqrt{m \wedge d}\|A\|, \quad \forall A \in \mathbb{R}^{m\times d},$$

where $m \wedge d = \min\{m, d\}$. We will see that in our later discussions, the norm $|\cdot|$ in $\mathbb{R}^{m\times d}$ induced by (2.1) is more convenient.

Next, we let $T > 0$ be fixed and $(\Omega, \mathcal{F}, \{\mathcal{F}_t\}_{t\geq 0}, \mathbf{P})$ be as assumed at the beginning of §1. We denote

- for any sub-σ-field $\mathcal{G}$ of $\mathcal{F}$, $L^2_{\mathcal{G}}(\Omega; \mathbb{R}^m)$ to be the set of all $\mathcal{G}$-measurable $\mathbb{R}^m$-valued square integrable random variables;

- $L^2_{\mathcal{F}}(\Omega; L^2(0, T; \mathbb{R}^n))$ to be the set of all $\{\mathcal{F}_t\}_{t\geq 0}$-progressively measurable processes $X(\cdot)$ valued in $\mathbb{R}^n$ such that $\int_0^T E|X(t)|^2 dt < \infty$. The notation $L^2_{\mathcal{F}}(0, T; \mathbb{R}^n)$ is often used for simplicity, when there is no danger of confusion.

- $L^2_{\mathcal{F}}(\Omega; C([0, T]; \mathbb{R}^n))$ to be the set of all $\{\mathcal{F}_t\}_{t\geq 0}$-progressively measurable continuous processes $X(\cdot)$ taking values in $\mathbb{R}^n$, such that $E \sup_{t\in[0,T]} |X(t)|^2 < \infty$.

Also, for any Euclidean spaces M and N, we let

- $L^2_{\mathcal{F}}(0, T; W^{1,\infty}(M; N))$ be the set of all functions $f : [0, T] \times M \times \Omega \to N$, such that for any fixed $\theta \in M$, $(t, \omega) \mapsto f(t, \theta; \omega)$ is $\{\mathcal{F}_t\}_{t\geq 0}$-progressively measurable with $f(t, 0; \omega) \in L^2_{\mathcal{F}}(0, T; N)$, and there exists a constant $L > 0$, such that

$$|f(t, \theta; \omega) - f(t, \overline{\theta}; \omega)| \leq L|\theta - \overline{\theta}|, \quad \forall \theta, \overline{\theta} \in M, \text{ a.e. } t \in [0, T], \text{ a.s.};$$

- $L^2_{\mathcal{F}_T}(\Omega; W^{1,\infty}(\mathbb{R}^n; \mathbb{R}^m))$ be the set of all functions $g : \mathbb{R}^n \times \Omega \to \mathbb{R}^m$, such that $\omega \mapsto g(x; \omega)$ is $\mathcal{F}_T$-measurable for all $x \in \mathbb{R}^n$ and $x \mapsto g(x; \omega)$ is uniformly Lipschitz in $x \in \mathbb{R}^n$ and $g(0; \omega) \in L^2_{\mathcal{F}}(\Omega; \mathbb{R}^m)$.

Further, we define

$$(2.3) \qquad \begin{aligned} \mathcal{M}[0, T] \stackrel{\Delta}{=} & L^2_{\mathcal{F}}(\Omega; C([0, T]; \mathbb{R}^n)) \times L^2_{\mathcal{F}}(\Omega; C([0, T]; \mathbb{R}^m)) \\ & \times L^2_{\mathcal{F}}(0, T; \mathbb{R}^\ell). \end{aligned}$$

The norm of this space is defined by

$$(2.4) \qquad \begin{aligned} \|(X(\cdot), Y(\cdot), Z(\cdot))\| = & \Big\{ E \sup_{t\in[0,T]} |X(t)|^2 + E \sup_{t\in[0,T]} |Y(t)|^2 \\ & + E \int_0^T |Z(t)|^2 dt \Big\}^{1/2}, \end{aligned}$$

for all $(X(\cdot), Y(\cdot), Z(\cdot)) \in \mathcal{M}[0, T]$. It is clear that $\mathcal{M}[0, T]$ is a Banach space under norm (2.4).

We are now ready to give the formal description of an FBSDE. Let us consider an FBSDE in its most general form:

$$(2.5) \quad \begin{cases} dX(t) = b(t, X(t), Y(t), Z(t))dt + \sigma(t, X(t), Y(t), Z(t))dW(t), \\ dY(t) = h(t, X(t), Y(t), Z(t))dt + \widehat{\sigma}(t, X(t), Y(t), Z(t))dW(t), \\ X(0) = x, \qquad Y(T) = g(X(T)). \end{cases}$$

Here, the initial value x of $X(\cdot)$ is in $\mathbb{R}^n$; and b, σ, h, $\widehat{\sigma}$ and g are some suitable functions which satisfy the following *Standing Assumptions*: denoting $M = \mathbb{R}^n \times \mathbb{R}^m \times \mathbb{R}^\ell$, one has

$$(2.6) \quad \begin{cases} b \in L^2_{\mathcal{F}}(0, T; W^{1,\infty}(M; \mathbb{R}^n)), \quad \sigma \in L^2_{\mathcal{F}}(0, T; W^{1,\infty}(M; \mathbb{R}^{n \times d})), \\ h \in L^2_{\mathcal{F}}(0, T; W^{1,\infty}(M; \mathbb{R}^m)), \quad \widehat{\sigma} \in L^2_{\mathcal{F}}(0, T; W^{1,\infty}(M; \mathbb{R}^{m \times d})), \\ g \in L^2_{\mathcal{F}_T}(\Omega; W^{1,\infty}(\mathbb{R}^n; \mathbb{R}^m)). \end{cases}$$

Definition 2.1. A process $(X(\cdot), Y(\cdot), Z(\cdot)) \in \mathcal{M}[0, T]$ is called an *adapted solution* of (2.5) if the following holds for any $t \in [0, T]$, almost surely:

$$(2.7) \quad \begin{cases} X(t) = x + \displaystyle\int_0^t b(s, X(s), Y(s), Z(s))ds \\ \qquad\qquad + \displaystyle\int_0^t \sigma(s, X(s), Y(s), Z(s))dW(s), \\ Y(t) = g(X(T)) - \displaystyle\int_t^T h(s, X(s), Y(s), Z(s))ds \\ \qquad\qquad - \displaystyle\int_t^T \widehat{\sigma}(s, X(s), Y(s), Z(s))dW(s). \end{cases}$$

Furthermore, we say that FBSDE (2.5) is *solvable* if it has an adapted solution. An FBSDE is said to be *nonsolvable* if it is not solvable.

In what follows we shall try to answer the the following natural question: for given $b, \sigma, h, \widehat{\sigma}$ and g satisfying (2.6) and for given $x \in \mathbb{R}^n$, is (2.5) always solvable? In fact, what makes this type of SDE interesting is that the answer to this question is not affirmative, although the standing assumption (2.6) is already quite strong from the standard SDE point of view.

§3. Some Nonsolvable FBSDEs

In this section we shall first present some nonsolvability results, and then give some necessary conditions for the solvability.

It is well-known that two-point boundary value problems for ordinary differential equations do not necessarily admit solutions. On the other hand, an FBSDE can be viewed as a two-point boundary value problem for stochastic differential equations, with extra requirement that its solution is adapted solely to the forward filtration. Therefore, we do not expect the general existence and uniqueness result, even under the conditions that are usually considered *strong* in the SDE literature; for instance, the uniform Lipschitz conditions.

The following result is closely related to the solvability of two-point boundary value problem for ordinary differential equations.

Proposition 3.1. *Suppose that the following two-point boundary value problem for a system of linear ordinary differential equations does not admit any solution:*

$$(3.1) \qquad \begin{cases} \begin{pmatrix} \dot{X}(t) \\ \dot{Y}(t) \end{pmatrix} = \mathcal{A}(t) \begin{pmatrix} X(t) \\ Y(t) \end{pmatrix}, & t \in [0, T], \\ X(0) = x, \qquad Y(T) = GX(T), \end{cases}$$

where $\mathcal{A}(\cdot) : [0, T] \to \mathbb{R}^{(n+m) \times (n+m)}$ *is a deterministic integrable function and* $G \in \mathbb{R}^{m \times n}$. *Then, for any properly defined* $\sigma(t, x, y, z)$ *and* $\hat{\sigma}(t, x, y, z)$, *the following FBSDE:*

$$(3.2) \qquad \begin{cases} d \begin{pmatrix} X(t) \\ Y(t) \end{pmatrix} = \mathcal{A}(t) \begin{pmatrix} X(t) \\ Y(t) \end{pmatrix} dt + \begin{pmatrix} \sigma(t, X(t), Y(t), Z(t)) \\ \hat{\sigma}(t, X(t), Y(t), Z(t)) \end{pmatrix} dW(t), \\ X(0) = x, \qquad Y(T) = GX(T), \end{cases}$$

does not admit any adapted solution.

Here, by properly defined σ, we mean that for any $(X, Y, Z) \in \mathcal{M}[0, T]$ the process $\sigma(t, X(t), Y(t), Z(t))$ is in $L^2_{\mathcal{F}}(0, T; \mathbb{R}^{n \times d})$. The similar holds for $\hat{\sigma}$.

Proof. Suppose (3.2) admits an adapted solution $(X, Y, Z) \in \mathcal{M}[0, T]$. Then, $(EX(\cdot), EY(\cdot))$ is a solution of (3.1), a contradiction. This proves the assertion. $\qquad \square$

There are many examples of systems like (3.1) which do not admit solutions. Here is a very simple one: $(n = m = 1)$

$$(3.3) \qquad \begin{cases} \dot{X} = Y, \\ \dot{Y} = -X, \\ X(0) = x, \quad Y(T) = -X(T). \end{cases}$$

We can easily show that for $T = k\pi + \frac{3\pi}{4}$ (k, nonnegative integer), the above two-point boundary value problem does not admit a solution for any $x \in \mathbb{R} \setminus \{0\}$ and it admits infinitely many solutions for $x = 0$.

Using (3.3) and time scaling, we can construct a nonsolvable two-point boundary value problem for a system of linear ordinary differential equations of (3.1) type over any given finite time duration $[0, T]$ with the unknowns X, Y taking values in $\mathbb{R}^n$ and $\mathbb{R}^m$, respectively. Then, by Proposition 3.1, we see that for any duration $T > 0$ and any dimensions n, m, ℓ and d for the processes X, Y, Z and the Brownian motion $W(t)$, nonsolvable FBSDEs exist.

The case that we have discussed in the above is a little special since the drift of the FBSDE is linear. Let us now look at some more general case. The following result gives a necessary condition for the solvability of FBSDE (2.1).

Proposition 3.2. *Assume that b, σ, h and $\widehat{\sigma}$ satisfy (2.6). Assume further that σ and $\widehat{\sigma}$ are continuous in (t, x, y) uniformly in z, for each $\omega \in \Omega$; and that $g \in C^2 \cap C_b^1(\mathbb{R}^n; \mathbb{R}^m)$ and is deterministic. Suppose for some $x \in \mathbb{R}^n$, there exists a $T > 0$, such that (2.5) admits an adapted solution $(X, Y, Z) \in \mathcal{M}[0, T]$ with*

$$(3.4) \qquad \text{tr} \{g_{xx}^i(X)(\sigma\sigma^T)(\cdot, X, Y, Z)\} \in L_{\mathcal{F}}^2(0, T; \mathbb{R}), \quad 1 \le i \le m.$$

Then,

$$(3.5) \qquad \begin{aligned} &\inf_{z \in \mathbb{R}^\ell} |\widehat{\sigma}(T, X(T), g(X(T)), z) \\ &\qquad - g_x(X(T))\sigma(T, X(T), g(X(T)), z)| = 0, \quad \text{a.s.} \end{aligned}$$

Furthermore, suppose there exists a $T_0 > 0$, such that for all $T \in (0, T_0]$, (2.5) admits an adapted solution (X, Y, Z) (depending on $T > 0$) satisfying the following:

$$(3.6) \qquad \int_0^T E\big\{|b(s, X(s), Y(s), Z(s))|^2 + |\sigma(s, X(s), Y(s), Z(s))|^\beta\big\} ds \le C,$$

for some constants $C > 0$ and $\beta > 2$, independent of $T \in (0, T_0]$. Then,

$$(3.7) \qquad E \inf_{z \in \mathbb{R}^\ell} |\widehat{\sigma}(0, x, g(x), z) - g_x(x)\sigma(0, x, g(x), z)| = 0.$$

Proof. Let $(X, Y, Z) \in \mathcal{M}[0, T]$ be an adapted solution of (2.5). We denote

$$\begin{cases} \widetilde{h}(s) = (\widetilde{h}^1(s), \cdots, \widetilde{h}^m(s))^T, \\ \widetilde{h}^i = h^i - \langle g_x^i, b \rangle - \dfrac{1}{2}\text{tr}\,(g_{xx}^i \sigma\sigma^T), \quad 1 \le i \le m. \end{cases}$$

Here, we have suppressed X, Y, Z and we will do so below for the notational simplicity. Clearly, $\widetilde{h} \in L_{\mathcal{F}}^2(0, T; \mathbb{R}^m)$. Next, for any $i = 1, 2, \cdots, m$, by Itô's formula

$$\begin{aligned} 0 &= E|Y^i(T) - g^i(X(T))|^2 \\ &= E|Y^i(t) - g^i(X(t))|^2 + E \int_t^T |\widehat{\sigma}^i - g_x^i \sigma|^2 ds \\ (3.8) \qquad &\quad + E \int_t^T 2\big[Y^i(s) - g^i(X(s))\big]\big[h^i - \langle g_x^i, b \rangle - \dfrac{1}{2}\text{tr}\,(g_{xx}^i \sigma\sigma^T)\big] ds \\ &= E|Y^i(t) - g^i(X(t))|^2 + E \int_t^T |\widehat{\sigma}^i - g_x^i \sigma|^2 ds \\ &\quad + E \int_t^T 2\big[Y^i(s) - g^i(X(s))\big]\widetilde{h}^i(s) ds. \end{aligned}$$

On the other hand, by (2.5) and Itô's formula, we have

$$
Y^i(s) - g^i(X(s)) = Y^i(s) - Y^i(T) + g^i(X(T)) - g^i(X(s))
$$

(3.9)
$$
= -\int_s^T \widetilde{h}^i(r)dr - \int_s^T (\widehat{\sigma}^i - g_x^i\sigma)dW(r).
$$

Combining (3.8) and (3.9), we obtain that

$$
E|Y(t) - g(X(t))|^2 + E\int_t^T |\widehat{\sigma} - g_x\sigma|^2 ds
$$

$$
= -2E\int_t^T \langle Y(s) - g(X(s)), \widetilde{h}(s) \rangle ds
$$

(3.10)
$$
= 2E\int_t^T \langle \int_s^T \widetilde{h}(r)dr + \int_s^T [\widehat{\sigma} - g_x\sigma]dW(r), \widetilde{h}(s) \rangle ds
$$

$$
= 2E\int_t^T \langle \int_s^T \widetilde{h}(r)dr, \widetilde{h}(s) \rangle ds
$$

$$
\le (T - t)\int_t^T E|\widetilde{h}(r)|^2 dr = o(T - t).
$$

In the above, we have used the fact that

$$
E\Big\{ \langle \int_s^T [\widehat{\sigma} - g_x\sigma]dW(r), \widetilde{h}(s) \rangle \Big\} = 0.
$$

Consequently, we have that

(3.11)
$$
E\int_t^T \inf_{z\in\mathbf{R}^\ell} \big|\widehat{\sigma}(s, X(s), Y(s), z) - g_x(X(s))\sigma(s, X(s), Y(s), z)\big|^2 ds
$$
$$
\le E\int_t^T |\widehat{\sigma} - g_x\sigma|^2 ds = o(T - t).
$$

Since σ and $\widehat{\sigma}$ are continuous in (t, x, y), uniformly in z, the process $F(s) \triangleq \inf_{z\in\mathbf{R}^\ell} |\widehat{\sigma}(s, X(s), Y(s), z) - g_x(X(s))\sigma(s, X(s), Y(s), z)|^2$ is continuous, and an easy application of Lebesgue's Dominated Convergence Theorem and Differentiation Theorem leads to that

$$
EF(T) = \lim_{T-t\to 0} E\Big\{ \frac{1}{T - t}\int_t^T F(s)ds \Big\} = 0,
$$

proving (3.5) since $F(T)$ is nonnegative. Finally, if (3.6) holds, then by the forward equation in (2.5) one has

(3.12)
$$
\lim_{T\to 0} E|X(T) - x|^2 = 0,
$$

uniformly (note that $(X(\cdot), Y(\cdot), Z(\cdot))$ depends on the time duration $[0, T]$ on which (2.5) is solved). Hence, (3.7) follows.   □

We note that (3.4) holds if both g_{xx}^i and σ are bounded, and (3.6) holds if both b and σ are bounded.

An interesting corollary of Proposition 3.2 is the following nonsolvability result for FBSDEs.

Corollary 3.3. *Suppose $\widehat{\sigma}$ is continuous in (t, x, y, z) and uniformly Lipschitz continuous in (x, y, z). Suppose there exists an $\varepsilon > 0$, such that*

$$(3.13) \qquad \{\widehat{\sigma}(0, x, y, z) \mid z \in \mathbb{R}^\ell\} \subset \mathbb{R}^{m \times d} \setminus B_\varepsilon(\widehat{\sigma}_0), \quad a.s.$$

for some $(x, y) \in \mathbb{R}^n \times \mathbb{R}^m$ and some $\widehat{\sigma}_0 \in \mathbb{R}^{m \times d}$, where $B_\varepsilon(\widehat{\sigma}_0)$ is the closed ball in $\mathbb{R}^{m \times d}$ centered at $\widehat{\sigma}_0$ with radius ε. Then there exist smooth functions b, σ, h and g, such that the corresponding FBSDE (2.1) does not have adapted solutions over all small enough time durations $[0, T]$.

Proof. In the present case, we may choose b, σ, h and g such that (3.6) holds but (3.7) does not hold. Then our claim follows. $\qquad \square$

Since we are mainly interested in the case that FBSDEs do have adapted solutions, we should prevent the situation (3.13) from happening. A natural way of doing that is to assume that

$$(3.14) \quad \{\widehat{\sigma}(0, x, y, z) \mid z \in \mathbb{R}^\ell\} = \mathbb{R}^{m \times d}, \qquad \forall (x, y) \in \mathbb{R}^n \times \mathbb{R}^m, \ a.s.$$

This implies that $\ell \geq md$. Further, (3.14) suggests us to simply take

$$(3.15) \qquad \widehat{\sigma}(t, x, y, z) \equiv z, \qquad \forall (t, x, y) \in [0, T] \times \mathbb{R}^n \times \mathbb{R}^m,$$

with $z \in \mathbb{R}^{m \times d}$. From now on, we will restrict ourselves to such a situation. Hence, (2.5) becomes

$$(3.16) \quad \begin{cases} dX(t) = b(t, X(t), Y(t), Z(t))dt + \sigma(t, X(t), Y(t), Z(t))dW(t), \\ dY(t) = h(t, X(t), Y(t), Z(t))dt + Z(t)dW(t), \\ X(0) = x, \qquad Y(T) = g(X(T)). \end{cases}$$

Also, (2.3) now should be changed to the following:

$$(3.17) \quad \begin{aligned} \mathcal{M}[0, T] \triangleq L^2_{\mathcal{F}}(\Omega; C([0, T]; \mathbb{R}^n)) &\times L^2_{\mathcal{F}}(\Omega; C([0, T]; \mathbb{R}^m)) \\ &\times L^2_{\mathcal{F}}(0, T; \mathbb{R}^{m \times d}). \end{aligned}$$

We keep (2.4) as the norm of $\mathcal{M}[0, T]$, but now $|Z(t)|^2 = \mathrm{tr}\{Z(t)Z(t)^T\}$.

§4. Well-posedness of BSDEs

We now briefly look at the well-posedness of BSDEs. The purpose of this section is to recall a natural technique used in proving the well-posedness of BSDEs, namely, the *method of contraction mapping*.

We consider the following BSDE (compare with (3.16)):

$$(4.1) \quad \begin{cases} dY(t) = h(t, Y(t), Z(t))dt + Z(t)dW(t), \quad t \in [0, T], \\ Y(T) = \xi, \end{cases}$$

where $\xi \in L^2_{\mathcal{F}_T}(\Omega; \mathbb{R}^m)$ and $h \in L^2_{\mathcal{F}}(0, T; W^{1, \infty}(\mathbb{R}^m \times \mathbb{R}^{m \times d}; \mathbb{R}^m))$ i.e., (recall from §2), $h : [0, T] \times \mathbb{R}^m \times \mathbb{R}^{m \times d} \times \Omega \to \mathbb{R}^m$, such that $(t, \omega) \mapsto$

$h(t, y, z; \omega)$ is $\{\mathcal{F}_t\}_{t\geq 0}$-progressively measurable for all $(y, z) \in \mathbb{R}^m \times \mathbb{R}^{m \times d}$ with $h(t, 0, 0; \omega) \in L^2_{\mathcal{F}}(0, T; \mathbb{R}^m)$ and for some constant $L > 0$,

(4.2)
$$|h(t, y, z) - h(t, \overline{y}, \overline{z})| \leq L\{|y - \overline{y}| + |z - \overline{z}|\},$$
$$\forall y, \overline{y} \in \mathbb{R}^m, \ z, \overline{z} \in \mathbb{R}^{m \times d}, \text{ a.e. } t \in [0, T], \text{ a.s.}$$

Denote

(4.3)
$$\mathcal{N}[0, T] \triangleq L^2_{\mathcal{F}}(\Omega; C([0, T]; \mathbb{R}^m)) \times L^2_{\mathcal{F}}(0, T; \mathbb{R}^{m \times d}),$$

and

(4.4)
$$\|(Y(\cdot), Z(\cdot))\|_{\mathcal{N}[0,T]} \triangleq \left\{ E \sup_{0 \leq t \leq T} |Y(t)|^2 + E \int_0^T |Z(t)|^2 dt \right\}^{1/2}.$$

Then, $\mathcal{N}[0, T]$ is a Banach space under norm (4.4). We can similarly define $\mathcal{N}[t, T]$, for $t \in [0, T)$.

Let us introduce the following definition (compare with Definition 2.1).

Definition 4.1. A processes $(Y(\cdot), Z(\cdot)) \in \mathcal{N}[0, T]$ is called an adapted solution of (4.1) if the following holds:

(4.5)
$$Y(t) = \xi - \int_t^T h(s, Y(s), Z(s))ds - \int_t^T Z(s)dW(s),$$
$$\forall t \in [0, T], \text{ a.s.}$$

The following result gives the existence and uniqueness of adapted solutions to BSDE (4.1).

Theorem 4.2. Let $h \in L^2_{\mathcal{F}}(0, T; W^{1,\infty}(\mathbb{R}^m \times \mathbb{R}^{m \times d}; \mathbb{R}^m))$. Then, for any $\xi \in L^2_{\mathcal{F}_T}(\Omega; \mathbb{R}^m)$, BSDE (4.1) admits a unique adapted solution $(Y(\cdot), Z(\cdot))$.

Proof. For any $(y(\cdot), z(\cdot)) \in \mathcal{N}[0, T]$, we know that

(4.6)
$$h(\cdot) \equiv h(\cdot, y(\cdot), z(\cdot)) \in L^2_{\mathcal{F}}(0, T; \mathbb{R}^m).$$

Now, we define

(4.7)
$$\begin{cases} M(t) = E\{\xi - \int_0^T h(s)ds | \mathcal{F}_t\}, \\ Y(t) = E\{\xi - \int_t^T h(s)ds | \mathcal{F}_t\}, \end{cases} \quad t \in [0, T].$$

Then $M(t)$ is an $\{\mathcal{F}_t\}_{t\geq 0}$-martingale (square integrable), and

(4.8)
$$M(0) = E\{\xi - \int_0^T h(s)ds\} = Y(0).$$

Therefore, by the Martingale Representation Theorem, we can find a $Z(\cdot) \in L^2_{\mathcal{F}}(0, T; \mathbb{R}^{m \times d})$, such that

(4.9)
$$M(t) = M(0) + \int_0^t Z(s)dW(s), \quad \forall t \in [0, T].$$

Since ξ is $\mathcal{F}_T$-measurable, we see that (note (4.7)–(4.8))

$$(4.10) \qquad \xi - \int_0^T h(s)ds = M(T) = Y(0) + \int_0^T Z(s)dW(s).$$

Consequently, by (4.7)–(4.10), we obtain

$$
\begin{aligned}
Y(t) &= M(t) + \int_0^t h(s)ds \\
&= Y(0) + \int_0^t Z(s)dW(s) + \int_0^t h(s)ds \\
(4.11) \qquad &= \xi - \int_0^T h(s)ds - \int_0^T Z(s)dW(s) \\
&\quad + \int_0^t h(s)ds + \int_0^t Z(s)dW(s) \\
&= \xi - \int_t^T h(s)ds - \int_t^T Z(s)dW(s).
\end{aligned}
$$

It is not very hard to show that actually $(Y(\cdot), Z(\cdot)) \in \mathcal{N}[0, T]$ (See below for a similar proof). Thus, we obtain an adapted solution $(Y(\cdot), Z(\cdot))$ to the following equation:

$$(4.12) \qquad \begin{cases} dY(t) = h(t, y(t), z(t))dt + Z(t)dW(t), \\ Y(T) = \xi. \end{cases}$$

Now, let $(\overline{y}(\cdot), \overline{z}(\cdot)) \in \mathcal{N}[0, T]$ and $(\overline{Y}(\cdot), \overline{Z}(\cdot)) \in \mathcal{N}[0, T]$ be the corresponding solution of (4.12). Then, by Itô's formula and (4.2), we have

$$
\begin{aligned}
(4.13) \qquad & E|Y(t) - \overline{Y}(t)|^2 + E\int_t^T |Z(s) - \overline{Z}(s)|^2 ds \\
& \le 2LE \int_t^T |Y(s) - \overline{Y}(s)|\big\{|y(s) - \overline{y}(s)| + |z(s) - \overline{z}(s)|\big\}ds.
\end{aligned}
$$

Next, we set

$$(4.14) \qquad \begin{cases} \varphi(t) = \big\{E|Y(t) - \overline{Y}(t)|^2\big\}^{1/2}, \\ \psi(t) = \big\{E|y(t) - \overline{y}(t)|^2\big\}^{1/2} + \big\{E|z(t) - \overline{z}(t)|^2\big\}^{1/2}. \end{cases}$$

Then, (4.13) implies

$$(4.15) \quad \varphi(t)^2 + E\int_t^T |Z(s) - \overline{Z}(s)|^2 ds \le 2L\int_t^T \varphi(s)\psi(s)ds, \quad t \in [0, T].$$

We have the following lemma.

Lemma 4.3. Let (4.15) hold. Then,

$$(4.16) \quad \varphi(t)^2 + E\int_t^T |Z(s) - \overline{Z}(s)|^2 ds \le L^2\Big\{\int_t^T \psi(s)ds\Big\}^2, \quad \forall t \in [0, T].$$

Proof. We call the right hand side of (4.15) $2L\theta(t)$. Then, by (4.15),

(4.17) $$\theta'(t) = -\varphi(t)\psi(t) \geq -\psi(t)\sqrt{2L\theta(t)},$$

which yields

(4.18) $$\{\sqrt{\theta(t)}\}' \geq -\sqrt{L/2}\,\psi(t).$$

Noting $\theta(T) = 0$, we have

(4.19) $$-\sqrt{\theta(t)} \geq -\sqrt{L/2} \int_t^T \psi(s)ds.$$

Consequently,

(4.20) $$\theta(t) \leq \frac{L}{2}\left\{\int_t^T \psi(s)ds\right\}^2, \qquad \forall t \in [0,T].$$

Hence, (4.16) follows from (4.15) and (4.20). $\qquad\qquad\qquad\qquad\qquad\square$

Now, applying the above result to (4.13), we obtain

(4.21)
$$E|Y(t) - \overline{Y}(t)|^2 + E \int_t^T |Z(s) - \overline{Z}(s)|^2 ds$$
$$\leq L^2\left\{\int_t^T \left\{\left(E|y(s) - \overline{y}(s)|^2\right)^{1/2} + \left(E|z(s) - \overline{z}(s)|^2\right)^{1/2}\right\}ds\right\}^2$$
$$\leq C(T-t)\|(y(\cdot), z(\cdot)) - (\overline{y}(\cdot), \overline{z}(\cdot))\|^2_{\mathcal{N}[t,T]}.$$

Then, by Doob's inequality, we further have

(4.22)
$$\|(Y(\cdot), Z(\cdot)) - (\overline{Y}(\cdot), \overline{Z}(\cdot))\|^2_{\mathcal{N}[t,T]}$$
$$\leq C(T-t)\|(y(\cdot), z(\cdot)) - (\overline{y}(\cdot), \overline{z}(\cdot))\|^2_{\mathcal{N}[t,T]}, \qquad \forall t \in [0,T].$$

Here $C > 0$ is a constant depending only on L. By taking $\delta = \frac{1}{2C}$, we see that the map $(y(\cdot), z(\cdot)) \mapsto (Y(\cdot), Z(\cdot))$ is a contraction on the Banach space $\mathcal{N}[T-\delta, T]$. Thus, it admits a unique fixed point, which is the adapted solution of (4.1) with $[0,T]$ replaced by $[T-\delta, T]$. By continuing this procedure, we obtain existence and uniqueness of the adapted solutions to (4.1). $\qquad\qquad\qquad\qquad\qquad\qquad\qquad\qquad\qquad\qquad\quad\square$

We now prove the continuous dependence of the solutions on the final data ξ and the function h.

Theorem 4.4. *Let* $h, \overline{h} \in L^2_\mathcal{F}(0, T; W^{1,\infty}(\mathbb{R}^m \times \mathbb{R}^{m\times d}; \mathbb{R}^m))$ *and* $\xi, \overline{\xi} \in L^2_{\mathcal{F}_T}(\Omega; \mathbb{R}^m)$. *Let* $(Y(\cdot), Z(\cdot)), (\overline{Y}(\cdot), \overline{Z}(\cdot)) \in \mathcal{N}[0, T]$ *be the adapted solutions of (4.1) corresponding to* (h, ξ) *and* $(\overline{h}, \overline{\xi})$, *respectively. Then*

(4.23)
$$\|(Y(\cdot) - \overline{Y}(\cdot), Z(\cdot) - \overline{Z}(\cdot))\|^2_{\mathcal{N}[0,T]}$$
$$\leq C\left\{E|\xi - \overline{\xi}|^2 + E \int_0^T |h(s, Y(s), Z(s)) - \overline{h}(s, Y(s), Z(s))|^2 ds\right\},$$

with $C > 0$ being a constant only depending on $T > 0$ and the Lipschitz constants of h and $\overline{h}$.

Proof. We denote

$$
(4.24) \qquad \begin{cases} \widehat{Y}(\cdot) = Y(\cdot) - \overline{Y}(\cdot), & \widehat{Z}(\cdot) = Z(\cdot) - \overline{Z}(\cdot), \\ \widehat{\xi} = \xi - \overline{\xi}, & \widehat{h}(\cdot) = h(\cdot, Y(\cdot), Z(\cdot)) - \overline{h}(\cdot, Y(\cdot), Z(\cdot)). \end{cases}
$$

Applying Itô's formula to $|\widehat{Y}(\cdot)|^2$, we obtain

$$
\begin{aligned}
|\widehat{Y}(t)|^2 &+ \int_t^T |\widehat{Z}(s)|^2 ds \\
&= |\widehat{\xi}|^2 - 2 \int_t^T \langle \widehat{Y}(s), h(s, Y(s), Z(s)) - \overline{h}(s, \overline{Y}(s), \overline{Z}(s)) \rangle \, ds \\
&\quad - 2 \int_t^T \langle \widehat{Y}(s), \widehat{Z}(s) dW(s) \rangle \\
(4.25) \quad &\le |\widehat{\xi}|^2 + 2 \int_t^T \{ |\widehat{Y}(s)||\widehat{h}(s)| + L|\widehat{Y}(s)|(|\widehat{Y}(s)| + |\widehat{Z}(s)|) \} ds \\
&\quad - 2 \int_t^T \langle \widehat{Y}(s), \widehat{Z}(s) dW(s) \rangle \\
&\le |\widehat{\xi}|^2 + \int_t^T \{ (1 + 2L + 2L^2)|\widehat{Y}(s)|^2 + \tfrac{1}{2}|\widehat{Z}(s)|^2 + |\widehat{h}(s)|^2 \} ds \\
&\quad - 2 \int_t^T \langle \widehat{Y}(s), \widehat{Z}(s) dW(s) \rangle .
\end{aligned}
$$

Taking expectation in the above, we have

$$
\begin{aligned}
(4.26) \quad E|\widehat{Y}(t)|^2 &+ \frac{1}{2} E \int_t^T |\widehat{Z}(s)|^2 ds \le E|\widehat{\xi}|^2 + E \int_0^T |\widehat{h}(s)|^2 ds \\
&+ (1 + 2L + 2L^2) E \int_t^T |\widehat{Y}(s)|^2 ds, \qquad t \in [0, T].
\end{aligned}
$$

Thus, it follows from Gronwall's inequality that

$$
\begin{aligned}
(4.27) \quad E|\widehat{Y}(t)|^2 &+ E \int_t^T |\widehat{Z}(s)|^2 ds \\
&\le C \Big\{ E|\widehat{\xi}|^2 + E \int_0^T |\widehat{h}(s)|^2 ds \Big\}, \qquad \forall t \in [0, T].
\end{aligned}
$$

On the other hand, by Burkholder-Davis-Gundy's inequality (see Karatzas-

Shreve [1]), we have from (4.25) that (note (4.27))

$$E\{ \sup_{t\in[0,T]} |\widehat{Y}(t)|^2 \} \le C\{ E|\widehat{\xi}|^2 + E \int_0^T |\widehat{h}(s)|^2 ds \}$$

$$+ 2E \sup_{t\in[0,T]} \left| \int_t^T \langle \widehat{Y}(s), \widehat{Z}(s)dW(s) \rangle \right|$$

(4.28)

$$\le C\{ E|\widehat{\xi}|^2 + E \int_0^T |\widehat{h}(s)|^2 ds \}$$

$$+ C_1 \left(E \sup_{t\in[0,T]} |\widehat{Y}(t)|^2 \right)^{1/2} \left(E \int_0^T |\widehat{Z}(s)|^2 ds \right)^{1/2}.$$

Now (4.23) follows easily from (4.28) and (4.27). □

We see that Theorems 4.2 and 4.4 give the well-posedness of BSDE (4.1). These results are satisfactory since the conditions that we have imposed are nothing more than uniform Lipschitz conditions as well as certain measurability conditions. These conditions seem to be indispensable, unless some other special structure conditions are assumed.

§5. Solvability of FBSDEs in Small Time Durations

In this section we try to adopt the method of contraction mapping used in the previous section to prove the solvability of FBSDE (3.16) in small time durations. The main result is the following.

Theorem 5.1. *Let b, σ, h and g satisfy (2.6). Moreover, we assume that*

(5.1)
$$\begin{cases} |\sigma(t,x,y,z;\omega) - \sigma(t,x,y,\bar{z};\omega)| \le L_0|z - \bar{z}|, \\ \qquad \forall(x,y) \in \mathbb{R}^n \times \mathbb{R}^m, \ z,\bar{z} \in \mathbb{R}^{m\times d}, \ \text{a.e.} \ t \ge 0, \ \text{a.s.} \\ |g(x;\omega) - g(\bar{x};\omega)| \le L_1|x - \bar{x}|, \qquad \forall x,\bar{x} \in \mathbb{R}^n, \ \text{a.s.} \end{cases}$$

with

(5.2)
$$L_0 L_1 < 1.$$

Then there exists a $T_0 > 0$, such that for any $T \in (0,T_0]$ and any $x \in \mathbb{R}^n$, (3.16) admits a unique adapted solution $(X,Y,Z) \in \mathcal{M}[0,T]$.

Note that condition (5.2) is almost necessary. Here is a simple example for which (5.2) does not hold and the corresponding FBSDE does not have adapted solutions over any small time durations.

Example 5.2. Let $n = m = d = 1$. Consider the following FBSDEs:

(5.3)
$$\begin{cases} dX(t) = Z(t)dW(t), \\ dY(t) = Z(t)dW(t), \\ X(0) = 0, \qquad Y(T) = X(T) + \xi, \end{cases}$$

where ξ is $\mathcal{F}_T$-measurable only (say, $\xi = W(T)$). Clearly, in the present case, $L_0 = L_1 = 1$. Thus, (5.2) fails. If (5.3) admitted an adapted solution

(X, Y, Z), then the process $\eta \overset{\Delta}{=} Y - X$ would be $\{\mathcal{F}_t\}_{t \geq 0}$-adapted and satisfy the following:

(5.4)
$$\begin{cases} d\eta(t) = 0, & t \in [0, T], \\ \eta(T) = \xi. \end{cases}$$

We know from §1 that (5.4) does not admit an adapted solution unless ξ is deterministic.

Proof of Theorem 5.1. Let $0 < T_0 \leq 1$ be undetermined and $T \in (0, T_0]$. Let $x \in \mathbb{R}^n$ be fixed. We introduce the following norm:

(5.5) $$\|(Y, Z)\|_{\overline{\mathcal{N}}[0,T]} \overset{\Delta}{=} \sup_{t \in [0,T]} \left\{ E|Y(t)|^2 + E \int_t^T |Z(s)|^2 ds \right\}^{1/2},$$

for all $(Y, Z) \in \mathcal{N}[0, T]$. It is clear that norm (5.5) is weaker than (4.4). We let $\overline{\mathcal{N}}[0, T]$ be the completion of $\mathcal{N}[0, T]$ in $L^2_{\mathcal{F}}(0, T; \mathbb{R}^m) \times L^2_{\mathcal{F}}(0, T; \mathbb{R}^{m \times d})$ under norm (5.5). Take any $(Y_i, Z_i) \in \overline{\mathcal{N}}[0, T]$, $i = 1, 2$. We solve the following FSDE for X_i:

(5.6)
$$\begin{cases} dX_i = b(t, X_i, Y_i, Z_i)dt + \sigma(t, X_i, Y_i, Z_i)dW(t), & t \in [0, T], \\ X_i(0) = x. \end{cases}$$

It is standard that under our conditions, (5.6) admits a unique (strong) solution $X_i \in L^2_{\mathcal{F}}(\Omega; C([0, T]; \mathbb{R}^n))$. By Itô's formula and the Lipschitz continuity of b and σ (note (5.1)), we obtain

$$E|X_1(t) - X_2(t)|^2$$

(5.7)
$$\leq E \int_0^t \left\{ 2L|X_1 - X_2| \Big(|X_1 - X_2| + |Y_1 - Y_2| + |Z_1 - Z_2| \Big) \right.$$
$$\left. + \Big(L(|X_1 - X_2| + |Y_1 - Y_2|) + L_0|Z_1 - Z_2| \Big)^2 \right\} ds$$
$$\leq E \int_0^t \left\{ C_\varepsilon \Big(|X_1 - X_2|^2 + |Y_1 - Y_2|^2 \Big) + (L_0^2 + \varepsilon)|Z_1 - Z_2|^2 \right\} ds,$$

where $C_\varepsilon > 0$ only depends on L, L_0 and $\varepsilon > 0$. Then, by Gronwall's inequality, we obtain

(5.8) $$E|X_1(t) - X_2(t)|^2 \leq e^{C_\varepsilon T} E \int_0^T \left\{ C_\varepsilon |Y_1 - Y_2|^2 + (L_0^2 + \varepsilon)|Z_1 - Z_2|^2 \right\} ds.$$

Next, we solve the following BSDEs: $(i = 1, 2)$

(5.9)
$$\begin{cases} d\overline{Y}_i = h(t, X_i, Y_i, Z_i)dt + \overline{Z}_i dW(t), & t \in [0, T], \\ \overline{Y}_i(T) = g(X_i(T)). \end{cases}$$

We see from Theorem 4.2 that (for $i = 1, 2$) (5.9) admits a unique adapted solution $(\overline{Y}_i, \overline{Z}_i) \in \mathcal{N}[0, T] \subseteq \overline{\mathcal{N}}[0, T]$. Thus, we have defined a map

$\mathcal{T} : \overline{\mathcal{N}}[0,T] \to \overline{\mathcal{N}}[0,T]$ by $(Y_i, Z_i) \mapsto (\overline{Y}_i, \overline{Z}_i)$. Applying Itô's formula to $|\overline{Y}_1(t) - \overline{Y}_2(t)|^2$, we have (note (5.1) and (5.8))

$$
\begin{aligned}
E|\overline{Y}_1(t) &- \overline{Y}_2(t)|^2 + E \int_t^T |\overline{Z}_1 - \overline{Z}_2|^2 ds \\
&\leq L_1^2 E |X_1(T) - X_2(T)|^2 \\
&\quad + 2LE \int_t^T |\overline{Y}_1 - \overline{Y}_2| (|X_1 - X_2| + |Y_1 - Y_2| + |Z_1 - Z_2|) ds \\
&\leq L_1^2 E |X_1(T) - X_2(T)|^2 + C_\varepsilon E \int_t^T |\overline{Y}_1 - \overline{Y}_2|^2 ds \\
&\quad + \varepsilon E \int_t^T |Z_1 - Z_2|^2 ds + E \int_t^T (|X_1 - X_2|^2 + |Y_1 - Y_2|^2) ds \\
&\leq (L_1^2 + T)e^{C_\varepsilon T} E \int_0^T [C_\varepsilon |Y_1 - Y_2|^2 + (L_0^2 + \varepsilon)|Z_1 - Z_2|^2] ds \\
&\quad + \varepsilon E \int_0^T |Z_1 - Z_2|^2 ds + E \int_0^T |Y_1 - Y_2|^2 ds \\
&\quad + C_\varepsilon E \int_t^T |\overline{Y}_1 - \overline{Y}_2|^2 ds.
\end{aligned}
$$

(5.10)

In the above, C_ε could be different from that appeared in (5.7)–(5.8). But C_ε is still independent of $T > 0$. Using Gronwall's inequality, we have

$$
\begin{aligned}
E|\overline{Y}_1(t) &- \overline{Y}_2(t)|^2 + E \int_t^T |\overline{Z}_1 - \overline{Z}_2|^2 ds \\
&\leq e^{C_\varepsilon T} \Big\{ \widetilde{C}_\varepsilon E \int_0^T |Y_1 - Y_2|^2 ds \\
&\quad + [\varepsilon + (L_1^2 + T)(L_0^2 + \varepsilon)e^{C_\varepsilon T}] E \int_0^T |Z_1 - Z_2|^2 ds \Big\} \\
&\leq e^{C_\varepsilon T} [\widetilde{C}_\varepsilon T + \varepsilon + (L_1^2 + T)(L_0^2 + \varepsilon)e^{C_\varepsilon T}] \\
&\quad \cdot \|(Y_1, Z_1) - (Y_2, Z_2)\|_{\overline{\mathcal{N}}[0,T]}^2,
\end{aligned}
$$

(5.11)

where $\widetilde{C}_\varepsilon > 0$ is again independent of $T > 0$. In the above, the last inequality follows from the fact that for any $(Y, Z) \in \overline{\mathcal{N}}[0,T]$,

(5.12)
$$
\begin{cases}
E|Y(t)|^2 \leq \|(Y, Z)\|_{\overline{\mathcal{N}}[0,T]}^2, \quad \forall t \in [0,T], \\
\int_0^T E|Z(t)|^2 dt \leq \|(Y, Z)\|_{\overline{\mathcal{N}}[0,T]}^2,
\end{cases}
$$

Since (5.2) holds, by choosing $\varepsilon > 0$ small enough then choosing $T > 0$ small enough, we obtain

(5.13) $\qquad \|(\overline{Y}_1, \overline{Z}_1) - (\overline{Y}_2, \overline{Z}_2)\|_{\overline{\mathcal{N}}[0,T]} \leq \alpha \|(Y_1, Z_1) - (Y_2, Z_2)\|_{\overline{\mathcal{N}}[0,T]},$

for some $0 < \alpha < 1$. This means that the map $\mathcal{T} : \overline{\mathcal{N}}[0, T] \to \overline{\mathcal{N}}[0, T]$ is contractive. By the Contraction Mapping Theorem, there exists a unique fixed point (Y, Z) for $\mathcal{T}$. Then, similar to the proof of Theorem 4.2 we can show that actually $(Y, Z) \in \mathcal{N}[0, T]$. Finally, we let X be the corresponding solution of (5.6). Then $(X, Y, Z) \in \mathcal{M}[0, T]$ is a unique adapted solution of (3.16). The above argument applies for all small enough $T > 0$. Thus, we obtain a $T_0 > 0$, such that for all $T \in (0, T_0]$ and all $x \in \mathbb{R}^n$, (3.16) is uniquely solvable.   □

In the above proof, it is crucial that the time duration is small enough, besides condition (5.2). This is the main disadvantage of applying the Contraction Mapping Theorem to two-point boundary value problems. Starting from the next chapter, we are going to use different methods to approach the solvability problem for the FBSDE (3.16).

§6. Comparison Theorems for BSDEs and FBSDEs

In this section we study an important tool in the theory of the BSDEs—Comparison Theorems. The main ingredients in the proof of the desired comparison results are "linearization of the equation" plus a change of probability measure. We should also note that in the coupled FBSDE case the situation becomes quite different. We shall give an example in the end of this section to show that the simple-minded generalization from BSDEs to FBSDEs fails in general.

To begin with, we consider two BSDEs: for $i = 1, 2$,

$$(6.1) \qquad Y^i(t) = \xi^i + \int_t^T h^i(s, Y^i(s), Z^i(s)) ds - \int_t^T Z^i(s) dW(s),$$

where W is a d-dimensional Brownian motion, and naturally the dimension of Y's and Z's are assumed to be 1 and d, respectively. Assume that

$$(6.2) \quad \xi^i \in L^2_{\mathcal{F}_T}(\Omega; \mathbb{R}); \qquad h^i \in L^2_{\mathcal{F}}(0, T; W^{1,\infty}(\mathbb{R}^{d+1}, \mathbb{R})), \qquad i = 1, 2,$$

where $L^2_{\mathcal{F}}(0, T; W^{1,\infty}(\mathbb{R}^{d+1}, \mathbb{R}))$ is defined in §2. Since under these conditions both BSDEs are well-posed, we denote by (Y^i, Z^i), $i = 1, 2$ the two adapted solutions respectively. We have

Theorem 6.1. *Suppose that assumption (6.2) holds, and suppose that* $\xi^1 \geq \xi^2$, *and* $h^1(t, y, z) \geq h^2(t, y, z)$, *for all* $(y, z) \in \mathbb{R}^{d+1}$, **P***-almost surely.* *Then it holds that* $Y^1(t) \geq Y^2(t)$, *for all* $t \in [0, T]$, **P***-a.s.*

Proof. Denote $\widehat{Y}(t) = Y^1(t) - Y^2(t)$, $\widehat{Z}(t) = Z^1(t) - Z^2(t)$, $\forall t \in [0, T]$; $\widehat{\xi} = \xi^1 - \xi^2$; and

$$\widehat{h}(t) = h^1(t, Y^2(t), Z^2(t)) - h^2(t, Y^2(t), Z^2(t)), \qquad t \in [0, T].$$

Clearly, $\widehat{h}$ is an $\{\mathcal{F}_t\}_{t \geq 0}$-adapted, non-negative process; and $\widehat{Y}$ satisfies the following (linear!) BSDE:

$$\widehat{Y}(t) = \widehat{\xi} + \int_t^T \{[h^1(s, Y^1(s), Z^1(s)) - h^1(s, Y^2(s), Z^2(s))] + \widehat{h}(s)\}ds$$

(6.3)
$$- \int_t^T \widehat{Z}(s)dW(s)$$

$$= \widehat{\xi} + \int_t^T \{\alpha(s)\widehat{Y}(s) + \beta(s)\widehat{Z}(s) + \widehat{h}(s)\}ds - \int_t^T \widehat{Z}(s)dW(s),$$

where

$$\alpha(s) = \int_0^1 h_y^1(s, Y^2(s) + \lambda\widehat{Y}(s), Z^2(s) + \lambda\widehat{Z}(s))d\lambda;$$

$$\beta(s) = \int_0^1 h_z^1(s, Y^2(s) + \lambda\widehat{Y}(s), Z^2(s) + \lambda\widehat{Z}(s))d\lambda.$$

Clearly, α and β are $\{\mathcal{F}_t\}_{t\geq0}$-adapted processes, and are both uniformly bounded, thanks to (6.2). In particular, β satisfies the so-called *Novikov condition*, and therefore the process

$$M(t) = \exp\left\{ \int_0^t \beta(s)dW(s) - \frac{1}{2}\int_0^t |\beta(s)|^2 ds \right\}, \quad t \in [0, T]$$

is an **P**-martingale. We now define a new probability measure $\widehat{\mathbf{P}}$ by

$$\frac{d\widehat{\mathbf{P}}}{d\mathbf{P}} = M(T).$$

Then by Girsanov's theorem, $\widehat{W}(t) \triangleq W(t) - \int_0^t \beta(s)ds$ is a $\widehat{\mathbf{P}}$-Brownian motion, and under $\widehat{\mathbf{P}}$, $\widehat{Y}$ satisfies

(6.4) $$\widehat{Y}(t) = \widehat{\xi} + \int_t^T [\alpha(s)\widehat{Y}(s) + \widehat{h}(s)]ds - \int_t^T \widehat{Z}(s)d\widehat{W}(s).$$

Now define $\Gamma(t) = \exp\{\int_0^t \alpha(s)ds\}$, then Itô's formula shows that

$$\Gamma(T)\widehat{\xi} - \Gamma(t)\widehat{Y}(t) = -\int_t^T \Gamma(s)\widehat{h}(s)ds + \int_t^T \widehat{Z}(s)d\widehat{W}(t).$$

Taking conditional expectation $E^{\widehat{\mathbf{P}}}\{\cdot|\mathcal{F}_t\}$ on both sides above, and noticing the adaptedness of $\Gamma(\cdot)\widehat{Y}(\cdot)$ we obtain that

$$\Gamma(t)\widehat{Y}(t) = E^{\widehat{\mathbf{P}}}\left\{\Gamma(T)\widehat{\xi} + \int_t^T \Gamma(s)\widehat{h}(s)ds\Big|\mathcal{F}_t\right\} \geq 0, \quad \forall t \in [0, T],$$

$\widehat{\mathbf{P}}$-almost surely, whence **P**-almost surely, proving the theorem.   □

An interesting as well as important observation is that the comparison theorem fails when the BSDE is coupled with a forward SDE. To be more precise, let us consider the following FBSDEs: for $i = 1, 2$,

(6.5)
$$
\begin{cases}
X^i(t) = x^i + \displaystyle\int_0^t b^i(s, X^i(s), Y^i(s), Z^i(s))ds \\
\qquad\qquad + \displaystyle\int_0^t \sigma^i(s, X^i(s), Y^i(s), Z^i(s))dW(s) \\
Y^i(t) = g^i(X^i(T)) + \displaystyle\int_t^T h^i(s, X^i(s), Y^i(s), Z^i(s))ds \\
\qquad\qquad - \displaystyle\int_t^T Z^i(s)dW(s).
\end{cases}
$$

We would like to know whether $g_1(x) \geq g_2(x)$, $\forall x$ would imply $Y_1(t) \geq Y_2(t)$, for all t? The following example shows that it is not true in general.

Example 6.2. Assume that $d = 1$. Consider the FBSDE:

(6.6)
$$
\begin{cases}
dX(t) = \dfrac{X(t)}{(Z(t) - Y(t))^2 + 1}dt + X(t)dW(t), \\
dY(t) = \dfrac{Z(t)}{(Z(t) - Y(t))^2 + 1}dt + Z(t)dW(t), \\
X(0) = x; \qquad Y(T) = g(X(T)).
\end{cases}
$$

We first assume that $g(x) = g^1(x) = x$. Then, one checks directly that $X^1(t) \equiv Y^1(t) \equiv Z^1(t) = x\exp\{W(t) + t/2\}$, $t \in [0, T]$ is an adapted solution to (6.6). (In fact, it can be shown by using Four Step Scheme of Chapter 6 that this is the unique adapted solution to (6.6)!)

Now let $g^2(x) = x + 1$. Then one checks that $X^2(t) \equiv Z^2(t)$ and $Y^2(t) \equiv X^2(t) + 1 = Z^2(t) + 1$, $\forall t \in [0, T]$ is the (unique) adapted solution to (6.6) with $g^2(x) = x + 1$. Moreover, solving (6.6) explicitly again we have $Y^2(t) = 1 + x\exp\{W(t)\}$.

Consequently, we see that $Y^1(t) - Y^2(t) = xe^{W(t)}[e^{t/2} - 1] - 1$, which can be both positive or negative with positive probability, for any $t > 0$, that is, the comparison theorem of the Theorem 6.1 type does not hold!

□

Finally we should note that despite the discouraging counterexample above, the comparison theorem for FBSDEs in a certain form can still be proved under appropriate conditionis on the coefficients. A special case will be presented in Chapter 8 (§8.3), when we study the applications of FBSDE in Finance.

Chapter 2

Linear Equations

In this chapter, we are going to study linear FBSDEs in any finite time duration. We will start with the most general case. By deriving a necessary condition of solvability, we obtain a reduction to a simple form of linear FBSDEs. Then we will concentrate on that to obtain some necessary and sufficient conditions for solvability. For simplicity, we will restrict ourselves to the case of one-dimensional Brownian motion in §§1–4. Some extensions to the case with multi-dimensional Brownian motion will be given in §5.

§1. Compatible Conditions for Solvability

Let $(\Omega, \mathcal{F}, \{\mathcal{F}_t\}_{t\geq0}, \mathbf{P})$ be a complete filtered probability space on which defined a one-dimensional standard Brownian motion $W(t)$, such that $\{\mathcal{F}_t\}_{t\geq0}$ is the natural filtration generated by $W(t)$, augmented by all the $\mathbf{P}$-null sets in $\mathcal{F}$. We consider the following system of coupled linear FBSDEs:

$$(1.1) \quad \begin{cases} dX(t) = \{AX(t) + BY(t) + CZ(t) + Db(t)\}dt \\ \qquad + \{A_1X(t) + B_1Y(t) + C_1Z(t) + D_1\sigma(t)\}dW(t), \\ dY(t) = \{\widehat{A}X(t) + \widehat{B}Y(t) + \widehat{C}Z(t) + \widehat{D}\widehat{b}(t)\}dt \\ \qquad + \{\widehat{A}_1X(t) + \widehat{B}_1Y(t) + \widehat{C}_1Z(t) + \widehat{D}_1\widehat{\sigma}(t)\}dW(s), \\ \qquad\qquad\qquad\qquad\qquad\qquad\qquad t \in [0,T], \\ X(0) = x, \qquad Y(T) = GX(T) + Fg. \end{cases}$$

In the above, A, B, C etc. are (deterministic) matrices of suitable sizes, b, σ, $\widehat{b}$ and $\widehat{\sigma}$ are stochastic processes and g is a random variable. We are looking for $\{\mathcal{F}_t\}_{t\geq0}$-adapted processes $X(\cdot)$, $Y(\cdot)$ and $Z(\cdot)$, valued in $\mathbb{R}^n$, $\mathbb{R}^m$ and $\mathbb{R}^\ell$, respectively, satisfying the above. More precisely, we recall the following definition (see Definition 2.1 of Chapter 1):

Definition 1.1. A triple $(X, Y, Z) \in \mathcal{M}[0,T]$ is called an *adapted solution* of (1.1) if the following holds for all $t \in [0,T]$, almost surely:

$$(1.2) \quad \begin{cases} X(t) = x + \displaystyle\int_0^t \{AX(s) + BY(s) + CZ(s) + Db(s)\}ds \\ \qquad + \displaystyle\int_0^t \{A_1X(s) + B_1Y(s) + C_1Z(s) + D_1\sigma(s)\}dW(s), \\ Y(t) = GX(T) + Fg - \displaystyle\int_t^T \{\widehat{A}X(s) + \widehat{B}Y(s) + \widehat{C}Z(s) + \widehat{D}\widehat{b}(s)\}ds \\ \qquad - \displaystyle\int_t^T \{\widehat{A}_1X(s) + \widehat{B}_1Y(s) + \widehat{C}_1Z(s) + \widehat{D}_1\widehat{\sigma}(s)\}dW(s). \end{cases}$$

When (1.1) admits an adapted solution, we say that (1.1) is solvable.

In what follows, we will let

(1.3)
$$\begin{cases}
A, A_1 \in \mathbb{R}^{n \times n}; \ B, B_1 \in \mathbb{R}^{n \times m}; \ C, C_1 \in \mathbb{R}^{n \times \ell}; \\
\widehat{A}, \widehat{A}_1, G \in \mathbb{R}^{m \times n}; \ \widehat{B}, \widehat{B}_1 \in \mathbb{R}^{m \times m}; \ \widehat{C}, \widehat{C}_1 \in \mathbb{R}^{m \times \ell}; \\
D \in \mathbb{R}^{n \times \bar{n}}; \ D_1 \in \mathbb{R}^{n \times \bar{n}_1}; \ \widehat{D} \in \mathbb{R}^{m \times \bar{m}}; \\
\widehat{D}_1 \in \mathbb{R}^{m \times \bar{m}_1}; \ F \in \mathbb{R}^{m \times k}; \\
b \in L_{\mathcal{F}}^2(0, T; \mathbb{R}^{\bar{n}}); \ \sigma \in L_{\mathcal{F}}^2(0, T; \mathbb{R}^{\bar{n}_1}); \\
\widehat{b} \in L_{\mathcal{F}}^2(0, T; \mathbb{R}^{\bar{m}}); \ \widehat{\sigma} \in L_{\mathcal{F}}^2(0, T; \mathbb{R}^{\bar{m}_1}); \\
g \in L_{\mathcal{F}_T}^2(\Omega; \mathbb{R}^k); \ x \in \mathbb{R}^n.
\end{cases}$$

Following result gives a compatibility condition among the coefficients of (1.1) for its solvability.

Theorem 1.2. *Suppose there exists a $T > 0$, such that for all b, σ, $\widehat{b}$, $\widehat{\sigma}$, g and x satisfying (1.3), (1.1) admits an adapted solution $(X, Y, Z) \in \mathcal{M}[0, T]$. Then*

(1.4) $$\mathcal{R}(\widehat{C}_1 - GC_1) \supseteq \mathcal{R}(F) + \mathcal{R}(\widehat{D}_1) + \mathcal{R}(GD_1),$$

where $\mathcal{R}(S)$ is the range of operator S. In particular, if

(1.5) $$\mathcal{R}(F) + \mathcal{R}(\widehat{D}_1) + \mathcal{R}(GD_1) = \mathbb{R}^m,$$

then $\widehat{C}_1 - GC_1 \in \mathbb{R}^{m \times \ell}$ is onto and thus $\ell \geq m$.

To prove the above result, we need the following lemma, which is interesting by itself.

Lemma 1.3. *Suppose that for any $\bar{\sigma} \in L_{\mathcal{F}}^2(0, T; \mathbb{R}^{\bar{k}})$ and any $g \in L_{\mathcal{F}_T}^2(\Omega; \mathbb{R}^k)$, there exist $h \in L_{\mathcal{F}}^2(0, T; \mathbb{R}^m)$ and $f \in L_{\mathcal{F}}^2(\Omega; C([0, T]; \mathbb{R}^m))$, such that the following BSDE admits an adapted solution $(\overline{Y}, Z) \in L_{\mathcal{F}}^2(\Omega; C([0, T]; \mathbb{R}^m)) \times L_{\mathcal{F}}^2(0, T; \mathbb{R}^\ell)$:*

(1.6)
$$\begin{cases}
d\overline{Y}(t) = h(t)dt + [f(t) + \overline{C}_1 Z(t) + \overline{D}\bar{\sigma}(t)]dW(t), \quad t \in [0, T], \\
\overline{Y}(T) = Fg.
\end{cases}$$

where $\overline{C}_1 \in \mathbb{R}^{m \times \ell}$ and $\overline{D} \in \mathbb{R}^{m \times \bar{k}}$. Then,

(1.7) $$\mathcal{R}(\overline{C}_1) \supseteq \mathcal{R}(F) + \mathcal{R}(\overline{D}).$$

Proof. We prove our lemma by contradiction. Suppose (1.7) does not hold. Then we can find an $\eta \in \mathbb{R}^m$ such that

(1.8) $$\eta^T \overline{C}_1 = 0, \qquad \text{but } \eta^T F \neq 0, \quad \text{or} \quad \eta^T \overline{D} \neq 0.$$

Let $\zeta(t) = \eta^T \overline{Y}(t)$. Then $\zeta(\cdot)$ satisfies

(1.9)
$$\begin{cases} d\zeta(t) = \bar{h}(t)dt + [\bar{f}(t) + \eta^T \overline{D}\bar{\sigma}(t)]dW(t), \\ \zeta(T) = \eta^T Fg, \end{cases}$$

where $\bar{h}(t) = \eta^T h(t)$, $\bar{f}(t) = \eta^T f(t)$. We claim that for some choice of g and $\bar{\sigma}(\cdot)$, (1.9) does not admit an adapted solution $\zeta(\cdot)$ for any $\bar{h} \in L^2_{\mathcal{F}}(0, T; \mathbb{R})$ and $\bar{f} \in L^2_{\mathcal{F}}(\Omega; C([0, T]; \mathbb{R}))$. To show this, we construct a deterministic Lebesgue measurable function β satisfying the following:

(1.10)
$$\begin{cases} \beta(s) = \pm 1, \qquad \forall s \in [0, T], \\ |\{s \in [T_i, T] \mid \beta(s) = 1\}| = \dfrac{T - T_i}{2}, \\ |\{s \in [T_i, T] \mid \beta(s) = -1\}| = \dfrac{T - T_i}{2}, \end{cases} \qquad i \geq 1,$$

for a sequence $T_i \uparrow T$, where $|\{\cdots\}|$ stands for the Lebesgue measure of $\{\cdots\}$. Such a function exists by some elementary construction. Now, we separate two cases.

   Case 1. $\eta^T F \neq 0$. We may assume that $|F^T \eta| = 1$.
   Let us choose

(1.11)
$$g = \left(\int_0^T \beta(s)dW(s) \right) F^T \eta, \qquad \bar{\sigma}(t) \equiv 0.$$

Then, by defining

(1.12)
$$\hat{\zeta}(t) = \left(\int_0^t \beta(s)dW(s) \right), \qquad t \in [0, T],$$

we have

(1.13)
$$\begin{cases} d[\zeta(t) - \hat{\zeta}(t)] = \bar{h}(t)dt + [\bar{f}(t) - \beta(t)]dW(t), \quad t \in [0, T], \\ \zeta(T) - \hat{\zeta}(T) = 0. \end{cases}$$

Applying Itô's formula to $|\zeta(t) - \hat{\zeta}(t)|^2$, we obtain

$$E|\zeta(t) - \hat{\zeta}(t)|^2 + E \int_t^T |\bar{f}(s) - \beta(s)|^2 ds$$

$$= -2E \int_t^T \langle \zeta(s) - \hat{\zeta}(s), \bar{h}(s) \rangle \, ds$$

(1.14)
$$= 2E \int_t^T \langle \int_s^T \bar{h}(r)dr + \int_s^T [\bar{f}(r) - \beta(r)]dW(r), \bar{h}(s) \rangle \, ds$$

$$= 2E \int_t^T \langle \int_s^T \bar{h}(r)dr, \bar{h}(s) \rangle \, ds$$

$$= E\left| \int_t^T \bar{h}(s)ds \right|^2 \leq (T - t) \int_t^T E|\bar{h}(s)|^2 ds.$$

Consequently, (note $\bar{h} \in L^2_{\mathcal{F}}(0, T; \mathbb{R})$ and $\bar{f} \in L^2_{\mathcal{F}}(\Omega; C([0, T]; \mathbb{R})))$

$$E \int_t^T |\bar{f}(T) - \beta(s)|^2 ds$$

(1.15)
$$\leq 2E \int_t^T |\bar{f}(s) - \beta(s)|^2 ds + 2E \int_t^T |\bar{f}(T) - \bar{f}(s)|^2 ds$$

$$\leq 2(T - t) \int_t^T E|\bar{h}(s)|^2 ds + 2E \int_t^T |\bar{f}(T) - \bar{f}(s)|^2 ds$$

$$= o(T - t).$$

On the other hand, by the definition of $\beta(\cdot)$, we have

(1.16)
$$E \int_{T_i}^T |\bar{f}(T) - \beta(s)|^2 ds$$

$$= \frac{T - T_i}{2} \Big(E|\bar{f}(T) - 1|^2 + E|\bar{f}(T) + 1|^2 \Big), \quad \forall i \geq 1.$$

Clearly, (1.16) contradicts (1.15), which means $\eta^T F \neq 0$ is not possible.

Case 2. $\eta^T F = 0$ and $\eta^T \overline{D} \neq 0$. We may assume that $|\overline{D}^T \eta| = 1$.

In this case, we choose $\bar{\sigma}(t) = \beta(t)\overline{D}^T \eta$ with $\beta(\cdot)$ satisfying (1.10). Thus, (1.9) becomes

(1.17)
$$\begin{cases} d\zeta(t) = \bar{h}(t)dt + [\bar{f}(t) + \beta(t)]dW(t), & t \in [0, T], \\ \zeta(T) = 0. \end{cases}$$

Then the argument used in Case 1 applies. Hence, $\eta^T \overline{D} \neq 0$ is impossible either, proving (1.7). $\qquad \qquad \qquad \qquad \qquad \qquad \qquad \qquad \qquad \qquad \qquad \Box$

Proof of Theorem 1.2. Let $(X, Y, Z) \in \mathcal{M}[0, T]$ be an adapted solution of (1.1). Set $\overline{Y}(t) = Y(t) - GX(t)$. Then $\overline{Y}(\cdot)$ satisfies the following BSDE:

(1.18)
$$\begin{cases} d\overline{Y} = \{(\widehat{A} - GA)X + (\widehat{B} - GB)Y \\ \qquad\quad + (\widehat{C} - GC)Z + \widehat{D}b - GDb\}dt \\ \qquad\quad + \{(\widehat{A}_1 - GA_1)X + (\widehat{B}_1 - GB_1)Y \\ \qquad\qquad + (\widehat{C}_1 - GC_1)Z + \widehat{D}_1\widehat{\sigma} - GD_1\sigma\}dW(t), \\ \overline{Y}(T) = Fg. \end{cases}$$

Denote

(1.19)
$$\begin{cases} h = (\widehat{A} - GA)X + (\widehat{B} - GB)Y + (\widehat{C} - GC)Z + \widehat{D}b - GDb, \\ f = (\widehat{A}_1 - GA_1)X + (\widehat{B}_1 - GB_1)Y. \end{cases}$$

We see that $h \in L^2_{\mathcal{F}}(0, T; \mathbb{R}^m)$ and $f \in L^2_{\mathcal{F}}(\Omega; C([0, T]; \mathbb{R}^m))$. One can rewrite (1.18) as follows:

(1.20)
$$\begin{cases} d\overline{Y} = hdt + \{f + (\widehat{C}_1 - GC_1)Z + \widehat{D}_1\widehat{\sigma} - GD_1\sigma\}dW(t), \\ \overline{Y}(T) = Fg. \end{cases}$$

Then, by Lemma 1.3, we obtain (1.4). The final conclusion is obvious.

□

To conclude this section, let us present the following further result, which might be less useful than Theorem 1.2, but still interesting.

Proposition 1.4. *Suppose that the assumption of Theorem 1.2 holds. For any b, σ, $\widehat{b}$, $\widehat{\sigma}$, g and x satisfying (1.3), let $(X, Y, Z) \in \mathcal{M}[0, T]$ be an adapted solution of (1.1). Then it holds*

(1.21)
$$
\begin{aligned}
[\widehat{A}_1 - GA_1 + (\widehat{B}_1 - GB_1)G]X(T) \\
+ (\widehat{B}_1 - GB_1)Fg \in \mathcal{R}(\widehat{C}_1 - GC_1), \qquad \text{a.s.}
\end{aligned}
$$

If, in addition, the following holds:

(1.22)
$$
\begin{cases}
\mathcal{R}(A + BG) + \mathcal{R}(BF) \subseteq \mathcal{R}(D), \\
\mathcal{R}(A_1 + B_1 G) + \mathcal{R}(B_1 F) \subseteq \mathcal{R}(D_1), \\
\mathcal{R}(\widehat{A} + \widehat{B}G) + \mathcal{R}(\widehat{B}F) \subseteq \mathcal{R}(\widehat{D}), \\
\mathcal{R}(\widehat{A}_1 + \widehat{B}_1 G) + \mathcal{R}(\widehat{B}_1 F) \subseteq \mathcal{R}(\widehat{D}_1),
\end{cases}
$$

then

(1.23)
$$
\begin{aligned}
\mathcal{R}\Big(\widehat{A}_1 - GA_1 + (\widehat{B}_1 - GB_1)G\Big) + \mathcal{R}\Big((\widehat{B}_1 - GB_1)F\Big) \\
\subseteq \mathcal{R}(\widehat{C}_1 - GC_1).
\end{aligned}
$$

Proof. Suppose $\eta \in \mathbb{R}^m$ such that

(1.24)
$$
\eta^T(\widehat{C}_1 - GC_1) = 0.
$$

Then, by (1.4), one has

(1.25)
$$
\eta^T F = 0, \quad \eta^T \widehat{D}_1 = 0, \quad \eta^T GD_1 = 0.
$$

Hence, from (1.20), we obtain

(1.26)
$$
\begin{cases}
d[\eta^T \overline{Y}(t)] = \eta^T h(t)dt + \eta^T f(t)dW(t), \quad t \in [0, T], \\
\eta^T \overline{Y}(T) = 0.
\end{cases}
$$

Applying Itô's formula to $|\eta^T \overline{Y}(t)|^2$, we have (similar to (1.14))

(1.27)
$$
\begin{aligned}
E|\eta^T \overline{Y}(t)|^2 + E \int_t^T |\eta^T f(s)|^2 ds &= -2E \int_t^T \eta^T \overline{Y}(s)\eta^T h(s)ds \\
&= 2E \int_t^T \Big[\int_s^T \eta^T h(r)dr + \int_s^T \eta^T f(r)dW(r)\Big]\eta^T h(s)ds \\
&= E\Big|\int_t^T \eta^T h(s)ds\Big|^2 \le (T - t) \int_t^T E|\eta^T h(s)|^2 ds.
\end{aligned}
$$

Dropping the first term on the left side of (1.27), then dividing both sides by $T - t$ and sending $t \to T$, we obtain

$$(1.28) \qquad\qquad\qquad E|\eta^T f(T)|^2 = 0.$$

By (1.19), and the relation $Y(T) = GX(T) + Fg$, we obtain

$$(1.29) \quad \eta^T[\widehat{A}_1 - GA_1 + (\widehat{B}_1 - GB_1)G]X(T) + \eta^T(\widehat{B}_1 - GB_1)Fg = 0, \quad \text{a.s.}$$

Thus, (1.21) follows. In the case (1.22) holds, for any $x \in \mathbb{R}^n$ and $g \in \mathbb{R}^m$ (deterministic), by some choice of b, σ, $\widehat{b}$ and $\widehat{\sigma}$, (1.1) admits an adapted solution $(X, Y, Z) \equiv (x, Gx + Fg, 0)$. Then, (1.21) implies (1.23). □

§2. Some Reductions

In this section, we are going to make some reductions under condition (1.5). We note that (1.5) is very general. It is true if, for example, $F = I \in \mathbb{R}^{m \times m}$, which is the case in many applications. Now, we assume (1.5). By Theorem 1.2, if we want (1.1) to be solvable for all given data, we must have $\widehat{C}_1 - GC_1$ to be onto (and thus $\ell \geq m$). Thus, it is reasonable to make the following assumption:

Assumption A. Let $\ell = m$ and $\widehat{C}_1 - GC_1 \in \mathbb{R}^{m \times m}$ be invertible.

Let us make some reductions under Assumption A. Set $\overline{Y} = Y - GX$. Then $\overline{Y}(T) = Fg$ and (see (1.18))

$$
\begin{aligned}
d\overline{Y} &= (\widehat{A}X + \widehat{B}Y + \widehat{C}Z + \widehat{D}b)dt \\
&\quad + (\widehat{A}_1 X + \widehat{B}_1 Y + \widehat{C}_1 Z + \widehat{D}_1 \widehat{\sigma})dW \\
&\quad - G(AX + BY + CZ + Db)dt \\
&\quad - G(A_1 X + B_1 Y + C_1 Z + D_1 \sigma)dW \\
(2.1) \qquad &= \Big\{ [\widehat{A} - GA + (\widehat{B} - GB)G]X + (\widehat{B} - GB)\overline{Y} \\
&\qquad + (\widehat{C} - GC)Z + \widehat{D}b - GDb \Big\}dt \\
&\quad + \Big\{ [\widehat{A}_1 - GA_1 + (\widehat{B}_1 - GB_1)G]X + (\widehat{B}_1 - GB_1)\overline{Y} \\
&\qquad + (\widehat{C}_1 - GC_1)Z + \widehat{D}_1 \widehat{\sigma} - GD_1 \sigma \Big\}dW.
\end{aligned}
$$

Define

$$
\begin{aligned}
(2.2) \qquad \overline{Z} &= [\widehat{A}_1 - GA_1 + (\widehat{B}_1 - GB_1)G]X + (\widehat{B}_1 - GB_1)\overline{Y} \\
&\quad + (\widehat{C}_1 - GC_1)Z + \widehat{D}_1 \widehat{\sigma} - GD_1 \sigma.
\end{aligned}
$$

Since $(\widehat{C}_1 - GC_1)$ is invertible, we have

$$
\begin{aligned}
(2.3) \qquad Z &= (\widehat{C}_1 - GC_1)^{-1}\{\overline{Z} - [\widehat{A}_1 - GA_1 + (\widehat{B}_1 - GB_1)G]X \\
&\quad - (\widehat{B}_1 - GB_1)\overline{Y} - (\widehat{D}_1 \widehat{\sigma} - GD_1 \sigma)\}.
\end{aligned}
$$

Then, it follows that

(2.4)
$$
\begin{cases}
dX = \left(\overline{A}X + \overline{B}\,\overline{Y} + \overline{C}\,\overline{Z} + \overline{b}\right)dt \\
\qquad\quad + \left(\overline{A}_1 X + \overline{B}_1 \overline{Y} + \overline{C}_1 \overline{Z} + \overline{\sigma}\right)dW, \\
d\overline{Y} = \left(\overline{A}_0 X + \overline{B}_0 \overline{Y} + \overline{C}_0 \overline{Z} + \overline{h}\right)dt + \overline{Z}dW, \\
X(0) = x, \qquad \overline{Y}(T) = Fg,
\end{cases}
$$

where

(2.5)
$$
\begin{cases}
\overline{A} = A + BG - C(\widehat{C}_1 - GC_1)^{-1}[\widehat{A}_1 - GA_1 + (\widehat{B}_1 - GB_1)G], \\
\overline{B} = B - C(\widehat{C}_1 - GC_1)^{-1}(\widehat{B}_1 - GB_1), \\
\overline{C} = C(\widehat{C}_1 - GC_1)^{-1}, \\
\overline{b} = Db - C(\widehat{C}_1 - GC_1)^{-1}(\widehat{D}_1\widehat{\sigma} - GD_1\sigma), \\
\overline{A}_1 = A_1 + B_1 G \\
\qquad\quad - C_1(\widehat{C}_1 - GC_1)^{-1}[\widehat{A}_1 - GA_1 + (\widehat{B}_1 - GB_1)G], \\
\overline{B}_1 = B_1 - C_1(\widehat{C}_1 - GC_1)^{-1}(\widehat{B}_1 - GB_1), \\
\overline{C}_1 = C_1(\widehat{C}_1 - GC_1)^{-1}, \\
\overline{\sigma} = D_1\sigma - C_1(\widehat{C}_1 - GC_1)^{-1}(\widehat{D}_1\widehat{\sigma} - GD_1\sigma), \\
\overline{A}_0 = \widehat{A} - GA + (\widehat{B} - GB)G \\
\qquad\quad - (\widehat{C} - GC)(\widehat{C}_1 - GC_1)^{-1}[\widehat{A}_1 - GA_1 + (\widehat{B}_1 - GB_1)G], \\
\overline{B}_0 = \widehat{B} - GB - (\widehat{C} - GC)(\widehat{C}_1 - GC_1)^{-1}(\widehat{B}_1 - GB_1), \\
\overline{C}_0 = (\widehat{C} - GC)(\widehat{C}_1 - GC_1)^{-1}, \\
\overline{h} = \widehat{D}b - GDb - (\widehat{C} - GC)(\widehat{C}_1 - GC_1)^{-1}(\widehat{D}_1\widehat{\sigma} - GD_1\sigma).
\end{cases}
$$

The above tells us that under Assumption A, (1.1) and (2.4) are equivalent. Next, we want to make a further reduction. To this end, let us denote

(2.6)
$$
\begin{cases}
\overline{\mathcal{A}} = \begin{pmatrix} \overline{A} & \overline{B} \\ \overline{A}_0 & \overline{B}_0 \end{pmatrix}, \quad \overline{\mathcal{C}} = \begin{pmatrix} \overline{C} \\ \overline{C}_0 \end{pmatrix}, \\
\overline{\mathcal{A}}_1 = \begin{pmatrix} \overline{A}_1 & \overline{B}_1 \\ 0 & 0 \end{pmatrix}, \quad \overline{\mathcal{C}}_1 = \begin{pmatrix} \overline{C}_1 \\ I \end{pmatrix}.
\end{cases}
$$

Let $\Psi(\cdot)$ be the solution of the following:

(2.7)
$$
\begin{cases}
d\Psi(t) = \overline{\mathcal{A}}\Psi(t)dt + \overline{\mathcal{A}}_1\Psi(t)dW(t), \qquad t \geq 0, \\
\Psi(0) = I.
\end{cases}
$$

Then (2.4) is equivalent to the following: For some $y \in \mathbb{R}^m$,

$$
(2.8) \quad
\begin{aligned}
\begin{pmatrix} X(t) \\ \overline{Y}(t) \end{pmatrix} = {} & \Psi(t) \begin{pmatrix} x \\ y \end{pmatrix} + \Psi(t) \int_0^t \Psi(s)^{-1} \Big[(\overline{C} - \overline{\mathcal{A}}_1 \overline{C}_1) \overline{Z}(s) \\
& + \begin{pmatrix} \overline{b}(s) \\ \overline{h}(s) \end{pmatrix} - \overline{\mathcal{A}}_1 \begin{pmatrix} \overline{\sigma}(s) \\ 0 \end{pmatrix} \Big] ds \\
& + \Psi(t) \int_0^t \Psi(s)^{-1} \Big[\overline{C}_1 \overline{Z}(s) + \begin{pmatrix} \overline{\sigma}(s) \\ 0 \end{pmatrix} \Big] dW(s), \\
& \hspace{3cm} t \in [0, T],
\end{aligned}
$$

with the property that

$$
(2.9) \quad
\begin{aligned}
Fg = {} & (0, I)\Psi(T) \begin{pmatrix} x \\ y \end{pmatrix} + (0, I)\Psi(T) \int_0^T \Psi(s)^{-1} \Big[(\overline{C} - \overline{\mathcal{A}}_1 \overline{C}_1) \overline{Z}(s) \\
& \hspace{1cm} + \begin{pmatrix} \overline{b}(s) \\ \overline{h}(s) \end{pmatrix} - \overline{\mathcal{A}}_1 \begin{pmatrix} \overline{\sigma}(s) \\ 0 \end{pmatrix} \Big] ds \\
& + (0, I)\Psi(T) \int_0^T \Psi(s)^{-1} \Big[\overline{C}_1 \overline{Z}(s) + \begin{pmatrix} \overline{\sigma}(s) \\ 0 \end{pmatrix} \Big] dW(s).
\end{aligned}
$$

Clearly, (2.9) is equivalent to the following: For some $y \in \mathbb{R}^m$ and $\overline{Z}(\cdot) \in L^2_{\mathcal{F}}(0, T; \mathbb{R}^m)$, it holds

$$
(2.10) \quad
\begin{aligned}
\eta \overset{\Delta}{=} {} & Fg - (0, I)\Psi(T) \begin{pmatrix} x \\ 0 \end{pmatrix} \\
& - (0, I)\Psi(T) \int_0^T \Psi(s)^{-1} \Big[\begin{pmatrix} \overline{b}(s) \\ \overline{h}(s) \end{pmatrix} ds - \overline{\mathcal{A}}_1 \begin{pmatrix} \overline{\sigma}(s) \\ 0 \end{pmatrix} \Big] \\
& - (0, I)\Psi(T) \int_0^T \Psi(s)^{-1} \begin{pmatrix} \overline{\sigma}(s) \\ 0 \end{pmatrix} dW(s) \\
= {} & (0, I)\Psi(T) \begin{pmatrix} 0 \\ y \end{pmatrix} + (0, I)\Psi(T) \int_0^T \Psi(s)^{-1} (\overline{C} - \overline{\mathcal{A}}_1 \overline{C}_1) \overline{Z}(s) ds \\
& + (0, I)\Psi(T) \int_0^T \Psi(s)^{-1} \overline{C}_1 \overline{Z}(s) dW(s).
\end{aligned}
$$

Thus, if we can solve the following:

$$
(2.11) \quad
\begin{cases}
d \begin{pmatrix} \widetilde{X} \\ \widetilde{Y} \end{pmatrix} = \Big\{ \overline{\mathcal{A}} \begin{pmatrix} \widetilde{X} \\ \widetilde{Y} \end{pmatrix} + \overline{C} \widetilde{Z} \Big\} dt + \Big\{ \overline{\mathcal{A}}_1 \begin{pmatrix} \widetilde{X} \\ \widetilde{Y} \end{pmatrix} + \overline{C}_1 \widetilde{Z} \Big\} dW, \\
\widetilde{X}(0) = 0, \qquad \widetilde{Y}(T) = \eta,
\end{cases}
$$

with η being given by (2.10), then for such a pair $y \equiv \widetilde{Y}(0)$ and $\overline{Z}(\cdot) \equiv \widetilde{Z}(\cdot)$, by setting $(X, \overline{Y})$ as (2.8), we obtain an adapted solution $(X, \overline{Y}, \overline{Z}) \in \mathcal{M}[0, T]$ of (2.4). The above procedure is reversible. Thus, by the equivalence between (2.4) and (1.1), we actually have the equivalence between the solvability of (1.1) and (2.11). Let us state this result as follows.

Theorem 2.1. *Let $F = I \in \mathbb{R}^{m \times m}$ and $\ell = m$. Then (1.1) is solvable for all b, σ, $\hat{b}$, $\hat{\sigma}$, x and g satisfying (1.3) if and only if (2.11) is solvable for all $\eta \in L^2_{\mathcal{F}_T}(\Omega; \mathbb{R}^m)$.*

We note that by Theorem 1.2, $F = I$ and $\ell = m$ imply Assumption A. Based on the above reduction, in what follows, we concentrate on the following FBSDE:

$$(2.12) \qquad \begin{cases} dX = \big(AX + BY + CZ\big)dt \\ \qquad\quad + \big(A_1 X + B_1 Y + C_1 Z\big)dW(t), \qquad t \in [0, T], \\ dY = \big(\hat{A}X + \hat{B}Y + \hat{C}Z\big)dt + ZdW(t), \\ X(0) = 0, \qquad Y(T) = g. \end{cases}$$

By denoting

$$(2.13) \qquad \begin{cases} \mathcal{A} = \begin{pmatrix} A & B \\ \hat{A} & \hat{B} \end{pmatrix}, \quad \mathcal{C} = \begin{pmatrix} C \\ \hat{C} \end{pmatrix}, \\ \mathcal{A}_1 = \begin{pmatrix} A_1 & B_1 \\ 0 & 0 \end{pmatrix}, \quad \mathcal{C}_1 = \begin{pmatrix} C_1 \\ I \end{pmatrix}, \end{cases}$$

we can write (2.12) as follows:

$$(2.14) \qquad \begin{cases} d\begin{pmatrix} X \\ Y \end{pmatrix} = \Big\{\mathcal{A}\begin{pmatrix} X \\ Y \end{pmatrix} + \mathcal{C}Z\Big\}dt + \Big\{\mathcal{A}_1\begin{pmatrix} X \\ Y \end{pmatrix} + \mathcal{C}_1 Z\Big\}dW, \\ X(0) = 0, \qquad Y(T) = \eta. \end{cases}$$

In what follows, we will not distinguish (2.12) and (2.14), and we will let

$$(2.15) \qquad \begin{cases} d\Phi(t) = \mathcal{A}\Phi(t)dt + \mathcal{A}_1\Phi(t)dW(t), \qquad t \in [0, T], \\ \Phi(0) = I. \end{cases}$$

If we call (X, Y) the *state* and Z the *control*, (2.12) is called a (linear) *stochastic control system*. Then, the solvability of (2.12) becomes the following *controllability* problem: For give $g \in L^2_{\mathcal{F}_T}(\Omega; \mathbb{R}^m)$, find a control $Z \in L^2_{\mathcal{F}}(0, T; \mathbb{R}^m)$, such that some initial state $(X(0), Y(0)) \in \{0\} \times \mathbb{R}^m$ can be steered to the final state $(X(T), Y(T)) \in L^2_{\mathcal{F}_T}(\Omega; \mathbb{R}^n) \times \{g\}$ at the moment $t = T$, almost surely. This is referred to as the controllability of the system (2.12) from $\{0\} \times \mathbb{R}^m$ to $L^2_{\mathcal{F}_T}(\Omega; \mathbb{R}^n) \times \{g\}$. We note that g is an $\mathcal{F}_T$-measurable square integrable random vector, and we need exactly control $Y(T)$ to g.

§3. Solvability of Linear FBSDEs

In this section, we are going to present some solvability results for linear FBSDE (2.12). The basic idea is adopted from the study of controllability in control theory. For convenience, we denote hereafter in this chapter that

$H = L^2_{\mathcal{F}_T}(\Omega; \mathbb{R}^m)$ and $\mathcal{H} = L^2_{\mathcal{F}}(0, T; \mathbb{R}^m)$ (which are Hilbert spaces to which the final datum g and the process $Z(\cdot)$ belong, respectively).

§3.1. Necessary conditions

First of all, we recall that if Φ is the solution of (2.15), then, Φ^{-1} exists and it satisfies the following linear SDE:

(3.1)
$$\begin{cases} d\Phi^{-1} = -\Phi^{-1}\big[\mathcal{A} - \mathcal{A}_1^2\big]dt - \Phi^{-1}\mathcal{A}_1 dW(t), & t \geq 0, \\ \Phi^{-1}(0) = I. \end{cases}$$

Moreover, $(X, Y, Z) \in \mathcal{M}[0, T]$ is an adapted solution of (2.12) if and only if the following variation of constant formula holds:

(3.2)
$$\begin{aligned} \begin{pmatrix} X(t) \\ Y(t) \end{pmatrix} &= \Phi(t) \begin{pmatrix} 0 \\ y \end{pmatrix} + \Phi(t) \int_0^t \Phi(s)^{-1}(\mathcal{C} - \mathcal{A}_1\mathcal{C}_1)Z(s)ds \\ &\quad + \Phi(t) \int_0^t \Phi(s)^{-1}\mathcal{C}_1 Z(s)dW(s), \quad t \in [0, T], \end{aligned}$$

for some $y \in \mathbb{R}^m$ with the property:

(3.3)
$$\begin{aligned} g = (0, I)\Big\{\Phi(T) \begin{pmatrix} 0 \\ y \end{pmatrix} + \Phi(T) \int_0^T \Phi(s)^{-1}(\mathcal{C} - \mathcal{A}_1\mathcal{C}_1)Z(s)ds \\ + \Phi(T) \int_0^T \Phi(s)^{-1}\mathcal{C}_1 Z(s)dW(s)\Big\}. \end{aligned}$$

Let us introduce an operator $\mathcal{K} : \mathcal{H} \to H$ as follows:

(3.4)
$$\begin{aligned} \mathcal{K}Z = (0, I)\Big\{\Phi(T) \int_0^T \Phi(s)^{-1}(\mathcal{C} - \mathcal{A}_1\mathcal{C}_1)Z(s)ds \\ + \Phi(T) \int_0^T \Phi(s)^{-1}\mathcal{C}_1 Z(s)dW(s)\Big\}. \end{aligned}$$

Then, for given $g \in H$, finding an adapted solution to (2.12) is equivalent to the following: Find $y \in \mathbb{R}^m$ and $Z \in \mathcal{H}$, such that

(3.5)
$$g = (0, I)\Phi(T) \begin{pmatrix} 0 \\ I \end{pmatrix} y + \mathcal{K}Z,$$

and define (X, Y) by (3.2). Then $(X, Y, Z) \in \mathcal{M}[0, T]$ is an adapted solution of (2.12). Hence, the study of operators $\Phi(T)$ and $\mathcal{K}$ is crucial to the solvability of linear FBSDE (2.12). We now make some investigations on $\Phi(\cdot)$ and $\mathcal{K}$. Let us first give the following lemma.

Lemma 3.1. *For any* $f \in L^1_{\mathcal{F}}(0, T; \mathbb{R}^{n+m})$ *and* $h \in L^2_{\mathcal{F}}(0, T; \mathbb{R}^{n+m})$, *it* *holds*

$$(3.6) \quad \begin{cases} E\Phi(t) = e^{\mathcal{A}t}, \\ E\left\{ \Phi(t) \int_0^t \Phi(s)^{-1} f(s) ds \right\} = \int_0^t e^{\mathcal{A}(t-s)} E f(s) ds, \\ E\left\{ \Phi(t) \int_0^t \Phi(s)^{-1} h(s) dW(s) \right\} \\ \qquad\qquad = \int_0^t e^{\mathcal{A}(t-s)} \mathcal{A}_1 E h(s) ds, \end{cases} \quad t \in [0, T].$$

Also, it holds that

$$(3.7) \quad E \sup_{0 \le t \le T} |\Phi(t)|^{2k} + E \sup_{0 \le t \le T} |\Phi(t)^{-1}|^{2k} < \infty, \qquad \forall k \ge 1.$$

Proof. Let us first prove the second equality in (3.6). The other two in (3.6) can be proved similarly. Set

$$(3.8) \quad \xi(t) = \Phi(t) \int_0^t \Phi(s)^{-1} f(s) ds, \qquad t \in [0, T].$$

Then $\xi(\cdot)$ satisfies the following SDE:

$$(3.9) \quad \begin{cases} d\xi(t) = [\mathcal{A}\xi(t) + f(t)]dt + \mathcal{A}_1 \xi(t) dW(t), & t \in [0, T], \\ \xi(0) = 0. \end{cases}$$

Taking expectation in (3.9), we obtain

$$(3.10) \quad \begin{cases} d[E\xi(t)] = [\mathcal{A}E\xi(t) + Ef(t)]dt, & t \in [0, T], \\ E\xi(0) = 0. \end{cases}$$

Thus,

$$(3.11) \quad E\xi(t) = \int_0^t e^{\mathcal{A}(t-s)} E f(s) ds, \qquad t \in [0, T],$$

proving our claim.

Now, we prove (3.7). For any $\xi_0 \in \mathbb{R}^{n+m}$, process $\xi(t) \overset{\Delta}{=} \Phi(t)\xi_0$ satisfies the following SDE:

$$(3.12) \quad \begin{cases} d\xi(t) = \mathcal{A}\xi(t)dt + \mathcal{A}_1 \xi(t) dW(t), & t \in [0, T], \\ \xi(0) = \xi_0. \end{cases}$$

Then, by Itô's formula, Burkholder-Davis-Gundy's inequality and Gronwall's inequality, we can show that

$$(3.13) \quad E \sup_{0 \le t \le T} |\xi(t)|^{2k} \le K |\xi_0|^{2k}, \qquad k \ge 1,$$

for some constant $K > 0$. Thus, the first term on the left hand side of (3.7) is finite. Similarly, one can prove that the second term is finite as well. $\qquad\qquad\qquad\qquad\qquad\qquad\qquad\qquad\qquad\qquad\qquad\qquad\qquad$ $\square$

From (3.7), we see that $\mathcal{K} : \mathcal{H} \to \mathcal{H}$ is a bounded linear operator. Now, if (2.12) admits an adapted solution, by taking expectation in (2.12), we obtain

$$(3.14) \qquad Eg = (0, I)\left\{ e^{\mathcal{A}T} \begin{pmatrix} 0 \\ I \end{pmatrix} y + \int_0^T e^{\mathcal{A}(T-s)} \mathcal{C}EZ(s)ds \right\},$$

for some $y \in \mathbb{R}^m$ and $EZ(\cdot) \in L^2(0, T; \mathbb{R}^m)$. This leads to the following necessary condition for the solvability of (2.12).

Theorem 3.2. *Suppose (2.12) is solvable for all $g \in H$. Then*

$$(3.15) \qquad \mathrm{rank}\left\{ (0, I)\left(e^{\mathcal{A}T} \begin{pmatrix} 0 \\ I \end{pmatrix}, \mathcal{C}, \mathcal{A}\mathcal{C}, \cdots, \mathcal{A}^{n+m-1}\mathcal{C} \right) \right\} = m.$$

Proof. Define

$$\mathcal{L}u(\cdot) \triangleq \int_0^T e^{\mathcal{A}(T-s)} \mathcal{C}u(s)ds, \qquad \forall u(\cdot) \in L^2(0, T; \mathbb{R}^m).$$

Then $\mathcal{L} : L^2(0, T; \mathbb{R}^m) \to \mathbb{R}^n$ is a linear bounded operator. We claim that

$$(3.16) \qquad \mathcal{R}(\mathcal{L}) = \mathcal{R}(\mathcal{C}) + \mathcal{R}(\mathcal{A}\mathcal{C}) + \cdots + \mathcal{R}(\mathcal{A}^{n+m-1}\mathcal{C}).$$

In fact, if $x \in \mathcal{R}(\mathcal{L})^\perp$, then, for any $u(\cdot) \in L^2(0, T; \mathbb{R}^m)$, it holds

$$0 = x^T \mathcal{L}u(\cdot) = \int_0^T x^T e^{\mathcal{A}(T-s)} \mathcal{C}u(s)ds,$$

which yields

$$x^T e^{\mathcal{A}s} \mathcal{C} = 0, \qquad \forall s \in [0, T].$$

Consequently,

$$x^T \mathcal{A}^k \mathcal{C} = \frac{d^k}{ds^k} \left[x^T e^{\mathcal{A}s} \mathcal{C} \right]\Big|_{s=0} = 0, \quad k \geq 0.$$

This implies that

$$x \in \left\{ \mathcal{R}(\mathcal{C}) + \mathcal{R}(\mathcal{A}\mathcal{C}) + \cdots + \mathcal{R}(\mathcal{A}^{n+m-1}\mathcal{C}) \right\}^\perp,$$

which results in

$$\mathcal{R}(\mathcal{C}) + \mathcal{R}(\mathcal{A}\mathcal{C}) + \cdots + \mathcal{R}(\mathcal{A}^{n+m-1}\mathcal{C}) \subseteq \mathcal{R}(\mathcal{L}).$$

The above proof is reversible with the add of Calay-Hamilton's theorem. Thus, we obtain the other inclusion, proving (3.16). Then (3.15) follows easily. $\qquad\qquad\qquad\qquad\qquad\qquad\qquad\qquad\qquad\qquad\qquad\qquad\qquad$ $\square$

We note that in the case $\mathcal{C} = 0$, (3.15) becomes

$$(3.17) \qquad \det \left\{ (0, I)e^{\mathcal{A}T} \begin{pmatrix} 0 \\ I \end{pmatrix} \right\} \neq 0.$$

This amounts to say that the FBSDEs (2.12) (with $\mathcal{C} = 0$) is solvable for all $g \in H$ implies that the corresponding two-point boundary value problem for ODEs:

$$(3.18) \qquad \begin{cases} \begin{pmatrix} \dot{X}(t) \\ \dot{Y}(t) \end{pmatrix} = \mathcal{A} \begin{pmatrix} X(t) \\ Y(t) \end{pmatrix}, & t \in [0, T], \\ X(0) = 0, & Y(T) = \bar{g}, \end{cases}$$

admits a solution for all $\bar{g} \in \mathbb{R}^m$.

Let us now present another necessary condition for the solvability of (2.12).

Theorem 3.3. *Let $\mathcal{C} = 0$. Suppose (2.12) is solvable for all $g \in H$. Then,*

$$(3.19) \qquad \det \left\{ (0, I)e^{\mathcal{A}t}\mathcal{C}_1 \right\} > 0, \qquad \forall t \in [0, T].$$

Consequently, if

$$(3.20) \qquad \widehat{T} = \inf\{T > 0 \mid \det \left[(0, I)e^{\mathcal{A}T}\mathcal{C}_1\right] = 0\} < \infty,$$

then, for any $T \geq \widehat{T}$, there exists a $g \in H$, such that (2.12) is not solvable.

Remark 3.4. The above result reveals a significant difference between the solvability of FBSDEs and that of two-point boundary value problems for ODEs. We note that for (3.18) to be solvable for all $\bar{g} \in \mathbb{R}^m$, if and only if (3.16) holds. Since the function $t \mapsto \det \left\{ (0, I)e^{\mathcal{A}t} \begin{pmatrix} 0 \\ I \end{pmatrix} \right\}$ is analytic (and it is equal to 1 at $t = 0$), except at most a discrete set of T's, (3.16) holds. That implies that for any $T_0 \in (0, \infty)$, if it happens that (3.18) is not solvable for $T = T_0$ with some $\bar{g} \in \mathbb{R}^m$, then, at some later time $T > T_0$, (3.18) will be solvable again for all $\bar{g} \in \mathbb{R}^m$. But, in the above FBSDEs case, if $\widehat{T} < \infty$, then for any $T \geq \widehat{T}$, we can always find a $g \in H$, such that (2.12) (with $\mathcal{C} = 0$) is not solvable. Thus, FBSDEs and the two-point boundary value problem for ODEs are significantly different as far as the solvable duration is concerned.

Proof of Theorem 3.3. Suppose there exists an $s_0 \in [0, T)$, such that

$$(3.21) \qquad \det \left\{ (0, I)e^{\mathcal{A}(T - s_0)}\mathcal{C}_1 \right\} = 0.$$

Note that $s_0 < T$ has to be true. Then there exists an $\eta \in \mathbb{R}^m$, $|\eta| = 1$, such that

$$(3.22) \qquad \eta^T (0, I)e^{\mathcal{A}(T - s_0)}\mathcal{C}_1 = 0.$$

We are going to prove that for any $\varepsilon > 0$ with $s_0 + \varepsilon < T$, there exists a $g \in L^2_{\mathcal{F}_{s_0+\varepsilon}}(\Omega; \mathbb{R}^m) \subseteq H$, such that (2.12) has no adapted solutions. To this end, we let $\beta : [0, T] \to \mathbb{R}$ be a Lebesgue measurable function such that

$$
(3.23) \quad
\begin{cases}
\beta(s) = \pm 1, \quad \forall s \in [0, s_0 + \varepsilon]; \quad \beta(s) = 0, \quad \forall s \in (s_0 + \varepsilon, T]; \\[2mm]
|\{s \in [s_0, s_k] \mid \beta(s) = 1\}| = \dfrac{s_k - s_0}{2}, \\[2mm]
\qquad\qquad\qquad\qquad\qquad\qquad\qquad\qquad\qquad k \geq 1, \\[2mm]
|\{s \in [s_0, s_k] \mid \beta(s) = -1\}| = \dfrac{s_k - s_0}{2},
\end{cases}
$$

for some sequence $s_k \downarrow s_0$ and $s_k \leq T - \varepsilon$. Next, we define

$$
(3.24) \qquad \zeta(t) = \int_0^t \beta(s) dW(s), \qquad t \in [0, T],
$$

and take $g = \zeta(T)\eta \in L^2_{\mathcal{F}_{s_0+\varepsilon}}(\Omega; \mathbb{R}^m) \subseteq H$. Suppose (2.12) admits an adapted solution $(X, Y, Z) \in \mathcal{M}[0, T]$ for this g. Then, for some $y \in \mathbb{R}^m$, we have (remember $\mathcal{C} = 0$)

$$
(3.25) \quad
\begin{aligned}
\zeta(T)\eta = (0, I)\Big\{ & e^{\mathcal{A}T} \begin{pmatrix} 0 \\ y \end{pmatrix} \\
& + \int_0^T e^{\mathcal{A}(T-s)} \Big[\mathcal{A}_1 \begin{pmatrix} X(s) \\ Y(s) \end{pmatrix} + \mathcal{C}_1 Z(s) \Big] dW(s) \Big\}.
\end{aligned}
$$

Applying η^T from left to (3.25) gives the following:

$$
(3.26) \qquad \zeta(T) = \alpha + \int_0^T \big\{ \gamma(s) + \langle \psi(s), Z(s) \rangle \big\} dW(s),
$$

where

$$
(3.27) \quad
\begin{cases}
\alpha = \eta^T (0, I) e^{\mathcal{A}T} \begin{pmatrix} 0 \\ y \end{pmatrix} \in \mathbb{R}, \\[3mm]
\gamma(\cdot) = \eta^T (0, I) e^{\mathcal{A}(T-\cdot)} \mathcal{A}_1 \begin{pmatrix} X(\cdot) \\ Y(\cdot) \end{pmatrix} \in L^2_{\mathcal{F}}(\Omega; C([0, T]; \mathbb{R})), \\[3mm]
\psi(\cdot) = \Big[\eta^T (0, I) e^{\mathcal{A}(T-\cdot)} \mathcal{C}_1 \Big]^T \text{ is analytic, } \psi(s_0) = 0.
\end{cases}
$$

Let us denote

$$
(3.28) \qquad \theta(t) = \alpha + \int_0^t [\gamma(s) + \langle \psi(s), Z(s) \rangle] dW(s), \qquad t \in [0, T].
$$

Then, it follows that

$$
(3.29) \quad
\begin{cases}
d[\theta(t) - \zeta(t)] = [\gamma(t) + \langle \psi(t), Z(t) \rangle - \beta(t)] dW(t), \quad t \in [0, T], \\
[\theta(T) - \zeta(T)] = 0.
\end{cases}
$$

By Itô's formula, we have

$$0 = E|\theta(t) - \zeta(t)|^2$$

(3.30)
$$+ E \int_t^T |\gamma(s) + \langle \psi(s), Z(s) \rangle - \beta(s)|^2 ds, \quad t \in [0, T].$$

Thus,

(3.31) $$\beta(s) - \gamma(s) = \langle \psi(s), Z(s) \rangle, \qquad \text{a.e. } s \in [0, T], \text{ a.s.}$$

which yields

(3.32) $$\int_{s_0}^{s_k} E|\beta(s) - \gamma(s)|^2 ds = \int_{s_0}^{s_k} E|\langle \psi(s), Z(s) \rangle|^2 ds, \quad \forall k \geq 1.$$

Now, we observe that (note $\gamma \in L_{\mathcal{F}}^2(\Omega; C([0, T]; \mathbb{R}))$) and (3.23))

$$\int_{s_0}^{s_k} E|\beta(s) - \gamma(s)|^2 ds$$

(3.33)
$$\geq \frac{1}{2} \int_{s_0}^{s_k} E|\beta(s) - \gamma(s_0)|^2 ds - \int_{s_0}^{s_k} E|\gamma(s) - \gamma(s_0)|^2 ds$$

$$\geq \frac{s_k - s_0}{4} E\left[|1 - \gamma(s_0)|^2 + |1 + \gamma(s_0)|^2\right] - o(s_k - s_0), \quad k \geq 1.$$

On the other hand, since $\psi(\cdot)$ is analytic with $\psi(s_0) = 0$, we must have

(3.34) $$\psi(s) = (s - s_0)\widetilde{\psi}(s), \qquad s \in [0, T],$$

for some $\widetilde{\psi}(\cdot)$ which is analytic and hence bounded on $[0, T]$. Consequently,

(3.35) $$\int_{s_0}^{s_k} E|\langle \psi(s), Z(s) \rangle|^2 ds \leq K(s_k - s_0)^2 \int_{s_0}^{s_k} E|Z(s)|^2 ds.$$

Hence, (3.32)–(3.33) and (3.35) imply

$$\frac{s_k - s_0}{4} E\left[|1 - \gamma(s_0)|^2 + |1 + \gamma(s_0)|^2\right] - o(s_k - s_0)$$

(3.36)
$$\leq K(s_k - s_0)^2 \int_{s_0}^{s_k} E|Z(s)|^2 ds, \quad \forall k \geq 1.$$

This is impossible. Finally, noting the fact that $\det\{(0, I)e^{At}\mathcal{C}_1\}\big|_{t=0} = 1$, we obtain (3.19). The final assertion is clear.   □

It is not clear if the above result holds for the case $\mathcal{C} \neq 0$ since the assumption $\mathcal{C} = 0$ is crucial in the proof.

§3.2. Criteria for solvability

Let us now present some results on the operator $\mathcal{K}$ (see (3.4) for definition) which will lead to some sufficient conditions for solvability of linear FBSDEs.

Lemma 3.5. *The range $\mathcal{R}(\mathcal{K})$ of $\mathcal{K}$ is closed in H.*

Proof. Let us denote $H_0 = L^2_{\mathcal{F}_T}(\Omega; \mathbb{R}^n)$ and $\widehat{H} = H_0 \times H \equiv L^2_{\mathcal{F}_T}(\Omega; \mathbb{R}^{n+m})$. Define

(3.37)
$$
\begin{aligned}
\widehat{\mathcal{K}}Z = {} & \Phi(T) \int_0^T \Phi(s)^{-1}(\mathcal{C} - \mathcal{A}_1\mathcal{C}_1)Z(s)ds \\
& + \Phi(T) \int_0^T \Phi(s)^{-1}\mathcal{C}_1 Z(s)dW(s), \quad Z \in \mathcal{H}.
\end{aligned}
$$

Then, by (3.7), $\widehat{\mathcal{K}}$ is a bounded linear operator and $\mathcal{K} = (0, I)\widehat{\mathcal{K}}$. We claim that the range $\mathcal{R}(\widehat{\mathcal{K}})$ of $\widehat{\mathcal{K}}$ is closed in $\widehat{H}$. To show this, let us take any convergence sequence

(3.38)
$$
\begin{pmatrix} X_k(T) \\ Y_k(T) \end{pmatrix} \equiv \widehat{\mathcal{K}}Z_k \to \zeta, \qquad \text{in } \widehat{H},
$$

where (X_k, Y_k) is the solution of the following:

(3.39)
$$
\begin{cases}
d\begin{pmatrix} X_k \\ Y_k \end{pmatrix} = \left\{\mathcal{A}\begin{pmatrix} X_k \\ Y_k \end{pmatrix} + \mathcal{C}Z_k\right\}dt + \left\{\mathcal{A}_1\begin{pmatrix} X_k \\ Y_k \end{pmatrix} + \mathcal{C}_1 Z_k\right\}dW(t), \\
\begin{pmatrix} X_k(0) \\ Y_k(0) \end{pmatrix} = 0.
\end{cases}
$$

Then, by Itô's formula, we have

(3.40)
$$
\begin{aligned}
& E\left\{|X_k(t)|^2 + |Y_k(t)|^2 + \int_t^T \left|\mathcal{A}_1\begin{pmatrix} X_k(s) \\ Y_k(s) \end{pmatrix} + \mathcal{C}_1 Z_k(s)\right|^2 ds\right\} \\
= {} & E\left\{|X_k(T)|^2 + |Y_k(T)|^2 \right. \\
& \left. - 2\int_t^T \left\langle \begin{pmatrix} X_k(s) \\ Y_k(s) \end{pmatrix}, \mathcal{A}\begin{pmatrix} X_k(s) \\ Y_k(s) \end{pmatrix} + \mathcal{C}Z_k(s) \right\rangle ds\right\}.
\end{aligned}
$$

We note that (recall $\mathcal{C}_1 = \begin{pmatrix} C_1 \\ I \end{pmatrix}$)

(3.41)
$$
\begin{aligned}
& \left|\mathcal{A}_1\begin{pmatrix} X_k \\ Y_k \end{pmatrix} + \mathcal{C}_1 Z_k\right|^2 \\
= {} & \langle (I + C_1^T C_1)Z_k, Z_k \rangle + \left|\mathcal{A}_1\begin{pmatrix} X_k \\ Y_k \end{pmatrix}\right|^2 + 2\langle C_1^T \mathcal{A}_1 \begin{pmatrix} X_k \\ Y_k \end{pmatrix}, Z_k \rangle \\
\geq {} & \frac{1}{2}|Z_k|^2 - C(|X_k|^2 + |Y_k|^2),
\end{aligned}
$$

for some constant $C > 0$. Thus, (3.40) implies

$$E\{|X_k(t)|^2 + |Y_k(t)|^2 + \int_t^T |Z_k(s)|^2 ds\}$$

(3.42)
$$\leq CE\{|X_k(T)|^2 + |Y_k(T)|^2$$

$$+ \int_t^T (|X_k(s)|^2 + |Y_k(s)|^2) ds\}, \qquad t \in [0, T].$$

Using Gronwall's inequality, we obtain

(3.43)
$$E\{|X_k(t)|^2 + |Y_k(t)|^2 + \int_t^T |Z_k(s)|^2 ds\}$$

$$\leq CE\{|X_k(T)|^2 + |Y_k(T)|^2\}, \qquad t \in [0, T].$$

From the convergence (3.38) and (3.41), we see that Z_k is bounded in $\mathcal{H}$. Thus, we may assume that $Z_k \to \tilde{Z}$ weakly in $\mathcal{H}$. Then it is easy to see that $\widehat{\mathcal{K}}\tilde{Z} = \zeta$, proving the closeness of $\mathcal{R}(\widehat{\mathcal{K}})$.

Now, $\mathcal{R}(\widehat{\mathcal{K}})$ is a Hilbert space with the induced inner product from that of $\widehat{H}$. In this space, we define an orthogonal projection $P_H : \widehat{H} \to \widehat{H}$ by the following:

(3.44)
$$P_H \begin{pmatrix} \xi \\ \eta \end{pmatrix} = \begin{pmatrix} 0 \\ \eta \end{pmatrix}, \qquad \forall \begin{pmatrix} \xi \\ \eta \end{pmatrix} \in \widehat{H} \equiv H_0 \times H.$$

Then the space

(3.45)
$$P_H(\mathcal{R}(\widehat{\mathcal{K}})) = \{0\} \times \mathcal{R}(\mathcal{K})$$

is closed in $\mathcal{R}(\widehat{\mathcal{K}})$ and so is in $\widehat{H}$. Hence, $\mathcal{R}(\mathcal{K})$ is closed in H. □

The following result gives some more information for the operator $\mathcal{K}$ when $\mathcal{C} = A_1 C_1 = 0$, which is equivalent to the conditions: $C = 0$, $\widehat{C} = 0$ and $A_1 C_1 + B_1 = 0$. Note that A_1, B_1 and C_1 are not necessarily zero.

Lemma 3.6. Let $\mathcal{C} = 0$ and let (3.19) hold. Then

(3.46)
$$\mathcal{R}(\mathcal{K}) = \{\eta \in H \mid E\eta = 0\} \overset{\Delta}{=} \mathcal{N}(E),$$

(3.47)
$$\mathcal{N}(\mathcal{K}) \overset{\Delta}{=} \{Z \in \mathcal{H} \mid \mathcal{K}Z = 0\} = \{0\}.$$

Proof. First of all, by Lemma 3.5, we see that $\mathcal{R}(\mathcal{K})$ is closed. Also, by (3.4) and Lemma 3.1, $\mathcal{R}(\mathcal{K}) \subseteq \mathcal{N}(E)$ (since $\mathcal{C} = A_1 C_1$). Thus, to show (3.46), it suffices to show that

(3.48)
$$\mathcal{N}(E) \bigcap \mathcal{R}(\mathcal{K})^\perp = \{0\}.$$

We now prove (3.48). Take $\eta \in \mathcal{N}(E)$. Suppose

(3.49)
$$0 = E\langle \eta, \mathcal{K}Z \rangle$$

$$= E\langle \eta, (0, I)\Phi(T) \int_0^T \Phi(s)^{-1} C_1 Z(s) dW(s) \rangle, \qquad \forall Z \in \mathcal{H}.$$

Denote

(3.50) $\left(\begin{array}{c}\overline{X}(t)\\ \overline{Y}(t)\end{array}\right) = \Phi(t) \int_0^t \Phi(s)^{-1} \mathcal{C}_1 Z(s) dW(s), \qquad t \in [0,T].$

Then, by $\mathcal{C} = \mathcal{A}_1 \mathcal{C}_1 = 0$, we have

(3.51)
$$\begin{cases} d\left(\dfrac{\overline{X}}{\overline{Y}}\right) = \mathcal{A}\left(\dfrac{\overline{X}}{\overline{Y}}\right) dt + \left\{\mathcal{A}_1\left(\dfrac{\overline{X}}{\overline{Y}}\right) + \mathcal{C}_1 Z\right\} dW(t), \\ \left(\dfrac{\overline{X}(0)}{\overline{Y}(0)}\right) = 0. \end{cases}$$

By Itô's formula and Gronwall's inequality, we obtain

(3.52) $E\{|\overline{X}(t)|^2 + |\overline{Y}(t)|^2\} \le K \int_0^t E|Z(s)|^2 ds, \qquad t \in [0,T].$

Also, we have

(3.53) $\left(\begin{array}{c}\overline{X}(t)\\ \overline{Y}(t)\end{array}\right) = \int_0^t e^{\mathcal{A}(t-s)}\left\{\mathcal{A}_1\left(\begin{array}{c}\overline{X}(s)\\ \overline{Y}(s)\end{array}\right) + \mathcal{C}_1 Z(s)\right\} dW(s), \quad t \in [0,T].$

Since $E\eta = 0$ and $\eta \in H$, by Martingale Representation Theorem, there exists a $\zeta \in \mathcal{H}$, such that

(3.54) $\eta = \int_0^T \zeta(s) dW(s).$

Then, from (3.49) and (3.53), we have

(3.55)
$$0 = E\langle \eta, \mathcal{K}Z \rangle = E\left\langle \eta, (0, I)\left(\begin{array}{c}\overline{X}(T)\\ \overline{Y}(T)\end{array}\right)\right\rangle$$
$$= \int_0^T E\left\langle \zeta(s), (0, I)e^{\mathcal{A}(T-s)}\left\{\mathcal{A}_1\left(\begin{array}{c}\overline{X}(s)\\ \overline{Y}(s)\end{array}\right) + \mathcal{C}_1 Z(s)\right\}\right\rangle ds.$$

This yields

(3.56)
$$\int_0^T E\left\langle \mathcal{C}_1^T e^{\mathcal{A}^T(T-s)}\left(\begin{array}{c}0\\ I\end{array}\right)\zeta(s), Z(s)\right\rangle ds$$
$$= -\int_0^T E\left\langle \mathcal{A}_1^T e^{\mathcal{A}^T(T-s)}\left(\begin{array}{c}0\\ I\end{array}\right)\zeta(s), \left(\begin{array}{c}\overline{X}(s)\\ \overline{Y}(s)\end{array}\right)\right\rangle ds.$$

By (3.49), the above holds for all $Z \in \mathcal{H}$. Now, let $0 < \delta < T$ and take

(3.57) $Z(s) = \mathcal{C}_1^T e^{\mathcal{A}^T(T-s)}\left(\begin{array}{c}0\\ I\end{array}\right)\zeta(s)\chi_{[T-\delta,T]}(s), \qquad s \in [0,T].$

Then $\overline{X}(s) = 0$, $\overline{Y}(s) = 0$ for all $s \in [0, T - \delta]$. Consequently, (3.56) and (3.52) result in

(3.58)
$$
\begin{aligned}
&\int_{T-\delta}^{T} E \left| C_1^T e^{\mathcal{A}^T (T-s)} \begin{pmatrix} 0 \\ I \end{pmatrix} \zeta(s) \right|^2 ds \\
&\leq K \int_{T-\delta}^{T} \left(E|\zeta(s)|^2 \right)^{1/2} \left(\int_{T-\delta}^{s} E|Z(r)|^2 dr \right)^{1/2} ds \\
&\leq K \int_{T-\delta}^{T} \left(E|\zeta(s)|^2 \right)^{1/2} \left(\int_{T-\delta}^{s} E|\zeta(r)|^2 dr \right)^{1/2} ds.
\end{aligned}
$$

By (3.19), we obtain

(3.59)
$$
\begin{aligned}
\int_{T-\delta}^{T} E|\zeta(s)|^2 ds &\leq K \int_{T-\delta}^{T} E \left| C_1^T e^{\mathcal{A}^T (T-s)} \begin{pmatrix} 0 \\ I \end{pmatrix} \zeta(s) \right|^2 ds \\
&\leq K \int_{T-\delta}^{T} \left(E|\zeta(s)|^2 \right)^{1/2} \left(\int_{T-\delta}^{s} E|\zeta(r)|^2 dr \right)^{1/2} ds \\
&\leq \frac{1}{2} \int_{T-\delta}^{T} E|\zeta(s)|^2 ds + K \int_{T-\delta}^{T} \int_{T-\delta}^{s} E|\zeta(r)|^2 dr ds.
\end{aligned}
$$

Thus, it follows that

(3.60)
$$
\int_{T-\delta}^{T} E|\zeta(s)|^2 ds \leq K\delta \int_{T-\delta}^{T} E|\zeta(s)|^2 ds,
$$

with $K > 0$ being an absolute constant (independent of δ). Therefore, for $\delta > 0$ small, we must have

(3.61)
$$
\zeta(s) = 0, \qquad \text{a.e. } s \in [T - \delta, T], \text{ a.s.}
$$

This together with (3.56) implies that

(3.62)
$$
\begin{aligned}
&\int_{0}^{T-\delta} E \left\langle C_1^T e^{\mathcal{A}^T (T-s)} \begin{pmatrix} 0 \\ I \end{pmatrix} \zeta(s), Z(s) \right\rangle ds \\
&= -\int_{0}^{T-\delta} E \left\langle A_1^T e^{\mathcal{A}^T (T-s)} \begin{pmatrix} 0 \\ I \end{pmatrix} \zeta(s), \begin{pmatrix} \overline{X}(s) \\ \overline{Y}(s) \end{pmatrix} \right\rangle ds.
\end{aligned}
$$

Then, thanks to (3.19), we can continue the above procedure to conclude that (3.61) holds over $[0, T]$ and hence it follows from (3.54) that $\eta = 0$. This proves (3.48).

We now prove (3.47). Suppose $\mathcal{K}Z = 0$. Again, we let $(\overline{X}(\cdot), \overline{Y}(\cdot))$ be defined by (3.50). Then, for any $\zeta \in \mathcal{H}$, by (3.53), we have

(3.63)
$$
\begin{aligned}
0 &= E \left\langle \int_0^T \zeta(s) dW(s), \mathcal{K}Z \right\rangle \\
&= E \int_0^T \left\langle \zeta(s), (0, I) e^{\mathcal{A}(T-s)} \left\{ A_1 \begin{pmatrix} \overline{X}(s) \\ \overline{Y}(s) \end{pmatrix} + C_1 Z(s) \right\} \right\rangle ds.
\end{aligned}
$$

This implies that

$$(3.64) \quad (0, I)e^{\mathcal{A}(T-s)}\left\{\mathcal{A}_1\begin{pmatrix}\overline{X}(s)\\\overline{Y}(s)\end{pmatrix} + \mathcal{C}_1 Z(s)\right\} = 0, \quad \text{a.e. } s \in [0, T], \text{ a.s.}$$

By (3.19), we easily see that

$$s \mapsto \mathcal{B}(s) \stackrel{\Delta}{=} \left\{(0, I)e^{\mathcal{A}(T-s)}\mathcal{C}_1\right\}^{-1}(0, I)e^{\mathcal{A}(T-s)}\mathcal{A}_1$$

is analytic and hence bounded over $[0, T]$. From (3.64), we obtain

$$(3.65) \qquad Z(s) = -\mathcal{B}(s)\begin{pmatrix}\overline{X}(s)\\\overline{Y}(s)\end{pmatrix}, \qquad \text{a.e. } s \in [0, T], \text{ a.s.}$$

Then, $(\overline{X}, \overline{Y})$ is the solution of

$$(3.66) \qquad \begin{cases} d\begin{pmatrix}\overline{X}\\\overline{Y}\end{pmatrix} = \mathcal{A}\begin{pmatrix}\overline{X}\\\overline{Y}\end{pmatrix}dt + [\mathcal{A}_1 - \mathcal{B}(t)]\begin{pmatrix}\overline{X}\\\overline{Y}\end{pmatrix}dW(t), \\ \begin{pmatrix}\overline{X}(0)\\\overline{Y}(0)\end{pmatrix} = 0. \end{cases}$$

Hence, we must have $(\overline{X}, \overline{Y}) = 0$, which yields $Z = 0$ due to (3.65). This proves (3.47). $\qquad\qquad\qquad\qquad\qquad\qquad\qquad\qquad\qquad\qquad\qquad\qquad\square$

A consequence of the above is the following.

Theorem 3.7. *Let* $\mathcal{C} = \mathcal{A}_1\mathcal{C}_1 = 0$. *Then, linear FBSDE (2.12) is solvable for all* $g \in H$ *if and only if (3.17) and (3.19) hold. In this case, the adapted solution to (2.12) is unique (for any given* $g \in H$).

Proof. Theorems 3.2 and 3.3 tell us that (3.17) and (3.19) are necessary. We now prove the sufficiency. First of all, for any $g \in H$, by (3.17), we can find $y \in \mathbb{R}^m$, such that (3.14) holds (note $\mathcal{C} = 0$). Then we have

$$(3.67) \qquad\qquad g - (0, I)\Phi(T)\begin{pmatrix}0\\I\end{pmatrix}y \in \mathcal{N}(E).$$

Next, by (3.46), there exists a $Z \in \mathcal{H}$, such that

$$(3.68) \qquad\qquad g - (0, I)\Phi(T)\begin{pmatrix}0\\I\end{pmatrix}y = \mathcal{K}Z.$$

For this pair $(y, Z) \in \mathbb{R}^m \times \mathcal{H}$, we define (X, Y) by (3.2). Then one can easily check that $(X, Y, Z) \in \mathcal{M}[0, T]$ is an adapted solution of (2.12). The uniqueness follows easily from (3.47) and (3.17). $\qquad\qquad\qquad\qquad\square$

The above result gives a complete solution to the solvability of linear FBSDE (2.12) with $\mathcal{C} = \mathcal{A}_1\mathcal{C}_1 = 0$. By Theorems 1.2, 2.1 and 3.7, we can obtain the solvability result for the original linear FBSDE (1.1). We omit the precise statement here.

§4. A Riccati Type Equation

In this section, we present another method. It will give a sufficient condition for the unique solvability of (2.12). We will obtain a Riccati type equation and a BSDE associated with (2.12). Let us now carry out a heuristic derivation.

Suppose $(X, Y, Z) \in \mathcal{M}[0, T]$ is an adapted solution of (2.12). We assume that X and Y are related by

$$(4.1) \qquad Y(t) = P(t)X(t) + p(t), \qquad \forall t \in [0, T], \text{ a.s.}$$

where $P : [0, T] \to \mathbb{R}^{m \times n}$ is a deterministic matrix-valued function and $p : [0, T] \times \Omega \to \mathbb{R}^m$ is an $\{\mathcal{F}_t\}_{t \geq 0}$-adapted process. We are going to derive the equations for $P(\cdot)$ and $p(\cdot)$. First of all, from (4.1) and the terminal condition in (2.12), we have

$$(4.2) \qquad g = P(T)X(T) + p(T).$$

Let us impose

$$(4.3) \qquad P(T) = 0, \qquad p(T) = g.$$

Since $g \in L^2_{\mathcal{F}_T}(\Omega; \mathbb{R}^m)$ and $p(\cdot)$ is required to be $\{\mathcal{F}_t\}_{t \geq 0}$-adapted, we should assume that $p(\cdot)$ satisfies a BSDE:

$$(4.4) \qquad \begin{cases} dp(t) = \alpha(t)dt + q(t)dW(t), & t \in [0, T], \\ p(T) = g, \end{cases}$$

with $\alpha(\cdot), q(\cdot) \in L^2_{\mathcal{F}}(0, T; \mathbb{R}^m)$ being undetermined. Next, by Itô's formula, we have (for simplicity, we suppress t below):

$$(4.5) \qquad \begin{aligned} dY &= \{\dot{P}X + P[AX + BY + CZ] + \alpha\}dt \\ &\quad + \{P[A_1X + B_1Y + C_1Z] + q\}dW \\ &= \{[\dot{P} + PA + PBP]X + PCZ + PBp + \alpha\}dt \\ &\quad + \{[PA_1 + PB_1P]X + PC_1Z + PB_1p + q\}dW, \end{aligned}$$

Now, compare (4.5) with the second equation in (2.12) (note (4.1)), we obtain that

$$(4.6) \quad [\dot{P} + PA + PBP]X + PCZ + PBp + \alpha = [\widehat{A} + \widehat{B}P]X + \widehat{C}Z + \widehat{B}p,$$

and

$$(4.7) \qquad (PA_1 + PB_1P)X + PC_1Z + PB_1p + q = Z.$$

By assuming $I - PC_1$ to be invertible, we have from (4.7) that

$$(4.8) \qquad Z = (I - PC_1)^{-1}\{(PA_1 + PB_1P)X + PB_1p + q\}.$$

Then, (4.6) can be written as

$$
\begin{aligned}
0 = \big[& \dot{P} + PA + PBP - \widehat{A} - \widehat{B}P \\
& + (PC - \widehat{C})(I - PC_1)^{-1}(PA_1 + PB_1P) \big] X \\
& + \big[PB - \widehat{B} + (PC - \widehat{C})(I - PC_1)^{-1}PB_1 \big] p \\
& + (PC - \widehat{C})(I - PC_1)^{-1} q + \alpha.
\end{aligned}
$$

(4.9)

Now, we introduce the following differential equation for $\mathbb{R}^{m \times n}$-valued function $P(\cdot)$:

(4.10)
$$
\begin{cases}
\dot{P} + PA + PBP - \widehat{A} - \widehat{B}P \\
\quad + (PC - \widehat{C})(I - PC_1)^{-1}(PA_1 + PB_1P) = 0, \quad t \in [0, T], \\
P(T) = 0.
\end{cases}
$$

We refer to (4.10) as a *Riccati type equation*. Suppose (4.10) admits a solution $P(\cdot)$ over $[0, T]$ such that

(4.11) $[I - P(t)C_1]^{-1}$ is bounded for $t \in [0, T]$.

Then, (4.9) gives

$$
\begin{aligned}
\alpha = - & \big[PB - \widehat{B} + (PC - \widehat{C})(I - PC_1)^{-1}PB_1 \big] p \\
& - (PC - \widehat{C})(I - PC_1)^{-1} q.
\end{aligned}
$$

Combining this with (4.4), we see that one should introduce the following BSDE:

(4.12)
$$
\begin{cases}
dp = - \Big\{ \big[PB - \widehat{B} + (PC - \widehat{C})(I - PC_1)^{-1}PB_1 \big] p \\
\qquad\quad + (PC - \widehat{C})(I - PC_1)^{-1} q \Big\} dt + q\, dW, \quad t \in [0, T], \\
p(T) = g.
\end{cases}
$$

When (4.10) admits a solution $P(\cdot)$ such that (4.11) holds, by Theorem 3.2 of Chapter 1, BSDE (4.12) admits a unique adapted solution $(p(\cdot), q(\cdot)) \in \mathcal{N}[0, T]$. Then we can define the following:

(4.13)
$$
\begin{cases}
\widetilde{A} = A + BP + C(I - PC_1)^{-1}(PA_1 + PB_1P), \\
\widetilde{A}_1 = A_1 + B_1P + C_1(I - PC_1)^{-1}(PA_1 + PB_1P), \\
\widetilde{b} = Bp + C(I - PC_1)^{-1}(PB_1p + q), \\
\widetilde{\sigma} = B_1p + C_1(I - PC_1)^{-1}(PB_1p + q).
\end{cases}
$$

It is clear that $\widetilde{A}$ and $\widetilde{A}_1$ are time-dependent matrix-valued functions and $\widetilde{b}$ and $\widetilde{\sigma}$ are $\{\mathcal{F}_t\}_{t \geq 0}$-adapted processes. Further, under (4.11), the following SDE admits a unique strong solution:

(4.14)
$$
\begin{cases}
dX = (\widetilde{A}X + \widetilde{b})dt + (\widetilde{A}_1X + \widetilde{\sigma})dW, \quad t \in [0, T], \\
X(0) = x.
\end{cases}
$$

The following theorem gives a representation of the adapted solution of FBSDE (2.12).

Theorem 4.1. *Let (4.10) admits a solution $P(\cdot)$ such that (4.11) holds. Then FBSDE (2.12) admits a unique adapted solution $(X, Y, Z) \in \mathcal{M}[0, T]$ which is determined by (4.14), (4.1) and (4.8).*

Proof. First of all, a direct computation shows that the process (X, Y, Z) determined by (4.14), (4.1) and (4.8) is an adapted solution of (2.12). We now prove the uniqueness. Let $(X, Y, Z) \in \mathcal{M}[0, T]$ be any adapted solution of (2.12). Set

$$(4.15) \qquad \begin{cases} \overline{Y} = PX + p, \\ \overline{Z} = (I - PC_1)^{-1}[(PA_1 + PB_1P)X + PB_1p + q], \end{cases}$$

where P and (p, q) are (adapted) solutions of (4.10) and (4.12), respectively. Denote $\widehat{Y} = Y - \overline{Y}$ and $\widehat{Z} = Z - \overline{Z}$. Then a direct computation shows that

$$(4.16) \qquad \begin{cases} d\widehat{Y} = \big[(PB - \widehat{B})\widehat{Y} \\ \qquad\quad + (PC - \widehat{C})\widehat{Z}\big]dt + \big[PB_1\widehat{Y} - (I - PC_1)\widehat{Z}\big]dW(t), \\ \widehat{Y}(T) = 0. \end{cases}$$

By (4.11), we may set

$$(4.17) \qquad \widetilde{Z} = PB_1\widehat{Y} - (I - PC_1)\widehat{Z},$$

to get the following equivalent BSDE (of (4.16)):

$$(4.18) \qquad \begin{cases} d\widehat{Y} = \big\{[PB - \widehat{B} + (PC - \widehat{C})(I - PC_1)^{-1}PB_1]\widehat{Y} \\ \qquad\quad - (PC - \widehat{C})(I - PC_1)^{-1}\widetilde{Z}\big\}dt + \widetilde{Z}dW(t), \\ \widehat{Y}(T) = 0. \end{cases}$$

It is clear that such a BSDE admits a unique adapted solution $(\widehat{Y}, \widetilde{Z}) = 0$ (see Chapter 1, §3). Consequently, $\widehat{Z} = 0$. Hence, by (4.15), we obtain

$$(4.19) \qquad \begin{cases} Y = PX + p, \\ Z = (I - PC_1)^{-1}[(PA_1 + PB_1P)X + PB_1p + q], \end{cases}$$

This means that any adapted solution (X, Y, Z) of (2.12) must satisfy (4.19). Then, similar to the heuristic derivation above, we have that X has to be the solution of (4.14). Hence, we obtain the uniqueness. $\qquad\square$

The following result tells us something more.

Proposition 4.2. *Let (4.10) admits a solution $P(\cdot)$ such that (4.11) holds for $t \in [T_0, T]$ (with some $T_0 \geq 0$). Then, for any $\widetilde{T} \in [0, T - T_0]$, linear FBSDE (2.12) is uniquely solvable on $[0, \widetilde{T}]$.*

Proof. Let

(4.20) $$\widetilde{P}(t) = P(t + T - \widetilde{T}), \qquad t \in [0, \widetilde{T}].$$

Then $\widetilde{P}(\cdot)$ satisfies (4.10) with $[0, T]$ replaced by $[0, \widetilde{T}]$ and

(4.21) $$[I - \widetilde{P}(t)C_1]^{-1} \text{ is bounded for } t \in [0, \widetilde{T}].$$

Thus, Theorem 4.1 applies.   □

The above proposition tells that if (4.10) admits a solution $P(\cdot)$ satisfying (4.11), FBSDE (2.12) is uniquely solvable over any $[0, \widetilde{T}]$ (with $\widetilde{T} \leq T$). Then in the case $\mathcal{C} = \mathcal{A}_1 \mathcal{C}_1$, by Theorem 3.2, the corresponding two-point boundary value problem (3.17) of ODE over $[0, \widetilde{T}]$ admits a solution for all $g \in \mathbb{R}^m$, of which a necessary and sufficient condition is

(4.22) $$\det \left\{ (0, I) e^{\mathcal{A}t} \begin{pmatrix} 0 \\ I \end{pmatrix} \right\} > 0, \qquad \forall t \in [0, T].$$

Therefore, by Theorem 3.7, compare (4.22) and (3.17), we see that the solvability of Riccati type equation (4.10) is *only* a sufficient condition for the solvability of (2.12) (at least for the case $\mathcal{C} = \mathcal{A}_1 \mathcal{C}_1 = 0$).

In the rest of this section, we concentrate on the case $\mathcal{C} = 0$. We do not assume that $\mathcal{A}_1 \mathcal{C}_1 = 0$. In this case, (4.10) becomes

(4.23) $$\begin{cases} \dot{P} + PA + PBP - \widehat{A} - \widehat{B}P = 0, & t \in [0, T], \\ P(T) = 0, \end{cases}$$

and the BSDE (4.12) is reduced to

(4.24) $$\begin{cases} dp = [\widehat{B} - PB]p\,dt + q\,dW(t), & t \in [0, T], \\ p(T) = g. \end{cases}$$

We have seen that (4.22) is a necessary condition for (4.23) having a solution $P(\cdot)$ satisfying (4.11). The following result gives the inverse of this.

Theorem 4.3. *Let* $C = 0$, $\widehat{C} = 0$. *Let (4.22) hold. Then (4.23) admits a unique solution* $P(\cdot)$ *which has the following representation:*

(4.25) $$P(t) = -\left[(0, I) e^{\mathcal{A}(T-t)} \begin{pmatrix} 0 \\ I \end{pmatrix} \right]^{-1} (0, I) e^{\mathcal{A}(T-t)} \begin{pmatrix} I \\ 0 \end{pmatrix}, \qquad t \in [0, T].$$

Moreover, it holds

(4.26)
$$I - P(t)C_1 = \left[(0, I) e^{\mathcal{A}(T-t)} \begin{pmatrix} 0 \\ I \end{pmatrix} \right]^{-1} \left[(0, I) e^{\mathcal{A}(T-t)} \begin{pmatrix} C_1 \\ I \end{pmatrix} \right],$$
$$t \in [0, T].$$

Consequently, if in addition to (4.22), (3.19) holds, then (4.11) holds and the linear FBSDE (2.12) (with $\mathcal{C} = 0$*) is uniquely solvable with the representation given by (4.14), (4.1) and (4.8).*

Proof. Let us first check that (4.25) is a solution of (4.23). To this end, we denote

$$(4.27) \qquad \Theta(t) = (0, I)e^{\mathcal{A}(T-t)} \begin{pmatrix} 0 \\ I \end{pmatrix}, \qquad t \in [0, T].$$

Then we have (recall (2.13) for the definition of $\mathcal{A}$)

$$(4.28) \qquad \dot{\Theta}(t) = -(0, I)e^{\mathcal{A}(T-t)} \begin{pmatrix} I \\ 0 \end{pmatrix} B - \Theta(t)\widehat{B}.$$

Hence,

$$
\begin{aligned}
\dot{P} &= \Theta^{-1}\dot{\Theta}\Theta^{-1}(0, I)e^{\mathcal{A}(T-t)} \begin{pmatrix} I \\ 0 \end{pmatrix} + \Theta^{-1}(0, I)e^{\mathcal{A}(T-t)}\mathcal{A} \begin{pmatrix} I \\ 0 \end{pmatrix} \\
&= \Theta^{-1}\Big\{ -(0, I)e^{\mathcal{A}(T-t)} \begin{pmatrix} I \\ 0 \end{pmatrix} B - \Theta\widehat{B}\Big\}(-P) \\
&\qquad + \Theta^{-1}(0, I)e^{\mathcal{A}(T-t)} \begin{pmatrix} A \\ \widehat{A} \end{pmatrix} \\
&= (PB - \widehat{B})(-P) + \Theta^{-1}(0, I)e^{\mathcal{A}(T-t)} \begin{pmatrix} I \\ 0 \end{pmatrix} A + \widehat{A} \\
&= -PBP + \widehat{B}P - PA + \widehat{A}.
\end{aligned}
$$

(4.29)

Thus, $P(\cdot)$ given by (4.25) is a solution of (4.23). Uniqueness is obvious since (4.23) is a terminal value problem with the right hand side of the equation being locally Lipschitz. Finally, an easy calculation shows (4.26) holds. Then we complete the proof. □

§5. Some Extensions

In this section, we briefly look at the case with multi-dimensional Brownian motion. Let $W(t) \equiv (W^1(t), \cdots, W^d(t))$ be a d-dimensional Brownian motion defined on $(\Omega, \mathcal{F}, \{\mathcal{F}_t\}_{t\geq 0}, \mathbf{P})$ with $\{\mathcal{F}_t\}_{t\geq 0}$ being the natural filtration of $W(\cdot)$ augmented by all the $\mathbf{P}$-null sets. Similar to the case of one-dimensional Brownian motion, we may also start with the most general case, by using some necessary conditions for solvability to obtain a reduced FBSDE. For simplicity, we skip this step and directly consider the following FBSDE:

$$(5.1) \quad
\begin{cases}
dX = (AX + BY)dt \\
\qquad + \displaystyle\sum_{i=1}^{d} (A_1^i X + B_1^i Y + C_1^i Z^i)dW^i(t), \qquad t \in [0, T], \\
dY = (\widehat{A}X + \widehat{B}Y)dt + \displaystyle\sum_{i=1}^{d} Z^i dW^i(t), \\
X(0) = 0, \qquad Y(T) = g,
\end{cases}
$$

where A, B, etc. are certain matrices of proper sizes. Note that we only consider the case that Z does not appear in the drift here since we have only completely solved such a case. We keep the notation $\mathcal{A}$ as in (2.13) and let

$$(5.2) \qquad \mathcal{A}_1^i = \begin{pmatrix} A_1^i & B_1^i \\ 0 & 0 \end{pmatrix}, \quad \mathcal{C}_1^i \begin{pmatrix} C_1^i \\ I \end{pmatrix}, \qquad 1 \le i \le d.$$

If we assume $X(\cdot)$ and $Y(\cdot)$ are related by (4.1), then, we can derive a Riccati type equation, which is exactly the same as (4.23). The associated BSDE is now replaced by the following:

$$(5.3) \qquad \begin{cases} dp = [\widehat{B} - PB]pdt + \sum_{i=1}^{d} q^i dW^i(t), & t \in [0, T], \\ p(T) = g. \end{cases}$$

Also, (4.13), (4.14) and (4.8) are now replaced by the following:

$$(5.4) \qquad \begin{cases} \widetilde{A} = A + BP, \qquad \widetilde{b} = Bp, \\ \widetilde{A}_1^i = A_1^i + B_1^i P \\ \qquad\quad + C_1^i (I - PC_1^i)^{-1}(PA_1^i + PB_1^i P), \qquad 1 \le i \le d, \\ \widetilde{\sigma}^i = B_1^i p + C_1^i (I - PC_1^i)^{-1}(PB_1^i p + q^i), \end{cases}$$

$$(5.5) \qquad \begin{cases} dX = (\widetilde{A}X + \widetilde{b})dt + \sum_{i=1}^{d} (\widetilde{A}_1^i X + \widetilde{\sigma}^i)dW^i(t), & t \in [0, T], \\ X(0) = 0, \end{cases}$$

$$(5.6) \quad Z^i = (I - PC_1^i)^{-1}\big\{(PA_1^i + PB_1^i P)X + PB_1^i p + q^i\big\}, \quad 1 \le i \le d.$$

Our main result is the following.

Theorem 5.1. *Let (4.22) hold and*

$$(5.7) \qquad \det\left\{(0, I)e^{\mathcal{A}t}\mathcal{C}_1^i\right\} > 0, \qquad \forall t \in [0, T], \ 1 \le i \le d.$$

Then (4.23) admits a unique solution $P(\cdot)$ given by (4.25) such that

$$(5.8) \qquad [I - P(t)C_1^i]^{-1} \text{ is bounded for } t \in [0, T], \ 1 \le i \le d,$$

and the FBSDE (5.1) admits a unique adapted solution $(X, Y, Z) \in \mathcal{M}[0, T]$ which can be represented by (5.5), (4.1) and (5.6).

The proof can be carried out similar to the case of one-dimensional Brownian motion. We leave the proof to the interested readers.

Chapter 3

Method of Optimal Control

In this chapter, we study the solvability of the following general nonlinear FBSDE: (the same form as (3.16) in Chapter 1)

(0.1) $\begin{cases} dX(t) = b(t, X(t), Y(t), Z(t))dt + \sigma(t, X(t), Y(t), Z(t))dW(t), \\ dY(t) = h(t, X(t), Y(t), Z(t))dt + Z(t)dW(t), \qquad t \in [0, T], \\ X(0) = x, \qquad Y(T) = g(X(T)). \end{cases}$

Here, we assume that functions b, σ, h and g are all deterministic, i.e., they are not explicitly depending on $\omega \in \Omega$; and $T > 0$ is any positive number. Thus, we have an FBSDE in a (possibly large) finite time duration. As we have seen in Chapter 1, §4, under certain Lipschitz conditions, (0.1) admits a unique adapted solution $(X(\cdot), Y(\cdot), Z(\cdot)) \in \mathcal{M}[0, T]$, provided $T > 0$ is relatively small. But, for general $T > 0$, we see from Chapter 2 that even if b, σ, h and g are all affine in the variables X, Y and Z, system (0.1) is not necessarily solvable. In what follows, we are going to introduce a method using optimal control theory to study the solvability of (0.1) in any finite time duration $[0, T]$. We refer to such an approach as the *method of optimal control*.

§1. Solvability and the Associated Optimal Control Problem

§1.1. An optimal control problem

Let us make an observation on solvability of (0.1) first. Suppose $(X(\cdot), Y(\cdot), Z(\cdot)) \in \mathcal{M}[0, T]$ is an adapted solution of (0.1). By letting $y = Y(0) \in \mathbb{R}^m$, we see that $(X(\cdot), Y(\cdot))$ satisfies the following FSDE:

(1.1) $\begin{cases} dX(t) = b(t, X(t), Y(t), Z(t))dt + \sigma(t, X(t), Y(t), Z(t))dW(t), \\ dY(t) = h(t, X(t), Y(t), Z(t))dt + Z(t)dW(t), \qquad t \in [0, T], \\ X(0) = x, \qquad Y(0) = y, \end{cases}$

with $Z(\cdot) \in \mathcal{Z}[0, T] \triangleq L_{\mathcal{F}}^2(0, T; \mathbb{R}^{m \times d})$ being a suitable process. We note that y and $Z(\cdot)$ have to be chosen so that the solution $(X(\cdot), Y(\cdot))$ of (1.1) satisfies the following terminal constraint:

(1.2) $$Y(T) = g(X(T)).$$

On the other hand, if we can find an $y \in \mathbb{R}^m$ and a $Z(\cdot) \in \mathcal{Z}[0, T]$, such that (1.1) admits a strong solution $(X(\cdot), Y(\cdot))$ with the terminal condition (1.2) being satisfied, then $(X(\cdot), Y(\cdot), Z(\cdot)) \in \mathcal{M}[0, T]$ is an adapted solution of (0.1). Hence, (0.1) is solvable if and only if one can find an $y \in \mathbb{R}^m$ and a $Z(\cdot) \in \mathcal{Z}[0, T]$, such that (1.1) admits a strong solution $(X(\cdot), Y(\cdot))$ satisfying (1.2).

The above observation can be viewed in a different way using the stochastic control theory. Let us call (1.1) a *stochastic control system* with $(X(\cdot), Y(\cdot))$ being the *state process*, $Z(\cdot)$ being the *control process*, and $(x, y) \in \mathbb{R}^n \times \mathbb{R}^m$ being the *initial state*. Then the solvability of (0.1) is equivalent to the following *controllability* problem for (1.1) with the *target*:

$$(1.3) \qquad\qquad \mathcal{T} = \{(x, g(x)) \mid x \in \mathbb{R}^n\}.$$

Problem (C). For any $x \in \mathbb{R}^n$, find an $y \in \mathbb{R}^m$ and a control $Z(\cdot) \in \mathcal{Z}[0, T]$, such that

$$(1.4) \qquad\qquad (X(T), Y(T)) \in \mathcal{T}, \qquad \text{a.s.}$$

Problem (C) having a solution means that the state $(X(t), Y(t))$ of system (1.1) can be *steered* from $\{x\} \times \mathbb{R}^m$ (at time $t = 0$) to the target $\mathcal{T}$, given by (1.3), at time $t = T$, almost surely, by choosing a suitable control $Z(\cdot) \in \mathcal{Z}[0, T]$.

In the previous chapter, we have presented some results related to this aspect for linear FBSDEs. We point out that the above controllability problem is very difficult for nonlinear case. However, the above formulation leads us to considering a related *optimal control problem*, which essentially decomposes the solvability problem of the original FBSDE into several relatively easier ones; and we can treat them separately. Let us now introduce the optimal control problem associated with (0.1).

Again, we consider the stochastic control system (1.1). Let us make the following assumption:

(H1) Functions $b(t, x, y, z)$, $\sigma(t, x, y, z)$, $h(t, x, y, z)$ and $g(x)$ are continuous and there exists a constant $L > 0$, such that for $\varphi = b, \sigma, h, g$, it holds that

$$(1.5) \quad \begin{cases} |\varphi(t, x, y, z) - \varphi(t, \overline{x}, \overline{y}, \overline{z})| \leq L(|x - \overline{x}| + |y - \overline{y}| + |z - \overline{z}|), \\ |\varphi(t, 0, 0, 0)|, \ |\sigma(t, x, y, 0)| \leq L, \\ \qquad \forall t \in [0, T], \ x, \overline{x} \in \mathbb{R}^n, \ y, \overline{y} \in \mathbb{R}^m, \ z, \overline{z} \in \mathbb{R}^{m \times d}. \end{cases}$$

Under the above (H1), we see that for any $(x, y) \in \mathbb{R}^n \times \mathbb{R}^m$, and $Z(\cdot) \in \mathcal{Z}[0, T]$, (1.1) admits a unique strong solution, denoted by, $(X(\cdot), Y(\cdot)) \equiv (X(\cdot; x, y, Z(\cdot)), Y(\cdot; x, y, Z(\cdot)))$, indicating the dependence on $(x, y, Z(\cdot))$. Next, we introduce a functional (called *cost functional*). The purpose is to impose certain kind of penalty on the difference $Y(T) - g(X(T))$ being large. To this end, we define

$$(1.6) \qquad f(x, y) = \sqrt{1 + |y - g(x)|^2} - 1, \qquad \forall (x, y) \in \mathbb{R}^n \times \mathbb{R}^m.$$

Clearly, f is as smooth as g and satisfying the following:

$$(1.7) \quad \begin{cases} f(x, y) \geq 0, \qquad \forall (x, y) \in \mathbb{R}^n \times \mathbb{R}^m, \\ f(x, y) = 0, \quad \text{if and only if} \quad y = g(x). \end{cases}$$

In the case that (H1) holds, we have

(1.8)
$$|f(x,y) - f(\overline{x},\overline{y})| \leq L|x - \overline{x}| + |y - \overline{y}|,$$
$$\forall (x,y), (\overline{x},\overline{y}) \in \mathbb{R}^n \times \mathbb{R}^m.$$

Now, we define the *cost functional* as follows:

(1.9) $J(x,y;Z(\cdot)) \overset{\Delta}{=} Ef(X(T;x,y,Z(\cdot)), Y(T;x,y,Z(\cdot))).$

The following is the *optimal control problem* associated with (0.1).

Problem (OC). For any given $(x,y) \in \mathbb{R}^n \times \mathbb{R}^m$, find a $\overline{Z}(\cdot) \in \mathcal{Z}[0,T]$, such that

(1.10) $\overline{V}(x,y) \overset{\Delta}{=} \inf_{Z(\cdot) \in \mathcal{Z}[0,T]} J(x,y;Z(\cdot)) = J(x,y;\overline{Z}(\cdot)).$

Any $\overline{Z}(\cdot) \in \mathcal{Z}[0,T]$ satisfying (1.10) is call an *optimal control*, the corresponding state process

$$(\overline{X}(\cdot), \overline{Y}(\cdot)) \overset{\Delta}{=} (X(\cdot;x,y,\overline{Z}(\cdot)), Y(\cdot;x,y,\overline{Z}(\cdot)))$$

is called an *optimal state process*. Sometimes, $(\overline{X}(\cdot), \overline{Y}(\cdot), \overline{Z}(\cdot))$ is referred to as an optimal triple of *Problem(OC)*.

We have seen that the optimality in *Problem(OC)* depends on the initial state (x,y). The number $\overline{V}(x,y)$ (which depends on (x,y)) in (1.10) is called the *optimal cost function* of *Problem(OC)*. By definition, we have

(1.11) $\overline{V}(x,y) \geq 0, \qquad \forall (x,y) \in \mathbb{R}^n \times \mathbb{R}^m.$

We point out that in the associated optimal control problem, it is possible to choose some other function f having similar properties as (1.7). For definiteness and some later convenience, we choose f of form (1.6).

Next, we introduce the following:

(1.12) $\mathcal{N}(\overline{V}) \overset{\Delta}{=} \{(x,y) \in \mathbb{R}^n \times \mathbb{R}^m \mid \overline{V}(x,y) = 0\}.$

This set is called the *nodal set* of function $\overline{V}$. We have the following simple result.

Proposition 1.1. *For $x \in \mathbb{R}^n$, FBSDE (0.1) admits an adapted solution if and only if*

(1.13) $\mathcal{N}(\overline{V}) \bigcap [\{x\} \times \mathbb{R}^m] \neq \phi,$

and for some $(x,y) \in \mathcal{N}(\overline{V})$, there exists an optimal control $Z(\cdot) \in \mathcal{Z}[0,T]$, such that

(1.14) $\overline{V}(x,y) = J(x,y;Z(\cdot)) = 0.$

Proof. Let $(X(\cdot), Y(\cdot), Z(\cdot)) \in \mathcal{M}[0,T]$ be an adapted solution of (0.1). Let $y = Y(0) \in \mathbb{R}^m$. Then (1.14) holds which gives $(x,y) \in \mathcal{N}(\overline{V})$ and (1.13) follows.

Conversely, if (1.14) holds with some $(x, y) \in \mathbb{R}^n \times \mathbb{R}^m$ and $Z(\cdot) \in \mathcal{Z}[0, T]$, then $(X(\cdot), Y(\cdot), Z(\cdot)) \in \mathcal{M}[0, T]$ is an adapted solution of (0.1).

□

In light of Proposition 1.1, we propose the following procedure to solve the FBSDE (0.1):

(i) Determine the function $\overline{V}(x, y)$.

(ii) Find the nodal set $\mathcal{N}(\overline{V})$ of $\overline{V}$; and restrict $x \in \mathbb{R}^n$ to satisfy (1.13).

(iii) For given $x \in \mathbb{R}^n$ satisfying (1.13), let $y \in \mathbb{R}^m$ such that $(x, y) \in \mathcal{N}(\overline{V})$. Find an optimal control $Z(\cdot) \in \mathcal{Z}[0, T]$ of $Problem(OC)$ with the initial state (x, y). Then the optimal triple $(X(\cdot), Y(\cdot), Z(\cdot)) \in \mathcal{M}[0, T]$ is an adapted solution of (0.1).

It is clear that in the above, (i) is a PDE problem; (ii) is a minimizing problem over $\mathbb{R}^m$; and (iii) is an existence of optimal control problem. Hence, the solvability of original FBSDE (0.1) has been decomposed into the above three major steps. We shall investigate these steps separately.

§1.2. Approximate solvability

We now introduce a notion which will be useful in practice and is related to condition (1.13).

Definition 1.2. For given $x \in \mathbb{R}^n$, (0.1) is said to be *approximately solvable* if for any $\varepsilon > 0$, there exists a triple $(X_\varepsilon(\cdot), Y_\varepsilon(\cdot), Z_\varepsilon(\cdot)) \in \mathcal{M}[0, T]$, such that (0.1) is satisfied except the last (terminal) condition, which is replaced by the following:

$$(1.15) \qquad\qquad E\big|Y_\varepsilon(T) - g(X_\varepsilon(T))\big| < \varepsilon.$$

We call $(X_\varepsilon(\cdot), Y_\varepsilon(\cdot), Z_\varepsilon(\cdot))$ an *approximate adapted solution* of (0.1) *with accuracy* ε.

It is clear that for given $x \in \mathbb{R}^n$, if (0.1) is solvable, then it is approximately solvable. We should note, however, even if all the coefficients of an FBSDE are uniformly Lipschitz, one still cannot guarantee its approximate solvability. Here is a simple example.

Example 1.3. Consider the following simple FBSDE:

$$(1.16) \qquad \begin{cases} dX(t) = Y(t)dt + dW(t), \\ dY(t) = -X(t)dt + Z(t)dW(t), \\ X(0) = x, \qquad Y(T) = -X(T), \end{cases}$$

with $T = \frac{3\pi}{4}$ and $x \neq 0$. It is obvious that the coefficients of this FBSDE are all uniformly Lipschitz. However, we claim that (1.16) is not approximately solvable. To see this, note that by the variation of constants formula with

$y = Y(0)$, we have

$$\begin{pmatrix} X(t) \\ Y(t) \end{pmatrix} = \begin{pmatrix} \cos t & \sin t \\ -\sin t & \cos t \end{pmatrix} \begin{pmatrix} x \\ y \end{pmatrix}$$

(1.17)

$$+ \int_0^t \begin{pmatrix} \cos(t-s) & \sin(t-s) \\ -\sin(t-s) & \cos(t-s) \end{pmatrix} \begin{pmatrix} 1 \\ Z(s) \end{pmatrix} dW(s).$$

Plugging $t = T = \frac{3\pi}{4}$ into (1.17), we obtain that

$$X(T) + Y(T) = -\sqrt{2}x + \int_0^T \eta(s) dW(s),$$

where η is some process in $L^2_{\mathcal{F}}(0, T; \mathbb{R})$. Consequently, by Jensen's inequality we have

$$E|Y(T) - g(X(T))| = E|X(T) + Y(T)| \geq |E[X(T) + Y(T)]| = \sqrt{2}|x| > 0,$$

for all $(y, Z) \in \mathbb{R}^m \times \mathcal{Z}[0, T]$. Thus, by Definition 1.2, FBSDE (1.16) is not approximately solvable (whence not solvable). □

The following result establishes the relationship between the approximate solvability of FBSDE (0.1) and the optimal cost function of the associated control problem.

Proposition 1.4. *Let (H1) hold. For a given $x \in \mathbb{R}^n$, the FBSDE (0.1) is approximately solvable if and only if the following holds:*

(1.18) $$\inf_{y \in \mathbb{R}^m} \overline{V}(x, y) = 0.$$

Proof. We first claim that the inequality (1.15) in Definition 1.2 can be replaced by

(1.19) $$Ef(X_\varepsilon(T), Y_\varepsilon(T)) < \varepsilon.$$

Indeed, by the following elementary inequalities:

(1.20) $$\frac{r \wedge r^2}{3} \leq \sqrt{1 + r^2} - 1 \leq r, \qquad \forall r \in [0, \infty),$$

we see that if (1.15) holds, so does (1.19). Conversely, (1.20) implies

$$Ef(X_\varepsilon(T), Y_\varepsilon(T)) \geq \frac{1}{3} E\Big(|Y_\varepsilon(T) - g(X_\varepsilon(T))|^2 I_{(|Y_\varepsilon(T) - g(X_\varepsilon(T))| \leq 1)} \Big)$$

$$+ \frac{1}{3} E\Big(|Y_\varepsilon(T) - g(X_\varepsilon(T))| I_{(|Y_\varepsilon(T) - g(X_\varepsilon(T))| > 1)} \Big).$$

Consequently, we have

(1.21) $$E|Y_\varepsilon(T) - g(X_\varepsilon(T))| \leq 3Ef(X_\varepsilon(T), Y_\varepsilon(T)) + \sqrt{3Ef(X_\varepsilon(T), Y_\varepsilon(T))}.$$

Thus (1.19) implies (1.15) with ε being replaced by $\varepsilon' = 3\varepsilon + \sqrt{3\varepsilon}$. Namely, (1.18) is equivalent to the approximately solvability, by Definition 1.2 and the definition of $\overline{V}$. □

Using Proposition 1.4, we can now claim the non-approximate solvability of the FBSDE (1.16) in a different way. By a direct computation using (1.21), one shows that

$$J(x, y; Z(\cdot)) = Ef(X(T), Y(T))$$
$$\geq \frac{1}{3}\Big[\sqrt{\sqrt{2}|x| + \frac{1}{4}} - \frac{1}{2}\Big]^2 > 0, \quad \forall Z(\cdot) \in \mathcal{Z}[0, T].$$

Thus,

$$\overline{V}(x, y) \geq \frac{1}{3}\Big[\sqrt{\sqrt{2}|x| + \frac{1}{4}} - \frac{1}{2}\Big]^2 > 0,$$

violating (1.18), whence not approximately solvable.

Next, we shall relate the approximate solvability to condition (1.13). To this end, let us introduce the following supplementary assumption.

(H2) There exists a constant $L > 0$, such that for all $(t, x, y, z) \in [0, T] \times \mathbb{R}^n \times \mathbb{R}^m \times \mathbb{R}^{m \times d}$, one of the following holds:

(1.22)
$$\begin{cases} |b(t, x, y, z)| + |\sigma(t, x, y, z)| \leq L(1 + |x|), \\ \langle h(t, x, y, z), y \rangle \geq -L(1 + |x|\,|y| + |y|^2), \end{cases}$$

(1.23)
$$\begin{cases} \langle h(t, x, y, z), y \rangle \geq -L(1 + |y|^2), \\ |g(x)| \leq L. \end{cases}$$

Proposition 1.5. *Let (H1) hold. Then (1.13) implies (1.18); conversely, if $\overline{V}(x, \cdot)$ is continuous, and (H2) holds, then (1.18) implies (1.13).*

Proof. That condition (1.13) implies (1.18) is obvious. We need only prove the converse. Let us first assume that $\overline{V}$ is continuous and (1.22) holds.

Since (1.18) implies the approximately solvability of (0.1), for every $\varepsilon \in (0, 1]$, we may let $(X_\varepsilon, Y_\varepsilon, Z_\varepsilon) \in \mathcal{M}[0, T]$ be an approximate adapted solution of (0.1) with accuracy ε. Some standard arguments using Itô's formula, Gronwall's inequality, and condition (1.22) will yield the following estimate

(1.24) $E|X_\varepsilon(t)|^2 \leq C(1 + |x|^2), \qquad \forall t \in [0, T], \ \varepsilon \in (0, 1].$

Here and in what follows, the constant $C > 0$ will be a generic one, depending only on L and T, and may change from line to line. By (1.24) and (1.15), we obtain

(1.25)
$$E|Y_\varepsilon(T)| \leq E|g(X_\varepsilon(T))| + E|Y_\varepsilon(T) - g(X_\varepsilon(T))|$$
$$\leq C(1 + |x|) + \varepsilon \leq C(1 + |x|).$$

Next, let $\langle x \rangle \overset{\Delta}{=} \sqrt{1+|x|^2}$. It is not hard to check that both $D\langle x \rangle$ and $D^2\langle x \rangle$ are uniformly bounded, thus applying Itô's formula to $\langle Y_\varepsilon(t) \rangle$, and note (1.22) and (1.24), we have

(1.26)
$$E\langle Y_\varepsilon(T)\rangle - E\langle Y_\varepsilon(t)\rangle$$
$$= E\int_t^T \frac{1}{\langle Y_\varepsilon(s)\rangle}\Big\{\langle Y_\varepsilon(s), h(s, X_\varepsilon(s), Y_\varepsilon(s), Z_\varepsilon(s))\rangle$$
$$+ \frac{1}{2}\Big[|Z_\varepsilon(s)|^2 - |Z_\varepsilon(s)^T\frac{Y_\varepsilon(s)}{\langle Y_\varepsilon(s)\rangle}|^2\Big]\Big\}ds$$
$$\geq -LE\int_t^T (1+|X_\varepsilon(s)| + \langle Y_\varepsilon(s)\rangle)ds$$
$$\geq -C(1+|x|) - LE\int_t^T \langle Y_\varepsilon(s)\rangle\,ds, \quad \forall t \in [0,T].$$

Now note that $|y| \leq \langle y \rangle \leq 1 + |y|$, we have by Gronwall's inequality and (1.25) that

(1.27) $E\langle Y_\varepsilon(t)\rangle \leq C(1+|x|), \qquad \forall t \in [0,T], \ \varepsilon \in (0,1].$

In particular, (1.27) leads to the boundedness of the set $\{|Y_\varepsilon(0)|\}_{\varepsilon>0}$. Thus, along a sequence we have $Y_{\varepsilon_k}(0) \to y$, as $k \to \infty$. The (1.13) will now follow easily from the continuity of $\overline{V}(x, \cdot)$ and the following equalities:

(1.28) $0 \leq \overline{V}(x, Y_{\varepsilon_k}(0)) \leq Ef(X_{\varepsilon_k}(T), Y_{\varepsilon_k}(T)) < \varepsilon_k.$

Finally, if (1.23) holds, then redoing (1.25) and (1.26), we see that (1.27) can be replaced by $E\langle Y_\varepsilon(t)\rangle \leq C$, $\forall t \in [0,T]$, $\varepsilon \in (0,1]$. Thus the same conclusion holds.   ☐

We will see in §3 that if (H1) holds, then $\overline{V}(\cdot, \cdot)$ is continuous.

§2. Dynamic Programming Method and the HJB Equation

We now study the optimal control problem associated with (0.1) via the Bellman's *dynamic programming method*. To this end, we let $s \in [0,T)$ and consider the following controlled system (compare with (1.1)):

(2.1) $\begin{cases} dX(t) = b(t, X(t), Y(t), Z(t))dt + \sigma(t, X(t), Y(t), Z(t))dW(t), \\ dY(t) = h(t, X(t), Y(t), Z(t))dt + Z(t)dW(t), \qquad t \in [s,T], \\ X(s) = x, \qquad Y(s) = y, \end{cases}$

Note that under assumption (H1) (see the paragraph containing (1.5)), for any $(s, x, y) \in [0,T) \times \mathbb{R}^n \times \mathbb{R}^m$ and $Z(\cdot) \in \mathcal{Z}[s,T] \overset{\Delta}{=} L^2_{\mathcal{F}}(s,T;\mathbb{R}^{m\times d})$, equation (2.1) admits a unique strong solution, denoted by, $(X(\cdot), Y(\cdot)) \equiv (X(\cdot;s,x,y,Z(\cdot)), Y(\cdot;s,x,y,Z(\cdot)))$. Next, we define the cost functional as follows:

(2.2) $J(s, x, y; Z(\cdot)) \overset{\Delta}{=} Ef(X(T;s,x,y,Z(\cdot)), Y(T;s,x,y,Z(\cdot))),$

with f defined by (1.6). Similar to *Problem(OC)*, we may pose the following optimal control problem.

Problem (OC)$_s$. For any given $(s, x, y) \in [0, T) \times \mathbb{R}^n \times \mathbb{R}^m$, find a $\overline{Z}(\cdot) \in \mathcal{Z}[s, T]$, such that

$$(2.3) \qquad V(s, x, y) \overset{\Delta}{=} \inf_{Z(\cdot) \in \mathcal{Z}[s,T]} J(s, x, y; Z(\cdot)) = J(s, x, y; \overline{Z}(\cdot)).$$

We also define

$$(2.4) \qquad V(T, x, y) = f(x, y), \qquad (x, y) \in \mathbb{R}^n \times \mathbb{R}^m.$$

Function $V(\cdot, \cdot, \cdot)$ defined by (2.3)–(2.4) is called the *value function* of the above family of optimal control problems (parameterized by $s \in [0, T)$). It is clear that when $s = 0$, *Problem(OC)$_s$* is reduced to *Problem(OC)* stated in the previous section. In other words, we have embedded *Problem(OC)* into a family of optimal control problems. We point out that this family of problems contains some very useful "dynamic" information due to allowing the initial moment $s \in [0, T)$ to vary. This is very crucial in the dynamic programming approach. From our definition, we see that

$$(2.5) \qquad V(0, x, y) = \overline{V}(x, y), \qquad \forall (x, y) \in \mathbb{R}^n \times \mathbb{R}^m.$$

Thus, if we can determine $V(s, x, y)$, we can do so for $\overline{V}(x, y)$. Recall that we called $\overline{V}(x, y)$ the *optimal cost function* of *Problem(OC)*, reserving the name *value function* for $V(s, x, y)$ for the conventional purpose. The following is the well-known *Bellman's principle of optimality*.

Theorem 2.1. For any $0 \le s \le \hat{s} \le T$, and $(x, y) \in \mathbb{R}^n \times \mathbb{R}^m$, it holds

$$(2.6) \quad V(s, x, y) = \inf_{Z(\cdot) \in \mathcal{Z}[s,T]} EV(\hat{s}, X(\hat{s}; s, x, y, Z(\cdot)), Y(\hat{s}; s, x, y, Z(\cdot))).$$

A rigorous proof of the above result is a little more involved. We present a sketch of the proof here.

Sketch of the proof. We denote the right hand side of (2.6) by $\widehat{V}(s, x, y)$. For any $Z(\cdot) \in \mathcal{Z}[s, T]$, by definition, we have

$$V(s, x, y) \le J(s, x, y; Z(\cdot))$$
$$= EJ(\hat{s}, X(\hat{s}; s, x, y, Z(\cdot)), Y(\hat{s}; s, x, y, Z(\cdot)); Z(\cdot)).$$

Thus, taking infimum over $Z(\cdot) \in \mathcal{Z}[s, T]$, we obtain

$$(2.7) \qquad\qquad V(s, x, y) \le \widehat{V}(s, x, y).$$

Conversely, for any $\varepsilon > 0$, there exists a $Z_\varepsilon(\cdot) \in \mathcal{Z}[s, T]$, such that

$$
\begin{aligned}
V(s, x, y) + \varepsilon &\ge J(s, x, y; Z_\varepsilon(\cdot)) \\
&= EJ(\hat{s}, X(\hat{s}; s, x, y, Z_\varepsilon(\cdot)), Y(\hat{s}; s, x, y, Z_\varepsilon(\cdot)); Z_\varepsilon(\cdot)) \\
&\ge EV(\hat{s}, X(\hat{s}; s, x, y, Z_\varepsilon(\cdot)), Y(\hat{s}; s, x, y, Z_\varepsilon(\cdot))) \\
&\ge \widehat{V}(s, x, y).
\end{aligned}
$$

(2.8)

Combining (2.7) and (2.8), we obtain (2.6).   □

Next, we introduce the *Hamiltonian* for the above optimal control problem:

(2.9)
$$\mathcal{H}(s, x, y, q, Q, z) \triangleq \left\{ \left\langle q, \begin{pmatrix} b(s, x, y, z) \\ h(s, x, y, z) \end{pmatrix} \right\rangle \right.$$
$$\left. + \frac{1}{2}\mathrm{tr}\left[Q \begin{pmatrix} \sigma(s, x, y, z) \\ z \end{pmatrix} \begin{pmatrix} \sigma(s, x, y, z) \\ z \end{pmatrix}^T \right] \right\},$$
$$\forall (s, x, y, q, Q, z) \in [0, T] \times \mathbb{R}^n \times \mathbb{R}^m$$
$$\times \mathbb{R}^{n+m} \times S^{n+m} \times \mathbb{R}^{m \times d},$$

and

(2.10)
$$H(s, x, y, q, Q) = \inf_{z \in \mathbb{R}^{m \times d}} \mathcal{H}(s, x, y, q, Q, z),$$
$$\forall (s, x, y, q, Q) \in [0, T] \times \mathbb{R}^n \times \mathbb{R}^m \times \mathbb{R}^{n+m} \times S^{n+m},$$

where S^{n+m} is the set of all $(n + m) \times (n + m)$ symmetric matrices. We see that since $\mathbb{R}^{m \times d}$ is not compact, the function H is not necessarily everywhere defined. We let

(2.11) $$\mathcal{D}(H) \triangleq \{(s, x, y, q, Q) \mid H(s, x, y, q, Q) > -\infty\}.$$

From above Theorem 2.1, we can obtain formally a PDE that the value function $V(\cdot, \cdot, \cdot)$ should satisfy.

Proposition 2.2. *Suppose $V(s, x, y)$ is smooth and H is continuous in* Int $\mathcal{D}(H)$. *Then*

(2.12) $$V_s(s, x, y) + H(s, x, y, DV(s, x, y), D^2V(s, x, y)) = 0,$$

for all $(s, x, y) \in [0, T] \times \mathbb{R}^n \times \mathbb{R}^m$, such that

(2.13) $$(s, x, y, DV(s, x, y), D^2V(s, x, y)) \in \mathrm{Int}\,\mathcal{D}(H),$$

where

$$DV = \begin{pmatrix} V_x \\ V_y \end{pmatrix}, \qquad D^2V = \begin{pmatrix} V_{xx} & V_{xy} \\ V_{xy}^T & V_{yy} \end{pmatrix}.$$

Proof. Let $(s, x, y) \in [0, T) \times \mathbb{R}^n \times \mathbb{R}^m$ such that (2.13) holds. For any $z \in \mathbb{R}^{m \times d}$, let $(X(\cdot), Y(\cdot))$ be the solution of (2.1) corresponding to (s, x, y) and $Z(\cdot) \equiv z$. Then, by (2.6) and Itô's formula, we have

(2.14)
$$0 \leq E\left\{ \frac{V(\hat{s}, X(\hat{s}), Y(\hat{s})) - V(s, x, y)}{\hat{s} - s} \right\}$$
$$\to V_s(s, x, y) + \mathcal{H}(s, x, y, DV(s, x, y), D^2V(s, x, y), z).$$

Taking infimum in $z \in \mathbb{R}^{m \times d}$, we see that

(2.15) $$V_s(s, x, y) + H(s, x, y, DV(s, x, y), D^2V(s, x, y)) \geq 0.$$

On the other hand, for any $\varepsilon > 0$ and $\hat{s} \in (s, T)$, by (2.6), there exists
a $Z(\cdot) \equiv Z_\varepsilon(\cdot) \in \mathcal{Z}[s, T]$, with the corresponding state being $(X(\cdot), Y(\cdot))$,
such that
(2.16)

$$
\varepsilon \geq E\left\{ \frac{V(\hat{s}, X(\hat{s}), Y(\hat{s})) - V(s, x, y)}{\hat{s} - s} \right\}
$$

$$
= \frac{1}{\hat{s} - s} E \int_s^{\hat{s}} \left\{ V_s(t, X(t), Y(t)) \right.
$$

$$
\left. + \mathcal{H}\big(t, X(t), Y(t), DV(t, X(t), Y(t)), D^2V(t, X(t), Y(t)), Z(t))\big) \right\} dt
$$

$$
\geq \frac{1}{\hat{s} - s} E \int_s^{\hat{s}} \left\{ V_s(t, X(t), Y(t)) \right.
$$

$$
\left. + H\big(t, X(t), Y(t), DV(t, X(t), Y(t)), D^2V(t, X(t), Y(t)))\big) \right\} dt
$$

$$
\to V_s(s, x, y) + H(s, x, y, DV(s, x, y), D^2V(s, x, y)).
$$

Here, we have used (2.13) and the assumption that H is continuous in
Int $\mathcal{D}(H)$. Combining (2.15)–(2.16), we obtain (2.12).   □

Equation (2.12) is called the *Hamilton-Jacobi-Bellman* (HJB for short)
equation associated with our optimal control problem. In principle, one
can determine the value function $V(\cdot, \cdot, \cdot)$ through solving (2.12)–(2.13)
together with the terminal condition (2.4). However, since $\mathcal{D}(H)$ might
be a very complicated set, solving (2.12)–(2.13) together with (2.4) is very
difficult. Thus, much more needs to be done in order to determine the value
function V.

§3. The Value Function

In this section, we are going to study the value function V introduced in
the previous section in some details.

§3.1. Continuity and semi-concavity

We first look at the continuity of the value function $V(s, x, y)$. Note that
since the control domain $\mathbb{R}^{m \times d}$ is not compact, we can only prove the
right-continuity of $V(s, x, y)$ in $s \in [0, T)$.

Proposition 3.1. *Let* (H1) *hold. Then* $V(s, x, y)$ *is right-continuous in*
$s \in [0, T)$ *and there exists a constant* $C > 0$, *such that*

(3.1)
$$
0 \leq V(s, x, y) \leq C(1 + |x| + |y|),
$$
$$
\forall (s, x, y) \in [0, T] \times \mathbb{R}^n \times \mathbb{R}^m,
$$

(3.2)
$$
|V(s, x, y) - V(s, \overline{x}, \overline{y})| \leq C(|x - \overline{x}| + |y - \overline{y}|),
$$
$$
\forall s \in [0, T], \ x, \overline{x} \in \mathbb{R}^n, \ y, \overline{y} \in \mathbb{R}^m.
$$

Proof. It is clear that for any $(s, x, y) \in [0, T] \times \mathbb{R}^n \times \mathbb{R}^m$, we have

$$
0 \leq V(s, x, y) \leq J(s, x, y; 0) \leq C(1 + |x| + |y|).
$$

This proves (3.1).

Next, let $s \in [0, T]$ and $(x, y), (\overline{x}, \overline{y}) \in \mathbb{R}^n \times \mathbb{R}^m$ be fixed. Then, for any $Z(\cdot) \in \mathcal{Z}[s, T]$, by Itô's formula and Gronwall's inequality, using (H1), we have

(3.3)
$$
\begin{aligned}
E &\left| X(t; s, x, y, Z(\cdot)) - X(t; s, \overline{x}, \overline{y}, Z(\cdot)) \right|^2 \\
&+ E \left| Y(t; s, x, y, Z(\cdot)) - Y(t; s, \overline{x}, \overline{y}, Z(\cdot)) \right|^2 \\
&\leq C \{ |x - \overline{x}|^2 + |y - \overline{y}|^2 \}, \qquad t \in [s, T],
\end{aligned}
$$

with $C > 0$ only depending on L and T. Then (3.2) follows from (1.8), which implies the (Lipschitz) continuity of $V(s, x, y)$ in (x, y).

We now prove the right-continuity of $V(s, x, y)$ in $s \in [0, T]$. First of all, it is clear that for any $Z(\cdot) \in \mathcal{Z}[0, T]$, the function

$$
(s, x, y) \mapsto J\big(s, x, y; Z\big|_{[s,T]}(\cdot)\big)
$$

is continuous. Thus, by the definition of V, it is necessary that $V(s, x, y)$ is upper semi-continuous. On the other hand, by (2.6) and (3.2), taking $Z(\cdot) = 0$, we have

(3.4) $V(s, x, y) \leq V(\widehat{s}, x, y) + C(\widehat{s} - s)^{1/2}, \qquad \forall 0 \leq s \leq \widehat{s} \leq T.$

Thus, by the upper semi-continuity of V, we must have

$$
\lim_{\widehat{s} \downarrow s} V(\widehat{s}, x, y) = V(s, x, y),
$$

which gives the right-continuity of V in $s \in [0, T)$.   □

From (2.5) and (3.2), we see that under (H1), the function $\overline{V}(x, y)$ is continuous, the assertion that we promised to prove in §1.

Next, we would like to establish another important property for the value function. To this end, we introduce the following definition.

Definition 3.2. A function $\varphi : \mathbb{R}^n \to \mathbb{R}$ is said to be *semi-concave* if there exists a constant $C > 0$, such that the function $\Phi(x) \equiv \varphi(x) - C|x|^2$ is concave on $\mathbb{R}^n$, i.e.,

(3.5) $\Phi(\lambda x + (1 - \lambda)\overline{x}) \geq \lambda \Phi(x) + (1 - \lambda)\Phi(\overline{x}), \quad \forall \lambda \in [0, 1], \ x, \overline{x} \in \mathbb{R}^n.$

A family of functions $\varphi_\varepsilon : \mathbb{R}^n \to \mathbb{R}$ is said to be semi-concave *uniformly* in ε if there exists a constant $C > 0$, independent of ε, such that $\varphi_\varepsilon(x) - C|x|^2$ is concave for all ε.

We have the following result.

Lemma 3.3. *Function* $\varphi : \mathbb{R}^n \to \mathbb{R}$ *is semiconcave if and only if*

(3.6)
$$
\begin{aligned}
\lambda \varphi(x) + (1 - \lambda)\varphi(\overline{x}) - \varphi(\lambda x + (1 - \lambda)\overline{x}) &\leq C\lambda(1 - \lambda)|x - \overline{x}|^2, \\
&\forall \lambda \in [0, 1], \ x, \overline{x} \in \mathbb{R}^n.
\end{aligned}
$$

In the case that $\varphi \in W^{2,1}_{loc}(\mathbb{R}^n)$, *it is semiconcave if and only if*

(3.7) $D^2 \varphi(x) \leq CI, \qquad \text{a.e. } x \in \mathbb{R}^n,$

where $D^2\varphi$ is the (generalized) Hessian of φ (i.e., it consists of second order weak derivatives of φ).

Proof. We have the following identity:

$$\lambda|x|^2 + (1-\lambda)|\bar{x}|^2 - |\lambda x + (1-\lambda)\bar{x}|^2 = \lambda(1-\lambda)|x-\bar{x}|^2,$$
$$\forall \lambda \in [0,1], \; x, \bar{x} \in \mathbb{R}^n.$$

Thus, we see immediately that (3.5) and (3.6) are equivalent.

Now, if φ is C^2, then, by the concavity of $\varphi(x) - C|x|^2$, we know that (3.7) holds. By Taylor expansion, we can prove the converse. For the general case, we may approximate φ using mollifier. $\qquad \square$

It is easy to see from the last conclusion of Lemma 3.3 that if φ has a bounded Hessian, i.e., φ_x is uniformly Lipschitz, then, it is semi-concave. This observation will be very useful below.

Let us make some further assumptions.

(H3) Functions b, σ, h and g are differentiable in (x,y) with the derivatives being uniformly Lipschitz continuous in $(x,y) \in \mathbb{R}^n \times \mathbb{R}^m$, uniformly in $(t,z) \in [0,T] \times \mathbb{R}^{m \times d}$.

We easily see that under (H3), the function f defined by (1.6) has a bounded Hessian, and thus it is semi-concave.

Now, we prove the following:

Theorem 3.4. *Let (H1) and (H3) hold. Then the value function $V(s,x,y)$ is semi-concave in $(x,y) \in \mathbb{R}^n \times \mathbb{R}^m$ uniformly in $s \in [0,T]$.*

Proof. Let $s \in [0,T)$, $x_0, x_1 \in \mathbb{R}^n$ and $y_0, y_1 \in \mathbb{R}^m$. Denote

$$(3.8) \qquad x_\lambda = \lambda x_1 + (1-\lambda)x_0, \; y_\lambda = \lambda y_1 + (1-\lambda)y_0, \qquad \lambda \in [0,1].$$

Then, for any $\varepsilon > 0$, there exists a $Z_\varepsilon(\cdot) \in \mathcal{Z}[s,T]$ (which is also depending on λ), such that

$$(3.9) \qquad J(s,x_\lambda,y_\lambda;Z_\varepsilon(\cdot)) < V(s,x_\lambda,y_\lambda) + \varepsilon.$$

We now fix the above $Z_\varepsilon(\cdot)$ and let $(X_\lambda(\cdot), Y_\lambda(\cdot))$ be the solution of (2.1) corresponding to $(s,x_\lambda,y_\lambda,Z_\varepsilon(\cdot))$. We denote

$$(3.10) \qquad \begin{cases} \eta_\lambda(r) = (X_\lambda(r), Y_\lambda(r)), \\ \xi_\lambda(r) = \lambda\eta_1(r) + (1-\lambda)\eta_0(r), \end{cases} \qquad \lambda \in [0,1], \; r \in [s,T].$$

Using Itô's formula and Gronwall's inequality, we have

$$(3.11) \; E|X_1(t) - X_0(t)|^4 + E|Y_1(t) - Y_0(t)|^4 \le C(|x_1-x_0|^4 + |y_1-y_0|^4).$$

Since $f(x, y)$ is semi-concave in (x, y) (by (H3)), we have

$$
(3.12) \quad
\begin{aligned}
&\lambda V(s, x_1, y_1) + (1 - \lambda)V(s, x_0, y_0) - V(s, x_\lambda, y_\lambda) - \varepsilon \\
&\leq \lambda J(s, x_1, y_1; Z_\varepsilon(\cdot)) + (1 - \lambda)J(s, x_0, y_0; Z_\varepsilon(\cdot)) \\
&\qquad\qquad\qquad\qquad - J(s, x_\lambda, y_\lambda; Z_\varepsilon(\cdot)) \\
&= E\Big\{\lambda f(\eta_1(T)) + (1 - \lambda)f(\eta_0(T)) - f(\eta_\lambda(T))\Big\} \\
&\leq CE\Big\{\lambda(1 - \lambda)|\eta_1(T) - \eta_0(T)|^2 + |\xi_\lambda(T) - \eta_\lambda(T)|^2\Big\}.
\end{aligned}
$$

Let us now estimate the right hand side of (3.12). For the first term, we have

$$
(3.13) \quad E|\eta_1(t) - \eta_0(t)|^2 \leq C(|x_1 - x_0|^2 + |y_1 - y_0|^2), \quad t \in [s, T].
$$

To estimate the second term on the right hand side of (3.12), let us denote

$$
(3.14) \quad
\begin{cases}
b_\lambda(r) = b(r, \eta_\lambda(r), Z_\varepsilon(r)), \\
h_\lambda(r) = h(r, \eta_\lambda(r), Z_\varepsilon(r)), \quad \lambda \in [0, 1], \ r \in [s, T]. \\
\sigma_\lambda(r) = \sigma(r, \eta_\lambda(r), Z_\varepsilon(r)),
\end{cases}
$$

Then, applying Itô's formula, one has (we suppress r in the integrand below)

$$
(3.15) \quad
\begin{aligned}
&E\big|\xi_\lambda(t) - \eta_\lambda(t)\big|^2 \\
&= 2E\int_s^t \langle \lambda X_1 + (1 - \lambda)X_0 - X_\lambda, \lambda b_1 + (1 - \lambda)b_0 - b_\lambda \rangle \, dr \\
&\quad + 2E\int_s^t \langle \lambda Y_1 + (1 - \lambda)Y_0 - Y_\lambda, \lambda h_1 + (1 - \lambda)h_0 - h_\lambda \rangle \, dr \\
&\quad + E\int_s^t |\lambda\sigma_1 + (1 - \lambda)\sigma_0 - \sigma_\lambda|^2 dr, \quad t \in [s, T].
\end{aligned}
$$

Note (we suppress r and $Z_\varepsilon(r)$ from the second line on)

$$
(3.16) \quad
\begin{aligned}
&|\lambda b_1(r) + (1 - \lambda)b_0(r) - b_\lambda(r)| \\
&= |\lambda b(\eta_1) + (1 - \lambda)b(\eta_0) - b(\eta_\lambda)| \\
&\leq |\lambda[b(\eta_1) - b(\xi_\lambda)] + (1 - \lambda)[b(\eta_0) - b(\xi_\lambda)]| + L|\xi_\lambda - \eta_\lambda| \\
&= \Big|\lambda \int_0^1 \langle b_\eta(\xi_\lambda + \alpha(1 - \lambda)(\eta_1 - \eta_0))d\alpha, (1 - \lambda)(\eta_1 - \eta_0) \rangle \\
&\quad + (1 - \lambda)\int_0^1 \langle b_\eta(\xi_\lambda - \alpha\lambda(\eta_1 - \eta_0))d\alpha, -\lambda(\eta_1 - \eta_0) \rangle\Big| \\
&\quad + L|\xi_\lambda - \eta_\lambda| \\
&\leq C\lambda(1 - \lambda)|\eta_1 - \eta_0| + L|\xi_\lambda - \eta_\lambda|.
\end{aligned}
$$

We have the similar estimates for the terms involving h and σ. Then it follows from (3.13) and (3.15) that

$$E|\xi_\lambda(t) - \eta_\lambda(t)|^2 \leq C\lambda^2(1-\lambda)^2(|x_1 - x_0|^2 + |y_1 - y_0|^2)$$
$$+ C\int_s^t E|\xi_\lambda(r) - \eta_\lambda(r)|^2 dr, \qquad \forall t \in [s, T].$$

By applying Gronwall's inequality, we obtain

(3.17) $$E|\xi_\lambda(t) - \eta_\lambda(t)|^2 \leq C\lambda^2(1-\lambda)^2(|x_1 - x_0|^2 + |y_1 - y_0|^2).$$

Combining (3.12), (3.13) and (3.17), we obtain the semi-concavity of $V(s, x, y)$ in (x, y), uniformly in $s \in [0, T]$.   □

§3.2. Approximation of the value function

We have seen that due to the noncompactness of the control domain $\mathbb{R}^{m \times d}$, it is not very easy to determine the value function V through a PDE (the HJB equation). In this subsection, we introduce some approximations of the value function, which will help us to determine the value function (approximately).

First of all, let $\widetilde{W}(t) \equiv (\widetilde{W}_1(t), \widetilde{W}_2(t))$ be an $(n+m)$-dimensional Brownian motion which is independent of $W(t)$ (embedded into an enlarged probability space, if necessary) and let $\{\widetilde{\mathcal{F}}_t\}_{t \geq 0}$ be the filtration generated by $W(t)$ and $\widetilde{W}(t)$, augmented by all the P-null sets in $\widetilde{\mathcal{F}}$. Define

(3.18)
$$\begin{cases} \mathcal{Z}_0[s, T] \overset{\Delta}{=} \mathcal{Z}[s, T], \\ \widetilde{\mathcal{Z}}_0[s, T] \overset{\Delta}{=} \{Z : [s, T] \times \Omega \to \mathbb{R}^{m \times d} \mid Z \text{ is } \{\widetilde{\mathcal{F}}_t\}_{t \geq 0}\text{-adapted}, \\ \qquad\qquad\qquad\qquad \int_0^T E|Z(t)|^2 dt < \infty \}. \end{cases}$$

Next, for any $\delta > 0$, we define

(3.19)
$$\begin{cases} \mathcal{Z}_\delta[s, T] \overset{\Delta}{=} \{Z \in \mathcal{Z}[s, T] \mid |Z(t)| \leq \frac{1}{\delta}, \quad \text{a.e. } t \in [s, T], \text{ a.s. }\}, \\ \widetilde{\mathcal{Z}}_\delta[s, T] \overset{\Delta}{=} \{Z \in \widetilde{\mathcal{Z}}_0[s, T] \mid |Z(t)| \leq \frac{1}{\delta}, \quad \text{a.e. } t \in [s, T], \text{ a.s. }\}. \end{cases}$$

The following inclusions are obvious.

(3.20)
$$\begin{array}{ccccc} \mathcal{Z}_0[s, T] & \supset & \mathcal{Z}_{\delta_1}[s, T] & \supset & \mathcal{Z}_{\delta_2}[s, T] \\ \cap & & \cap & & \cap \\ \widetilde{\mathcal{Z}}_0[s, T] & \supset & \widetilde{\mathcal{Z}}_{\delta_1}[s, T] & \supset & \widetilde{\mathcal{Z}}_{\delta_2}[s, T] \end{array} \qquad \forall \delta_2 \geq \delta_1 \geq 0.$$

In what follows, for any $Z \in \mathcal{Z}_0[s, T]$ (resp. $\widetilde{\mathcal{Z}}_0[s, T]$) and $\delta > 0$, we define the $\frac{1}{\delta}$-truncation of Z as follows:

(3.21) $$Z_\delta(t, \omega) = \begin{cases} Z(t, \omega), & \text{if } |Z(t, \omega)| \leq \frac{1}{\delta}, \\ \dfrac{Z(t, \omega)}{\delta|Z(t, \omega)|}, & \text{if } |Z(t, \omega)| > \frac{1}{\delta}. \end{cases}$$

Clearly, $Z_\delta \in \mathcal{Z}_\delta[s,T]$ (resp. $\widetilde{\mathcal{Z}}_\delta[s,T]$).

We now consider, for any $\varepsilon > 0$, the following *regularized state equation* (compare to (2.1)):

$$
(3.22) \quad
\begin{cases}
dX(t) = b(t, X(t), Y(t), Z(t))dt + \sigma(t, X(t), Y(t), Z(t))dW(t) \\
\qquad\qquad + \sqrt{2\varepsilon}d\widetilde{W}_1(t), \\
dY(t) = h(t, X(t), Y(t), Z(t))dt + Z(t)dW(t) \\
\qquad\qquad + \sqrt{2\varepsilon}d\widetilde{W}_2(t), \qquad\qquad t \in [s,T], \\
X(s) = x, \qquad Y(s) = y.
\end{cases}
$$

Define the cost functional by $J^{\delta,\varepsilon}(s,x,y;Z(\cdot))$ (resp. $\widetilde{J}^{\delta,\varepsilon}(s,x,y;Z(\cdot))$) which has the same form as (2.3) with the control being taken in $\mathcal{Z}_\delta[s,T]$ (resp. $\widetilde{\mathcal{Z}}_\delta[s,T]$) and the state satisfying (3.22), indicating the dependence on $\delta \geq 0$ and $\varepsilon \geq 0$. The corresponding optimal control problem is called *Problem* $(OC)_s^{\delta,\varepsilon}$ (resp. *Problem* $\widetilde{(OC)}_s^{\delta,\varepsilon}$). The corresponding (approximate) value functions are then defined as, respectively,

$$
(3.23) \quad
\begin{cases}
\widetilde{V}^{\delta,\varepsilon}(s,x,y) = \inf_{Z(\cdot) \in \widetilde{\mathcal{Z}}_\delta[s,T]} \widetilde{J}^{\delta,\varepsilon}(s,x,y;Z(\cdot)), \\
V^{\delta,\varepsilon}(s,x,y) = \inf_{Z(\cdot) \in \mathcal{Z}_\delta[s,T]} J^{\delta,\varepsilon}(s,x,y;Z(\cdot)).
\end{cases}
$$

Due to the inclusions in (3.20), we see that for any $(s,x,y) \in [0,T] \times \mathbb{R}^n \times \mathbb{R}^m$,

$$
(3.24) \quad
\begin{cases}
V^{\delta,\varepsilon}(s,x,y) \geq \widetilde{V}^{\delta,\varepsilon}(s,x,y) \geq 0, \quad \forall \delta, \varepsilon \geq 0, \\
\widetilde{V}^{\delta_2,\varepsilon}(s,x,y) \geq \widetilde{V}^{\delta_1,\varepsilon}(s,x,y), \quad \forall \delta_2 \geq \delta_1 \geq 0, \ \varepsilon \geq 0, \\
V^{\delta_2,\varepsilon}(s,x,y) \geq V^{\delta_1,\varepsilon}(s,x,y), \quad \forall \delta_2 \geq \delta_1 \geq 0, \ \varepsilon \geq 0.
\end{cases}
$$

Also, it is an easy observation that $V^{0,0}(s,x,y) = V(s,x,y)$, $\forall(s,x,y)$. Note that for $\delta > 0$ and $\varepsilon \geq 0$, the corresponding HJB equation for the value function $\widetilde{V}^{\delta,\varepsilon}(s,x,y)$ takes the following form:

$$
(3.25) \quad
\begin{cases}
\widetilde{V}_s^{\delta,\varepsilon} + \varepsilon\Delta\widetilde{V}^{\delta,\varepsilon} + H^\delta(s,x,y,D\widetilde{V}^{\delta,\varepsilon}, D^2\widetilde{V}^{\delta,\varepsilon}) = 0, \\
\qquad\qquad (s,x,y) \in (0,T) \times \mathbb{R}^n \times \mathbb{R}^m; \\
\widetilde{V}^{\delta,\varepsilon}(T,x,y) = f(x,y), \qquad (x,y) \in \mathbb{R}^n \times \mathbb{R}^m,
\end{cases}
$$

where Δ is the Laplacian operator in $\mathbb{R}^{n+m}$, and H^δ is defined by the following:

$$
H^\delta(s,x,y,q,Q) \triangleq \inf_{\substack{z \in \mathbb{R}^{m \times d} \\ |z| \leq 1/\delta}} \left\{ \left\langle q, \begin{pmatrix} b(s,x,y,z) \\ h(s,x,y,z) \end{pmatrix} \right\rangle \right.
$$
$$
\left. + \frac{1}{2}\mathrm{tr}\left[Q \begin{pmatrix} \sigma(s,x,y,z) \\ z \end{pmatrix} \begin{pmatrix} \sigma(s,x,y,z) \\ z \end{pmatrix}^T \right] \right\},
$$

for $(s, x, y, q, Q) \in [0, T] \times \mathbb{R}^n \times \mathbb{R}^m \times \mathbb{R}^{n+m} \times S^{n+m}$, where S^{n+m} is the set of all $(n+m) \times (n+m)$ symmetric matrices. We observe that for $\varepsilon > 0$, (3.25) is a *nondegenerate* nonlinear parabolic PDE; and for $\varepsilon = 0$, however, (3.25) is a *degenerate* nonlinear parabolic PDE. The following notion will be necessary for us to proceed further.

Definition 3.5. A continuous function $v : [0, T] \times \mathbb{R}^n \times \mathbb{R}^m \to \mathbb{R}$ is called a *viscosity subsolution* (resp. *viscosity supersolution*) of (3.25), if

$$
(3.26) \quad
\begin{aligned}
& v(T, x, y) \leq f(x, y), \qquad \forall (x, y) \in \mathbb{R}^n \times \mathbb{R}^m, \\
& (\text{resp. } v(T, x, y) \geq f(x, y), \qquad \forall (x, y) \in \mathbb{R}^n \times \mathbb{R}^m),
\end{aligned}
$$

and for any smooth function $\varphi(s, x, y)$ whenever the map $v - \varphi$ attains a local maximum (resp. minimum) at $(s, x, y) \in [0, T) \times \mathbb{R}^n \times \mathbb{R}^m$, it holds:

$$
(3.27) \quad
\begin{aligned}
& \varphi_s(s, x, y) + \varepsilon \Delta \varphi(s, x, y) \\
& \qquad + H^\delta(s, x, y, D\varphi(s, x, y), D^2\varphi(s, x, y)) \geq 0 \quad (\text{resp. } \leq 0).
\end{aligned}
$$

If v is both viscosity subsolution and viscosity supersolution of (3.25), we call it a *viscosity solution* of (3.25).

We note that in the above definition, v being continuous is enough. Thus, by this, we can talk about a solution of differential equations *without* its differentiability. Furthermore, such a notion admits the uniqueness.

The following proposition collects some basic properties of the approximate value functions.

Proposition 3.6. *Let (H1) hold. Then*

(i) $\widetilde{V}^{\delta,\varepsilon}(s, x, y)$ *and* $V^{\delta,\varepsilon}(s, x, y)$ *are continuous in* $(x, y) \in \mathbb{R}^n \times \mathbb{R}^m$, *uniformly in* $s \in [0, T]$ *and* $\delta, \varepsilon \geq 0$; *For fixed* $\delta > 0$ *and* $\varepsilon \geq 0$, $\widetilde{V}^{\delta,\varepsilon}(s, x, y)$ *and* $V^{\delta,\varepsilon}(s, x, y)$ *are continuous in* $(s, x, y) \in [0, T] \times \mathbb{R}^n \times \mathbb{R}^m$.

(ii) *For* $\delta > 0$ *and* $\varepsilon \geq 0$, $\widetilde{V}^{\delta,\varepsilon}(s, x, y)$ *is the unique viscosity solution of* (3.25), *and for* $\delta, \varepsilon > 0$, $\widetilde{V}^{\delta,\varepsilon}(s, x, y)$ *is the unique strong solution of (3.25).*

(iii) *For* $\delta > 0$ *and* $\varepsilon \geq 0$, $V^{\delta,\varepsilon}(s, x, y)$ *is a viscosity super solution of* (3.25), $V^{\delta,0}(s, x, y)$ *is the unique viscosity solution of (3.25) (with* $\varepsilon = 0$).

The proof of (i) is similar to that of Proposition 3.1 and the proof of (ii) and (iii) are by now standard, which we omit here for simplicity of presentation (see Yong-Zhou [1] and Fleming-Soner [1], for details).

The following result gives the continuous dependence of the approximate value functions on the parameters δ and ε.

Theorem 3.7. *Let (H1) hold. Then, for any* $s \in [0, T]$, *there exists a continuous function* $\eta_s : [0, \infty) \times [0, \infty) \to [0, \infty)$, *with* $\eta_s(0, r) = 0$ *for all* $r \geq 0$, *such that*

$$
(3.28) \quad
\begin{aligned}
& |\widetilde{V}^{\delta,\varepsilon}(s, x, y) - \widetilde{V}^{\hat\delta,\hat\varepsilon}(s, x, y)| \leq \eta_s(|\delta - \hat\delta| + |\varepsilon - \hat\varepsilon|, |x| + |y|), \\
& |V^{\delta,\varepsilon}(s, x, y) - V^{\hat\delta,\hat\varepsilon}(s, x, y)| \leq \eta_s(|\delta - \hat\delta| + |\varepsilon - \hat\varepsilon|, |x| + |y|), \\
& \forall (s, x, y) \in [0, T] \times \mathbb{R}^n \times \mathbb{R}^m, \ \delta, \hat\delta, \varepsilon, \hat\varepsilon \in [0, 1].
\end{aligned}
$$

Proof. Fix $(s, x, y) \in [0, T] \times \mathbb{R}^n \times \mathbb{R}^m$, $\delta, \hat{\delta}, \varepsilon, \hat{\varepsilon} \geq 0$, and $Z \in \mathcal{Z}[s, T]$. Let Z_δ (resp. $Z_{\hat{\delta}}$) be the $1/\delta$- (resp. $1/\hat{\delta}$-) truncation of Z; and (X, Y) (resp. $(\widehat{X}, \widehat{Y})$) the solution of (3.22) corresponding to (ε, Z_δ) (resp. $(\hat{\varepsilon}, Z_{\hat{\delta}})$). By Itô's formula and Gronwall's inequality,

(3.29)
$$E\{|X(T) - \widehat{X}(T)|^2 + |Y(T) - \widehat{Y}(T)|^2\}$$
$$\leq C\{E \int_s^T |Z_\delta(t) - Z_{\hat{\delta}}(t)|^2 dt + |\sqrt{\varepsilon} - \sqrt{\hat{\varepsilon}}|^2\},$$

where $C > 0$ depends only on L and T. Thus, we obtain

(3.30)
$$|V^{\delta, \varepsilon}(s, x, y) - V^{\delta, \hat{\varepsilon}}(s, x, y)|$$
$$\leq C|\sqrt{\varepsilon} - \sqrt{\hat{\varepsilon}}|, \quad \forall (s, x, y), \ \delta, \varepsilon, \hat{\varepsilon} \geq 0.$$

Combining with Proposition 3.6, we see that $V^{\delta, \varepsilon}(s, x, y)$ is continuous in $(\varepsilon, x, y) \in [0, \infty) \times \mathbb{R}^n \times \mathbb{R}^m$ uniformly in $\delta \geq 0$ and $s \in [0, T]$.

Next, for fixed $(s, x, y) \in [0, T] \times \mathbb{R}^n \times \mathbb{R}^m$, $\varepsilon \geq 0$, and $\hat{\delta} \geq \delta \geq 0$, by (3.24), we have

(3.31)
$$0 \leq V^{\hat{\delta}, \varepsilon}(s, x, y) - V^{\delta, \varepsilon}(s, x, y).$$

On the other hand, for any $\delta > 0$, and $\varepsilon_0 > 0$, we can choose $Z^{\varepsilon_0} \in \mathcal{Z}_\delta[s, T]$ so that

(3.32)
$$V^{\delta, \varepsilon}(s, x, y) + \varepsilon_0 > J^{\delta, \varepsilon}(s, x, y; Z^{\varepsilon_0}).$$

Let $Z^{\varepsilon_0}_{\hat{\delta}}$ be the $\frac{1}{\hat{\delta}}$-truncation of Z^{ε_0}, and denote the corresponding solution of (3.22) with Z^{ε_0} (resp. $Z^{\varepsilon_0}_{\hat{\delta}}$) by $(X^{\varepsilon_0}, Y^{\varepsilon_0})$ (resp. $(\widehat{X}^{\varepsilon_0}, \widehat{Y}^{\varepsilon_0})$). Setting $(X, Y) = (X^{\varepsilon_0}, Y^{\varepsilon_0})$, $(\widehat{X}, \widehat{Y}) = (\widehat{X}^{\varepsilon_0}, \widehat{Y}^{\varepsilon_0})$, $\varepsilon = \hat{\varepsilon}$, $Z_\delta = Z^{\varepsilon_0}$, and $Z_{\hat{\delta}} = Z^{\varepsilon_0}_{\hat{\delta}}$ in (3.29), we obtain

(3.33)
$$E\{|X^{\varepsilon_0}(T) - \widehat{X}^{\varepsilon_0}(T)|^2 + |Y^{\varepsilon_0}(T) - \widehat{Y}^{\varepsilon_0}(T)|^2\}$$
$$\leq CE \int_s^T |Z^{\varepsilon_0}(t) - Z^{\varepsilon_0}_{\hat{\delta}}(t)|^2 dt.$$

We consider the following two cases:

Case 1. $\delta > 0$. In this case, note that $|Z^{\varepsilon_0}(t) - Z^{\varepsilon_0}_{\hat{\delta}}(t)| \leq |1/\delta - 1/\hat{\delta}|$, a.e. $t \in [s, T]$, a.s. By (1.8) and (H1), one easily checks that

(3.34)
$$J^{\delta, \varepsilon}(s, x, y; Z^{\varepsilon_0}) \geq J^{\hat{\delta}, \varepsilon}(s, x, y; Z^{\varepsilon_0}_{\hat{\delta}}) - C\left|\frac{1}{\delta} - \frac{1}{\hat{\delta}}\right|$$
$$\geq V^{\hat{\delta}, \varepsilon}(s, x, y) - C\left|\frac{1}{\delta} - \frac{1}{\hat{\delta}}\right|.$$

Combining (3.31), (3.32) and (3.34), we obtain (note $\varepsilon_0 > 0$ is arbitrary)

(3.35)
$$0 \leq V^{\hat{\delta}, \varepsilon}(s, x, y) - V^{\delta, \varepsilon}(s, x, y)| \leq C\left|\frac{1}{\delta} - \frac{1}{\hat{\delta}}\right|,$$
$$\forall (s, x, y), \ \delta, \hat{\delta} > 0, \ \varepsilon \geq 0,$$

where C is again an absolute constant.

Case 2. $\delta = 0$. Now let $\hat{\delta} > 0$ be small enough so that the right side of (3.33) is no greater than ε_0^2. Then, similar to (3.36), we have

$$(3.36) \qquad J^{0,\varepsilon}(s,x,y;Z^{\varepsilon_0}) \geq V^{\hat{\delta},\varepsilon}(s,x,y) - \varepsilon_0.$$

Combing (3.31), (3.32) and (3.36), one has $0 \leq V^{\hat{\delta},\varepsilon}(s,x,y) - V^{0,\varepsilon}(s,x,y) \leq 2\varepsilon_0$, which shows that

$$(3.37) \qquad V^{\hat{\delta},\varepsilon}(s,x,y) \downarrow V^{0,\varepsilon}(s,x,y), \qquad \hat{\delta} \downarrow 0.$$

Since $V^{0,\varepsilon}(s,x,y)$ is continuous in (ε,x,y) (see (3.30) and Proposition 3.6-(i)), by Dini's theorem, we obtain that the convergence in (3.37) is uniform in (ε,x,y) on compact sets. Thus, for some continuous function $\eta_s : [0,\infty) \times [0,\infty) \to [0,\infty)$ with $\eta_s(0,r) = 0$ for all $r \geq 0$, one has

$$(3.38) \qquad \begin{aligned} &0 \leq V^{\hat{\delta},\varepsilon}(s,x,y) - V^{0,\varepsilon}(s,x,y) \leq \eta_s(\hat{\delta}, |x| + |y|), \\ &\forall (s,x,y), \ \varepsilon \in [0,1], \ \hat{\delta} \geq 0. \end{aligned}$$

Combining (3.30), (3.35) and (3.38), we have that $V^{\delta,\varepsilon}(s,x,y)$ is continuous in $(\delta,\varepsilon,x,y) \in [0,\infty) \times [0,\infty) \times \mathbb{R}^n \times \mathbb{R}^m$. The proof for $\widetilde{V}^{\delta,\varepsilon}$ is exactly the same. $\qquad\qquad\square$

Corollary 3.8. *Let (H1) hold. Then*

$$(3.39) \quad \widetilde{V}^{\delta,0}(s,x,y) = V^{\delta,0}(s,x,y), \quad \forall (s,x,y) \in [0,T] \times \mathbb{R}^n \times \mathbb{R}^m, \ \delta \geq 0.$$

Proof. If $\delta > 0$, then both $\widetilde{V}^{\delta,0}$ and $V^{\delta,0}$ are the viscosity solutions of the HJB equation (3.25). Thus, (3.39) follows from the uniqueness. By the continuity of $\widetilde{V}^{\delta,0}$ and $V^{\delta,0}$ in $\delta \geq 0$, we obtain (3.39) for $\delta = 0$. $\qquad\square$

Corollary 3.9. *Let $V(0,x,y) = 0$. Then, for any $\hat{\varepsilon} > 0$, there exist $\delta, \varepsilon > 0$ and $Z^{\delta,\varepsilon}(\cdot) \in \mathcal{Z}_\delta[0,T]$ satisfying*

$$(3.40) \qquad J^{\delta,\varepsilon}(0,x,y;Z^{\delta,\varepsilon}(\cdot)) < \hat{\varepsilon},$$

such that, if $(X^{\delta,\varepsilon}(\cdot), Y^{\delta,\varepsilon}(\cdot))$ is the solution of (2.1) with $Z(\cdot) = Z^{\delta,\varepsilon}(\cdot)$, then the triplet $(X^{\delta,\varepsilon}, Y^{\delta,\varepsilon}, Z^{\delta,\varepsilon})$ is an approximate solution of (1.1) with accuracy $3\hat{\varepsilon} + \sqrt{3\hat{\varepsilon}}$.

Proof. Let $V(0,x,y) = 0$. Since $V = V^{0,0}$, by Theorem 3.7, there exist $\delta, \varepsilon > 0$, such that $V^{\delta,\varepsilon}(0,x,y) < \hat{\varepsilon}$. Now by (3.9) we can find a $Z^{\varepsilon,\delta} \in \mathcal{Z}_\delta[0,T]$ such that (3.40) is satisfied. Let $(X^{\delta,\varepsilon}, Y^{\delta,\varepsilon})$ be the solutions of (2.1) with $s = 0$, and $Z = Z^{\delta,\varepsilon}$. Then we have (see (1.21))

$$\begin{aligned} E|Y^{\delta,\varepsilon}(T) &- g(X^{\delta,\varepsilon}(T))| \\ &\leq 3Ef(X^{\delta,\varepsilon}(T), Y^{\delta,\varepsilon}(T)) + \sqrt{3Ef(X^{\delta,\varepsilon}(T), Y^{\delta,\varepsilon}(T))} \\ &= 3J^{\delta,\varepsilon}(0,x,y;Z^{\delta,\varepsilon}(\cdot)) + \sqrt{3J^{\delta,\varepsilon}(0,x,y;Z^{\delta,\varepsilon}(\cdot))} \\ &\leq 3\hat{\varepsilon} + \sqrt{3\hat{\varepsilon}}. \end{aligned}$$

This proves our assertion. □

To conclude this section, we present the following result.

Proposition 3.10. *Let (H1) and (H3) hold. Then* $\widetilde{V}^{\delta,\varepsilon}(s,x,y)$ *is semi-concave uniformly in* $s \in [0,T]$, $\delta \in (0,1]$ *and* $\varepsilon \in [0,1]$. *In particular, there exists a constant* $C > 0$, *such that*

$$(3.41) \quad \Delta_y \widetilde{V}^{\delta,\varepsilon}(s,x,y) \leq C, \quad \forall (s,x,y) \in [0,T] \times \mathbb{R}^n \times \mathbb{R}^m, \ \delta,\varepsilon \in (0,1],$$

where $\Delta_y = \sum_{j=1}^m \partial_{y_j}^2$.

Proof. The proof of the first claim is similar to that of Theorem 3.4. To show the second one, we need only to note that by (3.7),

$$\Delta_y \widetilde{V}^{\delta,\varepsilon} = \mathrm{tr}\left\{ \begin{pmatrix} 0 & 0 \\ 0 & I \end{pmatrix} D^2 \widetilde{V}^{\delta,\varepsilon} \begin{pmatrix} 0 & 0 \\ 0 & I \end{pmatrix} \right\} \leq C,$$

which gives (3.41). □

§4. A Class of Approximately Solvable FBSDEs

We have seen from Proposition 1.4 that if (1.13) holds, then (0.1) is approximate solvable. Further, from §3 we see that if $(x,y) \in \mathbb{R}^n \times \mathbb{R}^m$ is such that $V(0,x,y) = 0$, then one can actually construct a sequence of approximate solutions to (0.1). Finally, (1.13) is also an important step of solving (0.1) (see Proposition 1.1). In this section, we look for conditions under which (1.13) holds. Moreover, we would like to construct the nodal set $\mathcal{N}(\overline{V})$ for some special and interesting cases.

In what follows, we restrict ourselves to the following FBSDE:

$$(4.1) \quad \begin{cases} dX(t) = b(t, X(t), Y(t))dt + \sigma(t, X(t), Y(t))dW(t), \\ dY(t) = h(t, X(t), Y(t))dt + Z(t)dW(t), \quad t \in [0,T], \\ X(0) = x, \quad Y(T) = g(X(T)). \end{cases}$$

The difference between (0.1) and (4.1) is that in (4.1), the functions b, σ and h are all independent of Z. To study the set $\mathcal{N}(\overline{V})$, we introduce the nodal set of value function $V(s,x,y)$:

$$(4.2) \quad \mathcal{N}(V) = \{ (s,x,y) \in [0,T] \times \mathbb{R}^n \times \mathbb{R}^m \mid V(s,x,y) = 0 \}.$$

Clearly,

$$(4.3) \quad \{0\} \times \mathcal{N}(\overline{V}) = \mathcal{N}(V) \bigcap [\{0\} \times \mathbb{R}^n \times \mathbb{R}^m].$$

We will study $\mathcal{N}(V)$ below, which will automatically give the information on $\mathcal{N}(\overline{V})$ that we are looking for.

Let us now first make an observation. Suppose there exists a function $\theta : [0,T] \times \mathbb{R}^n \to \mathbb{R}^m$, such that

$$(4.4) \quad V(s,x,\theta(s,x)) = 0, \quad \forall (s,x) \in [0,T] \times \mathbb{R}^n,$$

then it holds

(4.5) $\{(s, x, \theta(s, x)) \mid (s, x) \in [0, T] \times \mathbb{R}^n\} \subseteq \mathcal{N}(V).$

In particular,

(4.6) $(x, \theta(0, x)) \in \mathcal{N}(\overline{V}), \qquad \forall x \in \mathbb{R}^n.$

This gives the nonemptiness of the nodal set $\mathcal{N}(\overline{V})$. Thus, finding some way of determining $\theta(s, x)$ is very useful. Now, let us assume that both V and θ are smooth and we find an equation that is satisfied by θ (so that (4.4) holds). To this end, we define

(4.7) $w(s, x) = V(s, x, \theta(s, x)), \qquad \forall (s, x) \in [0, T] \times \mathbb{R}^n.$

Differentiating the above, we obtain

(4.8) $\begin{cases} w_s = V_s + \langle V_y, \theta_s \rangle, \\ w_{x_i} = V_{x_i} + \langle V_y, \theta_{x_i} \rangle, \qquad 1 \leq i \leq n, \\ w_{x_i x_j} = V_{x_i x_j} + \langle V_{x_i y}, \theta_{x_j} \rangle + \langle V_{y x_j}, \theta_{x_i} \rangle \\ \qquad\qquad + \langle V_y, \theta_{x_i x_j} \rangle + \langle V_{yy} \theta_{x_j}, \theta_{x_i} \rangle, \quad 1 \leq i, j \leq n. \end{cases}$

Clearly,

(4.9) $\operatorname{tr}[\sigma \sigma^T w_{xx}] = \operatorname{tr}\{\sigma \sigma^T [V_{xx} + 2V_{xy}\theta_x + \theta_x^T V_{yy}\theta_x + \langle V_y, \theta_{xx} \rangle]\},$

where we note that V_{xy} is an $(n \times m)$ matrix and θ_x is $(m \times n)$ matrix. Then it follows from (2.12) that (recall (4.1) for the form of functions b, σ and h)

$$0 = V_s + \frac{1}{2}\operatorname{tr}[\sigma \sigma^T V_{xx}] + \langle b, V_x \rangle + \langle h, V_y \rangle$$
$$+ \frac{1}{2} \inf_{z \in \mathbb{R}^{m \times d}} \operatorname{tr}[V_{xy}^T \sigma z^T + V_{xy} z \sigma^T + V_{yy} z z^T]$$
$$= w_s - \langle V_y, \theta_s \rangle + \frac{1}{2}\operatorname{tr}[\sigma \sigma^T (w_{xx} - 2V_{xy}\theta_x - \theta_x^T V_{yy}\theta_x)]$$
$$+ \langle b, w_x - \theta_x^T V_y \rangle + \langle h, V_y \rangle - \frac{1}{2} \langle \operatorname{tr}\sigma\sigma^T \theta_{xx}, V_y \rangle$$
(4.10)
$$+ \frac{1}{2} \inf_{z \in \mathbb{R}^{m \times d}} \operatorname{tr}[V_{xy}^T \sigma z^T + V_{xy} z \sigma^T + V_{yy} z z^T]$$
$$= \{w_s + \frac{1}{2}\operatorname{tr}[\sigma \sigma^T w_{xx}] + \langle b, w_x \rangle\}$$
$$- \langle V_y, \theta_s + \frac{1}{2}\operatorname{tr}[\sigma \sigma^T \theta_{xx}] + \theta_x b - h \rangle\}$$
$$+ \frac{1}{2} \inf_{z \in \mathbb{R}^{m \times d}} \operatorname{tr}[2(z - \theta_x \sigma)\sigma^T V_{xy} + (z z^T - \theta_x \sigma \sigma^T \theta_x^T) V_{yy}].$$

Thus, if we suppose θ to be a solution of the following system:

(4.11) $\begin{cases} \theta_s + \frac{1}{2}\operatorname{tr}[\sigma \sigma^T \theta_{xx}] + \theta_x b - h = 0, \quad (s, x) \in [0, T) \times \mathbb{R}^n, \\ \theta\big|_{s=T} = g. \end{cases}$

Then we have

$$(4.12) \qquad \begin{cases} w_s + \dfrac{1}{2} \text{tr}\,[\sigma \sigma^T w_{xx}] + \langle\, b, w_x \,\rangle \geq 0, \\[2mm] w\big|_{s=T} = 0. \end{cases}$$

Hence, by maximum principle, we obtain

$$(4.13) \qquad 0 \geq w(s,x) \equiv V(s,x,\theta(s,x)) \geq 0, \qquad \forall (s,x) \in [0,T] \times \mathbb{R}^n.$$

This gives (4.4). The above gives a proof of the following proposition.

Proposition 4.1. *Suppose the value function V is smooth and θ is a classical solution of (4.11). Then (4.4) holds.*

We know that $V(s,x,y)$ is not necessarily smooth. Also since $\sigma \sigma^T$ could be degenerate, (4.11) might have no classical solutions. Thus, the assumptions of Proposition 4.1 are rather restrictive. The goal of the rest of the section is to prove a result similar to the above without assuming the smoothness of V and the nondegeneracy of $\sigma \sigma^T$. To this end, we need the following assumption.

(H4) Function $g(x)$ is bounded in $C^{2+\alpha}(\mathbb{R}^n)$ for some $\alpha \in (0,1)$ and there exists a constant $L > 0$, such that

$$(4.14) \quad |b(s,x,0)| + |\sigma(s,x,0)| + |h(s,x,0)| \leq L, \quad \forall (s,x) \in [0,T] \times \mathbb{R}^n.$$

Our main result of this section is the following.

Theorem 4.3. *Let (H1)–(H3) hold. Then, for any $x \in \mathbb{R}^n$, (1.13) holds, and thus, (4.1) is approximately solvable.*

To prove this theorem we need some lemmas.

Lemma 4.4. *Let (H1)–(H3) hold. Then, for any $\varepsilon > 0$, there exists a unique classical solution $\theta^\varepsilon : [0,T] \times \mathbb{R}^n \to \mathbb{R}^m$ of the following (nondegenerate) parabolic system:*

$$(4.15) \qquad \begin{cases} \theta^\varepsilon_s + \varepsilon \Delta \theta^\varepsilon + \dfrac{1}{2} \text{tr}\,[\sigma \sigma^T \theta^\varepsilon_{xx}] + \theta^\varepsilon_x b - h = 0, \quad (s,x) \in [0,T) \times \mathbb{R}^n, \\[2mm] \theta^\varepsilon\big|_{s=T} = g, \end{cases}$$

with θ^ε, $\theta^\varepsilon_{x_i}$ and $\theta^\varepsilon_{x_i x_j}$ all being bounded (with the bounds depending on $\varepsilon > 0$, in general). Moreover, there exists a constant $C > 0$, independent of $\varepsilon \in (0,1]$, such that

$$(4.16) \qquad |\theta^\varepsilon(s,x)| \leq C, \quad \forall (s,x) \in [0,T] \times \mathbb{R}^n, \ \varepsilon \in (0,1].$$

Proof. We note that under (H1)–(H3), following hold:

(4.17)
$$\begin{cases} 0 \leq (\sigma\sigma^T)(s,x,y) \leq C(1+|y|^2)I, \\ |(\sigma_{x_i}\sigma^T)(s,x,y)| + |(\sigma_{y_k}\sigma^T)(s,x,y)| \leq C(1+|y|), \\ \qquad\qquad\qquad\qquad\qquad 1 \leq i \leq n,\ 1 \leq k \leq m, \\ |b(s,x,y)| \leq L(1+|y|), \\ -\langle h(s,x,y), y \rangle \leq L(1+|y|^2). \end{cases}$$

Thus, by Ladyzenskaja, et al [1], we know that for any $\varepsilon > 0$, there exists a unique classical solution θ^ε to (4.15) with θ^ε, $\theta^\varepsilon_{x_i}$ and $\theta^\varepsilon_{x_i x_j}$ all being bounded (with the bounds depending on $\varepsilon > 0$). Next, we prove (4.16). To this end, we fix an $\varepsilon \in (0,1]$ and denote

(4.18)
$$\mathcal{A}_\varepsilon w \stackrel{\Delta}{=} \varepsilon\Delta w + \frac{1}{2}\mathrm{tr}\,[\sigma\sigma^T(s,x,\theta^\varepsilon(s,x))w_{xx}] + \langle b(s,x,\theta^\varepsilon(s,x)), w_x \rangle$$
$$\equiv \sum_{i,j=1}^n a^\varepsilon_{ij} w_{x_i x_j} + \sum_{i=1}^n b^\varepsilon_i w_{x_i}.$$

Set

(4.19)
$$\overline{w}(s,x) \stackrel{\Delta}{=} \frac{1}{2}|\theta^\varepsilon(s,x)|^2 \equiv \frac{1}{2}\sum_{i=1}^m \theta^{\varepsilon,k}(s,x)^2.$$

Then it holds that (note (4.17))

$$\overline{w}_s = \sum_{k=1}^m \theta^{\varepsilon,k}\theta^{\varepsilon,k}_s = \sum_{k=1}^m \theta^{\varepsilon,k}[-\mathcal{A}_\varepsilon\theta^{\varepsilon,k} + h^k(s,x,\theta^\varepsilon)]$$
$$= \sum_{k=1}^m \theta^{\varepsilon,k}\Big[-\sum_{i,j=1}^n a^\varepsilon_{ij}\theta^{\varepsilon,k}_{x_i x_j} - \sum_{i=1}^n b^\varepsilon_i\theta^{\varepsilon,k}_{x_i} + h^k(s,x,\theta^\varepsilon) \Big]$$
$$= -\sum_{k=1}^m \sum_{i,j=1}^n a^\varepsilon_{ij}\{[(\tfrac{1}{2}\theta^{\varepsilon,k})^2]_{x_i x_j} - \theta^{\varepsilon,k}_{x_i}\theta^{\varepsilon,k}_{x_j}\}$$
$$\quad - \sum_{k=1}^m \sum_{i=1}^n b^\varepsilon_i[(\tfrac{1}{2}\theta^{\varepsilon,k})^2]_{x_i} + \sum_{k=1}^m \theta^{\varepsilon,k}h^k(s,x,\theta^\varepsilon)$$
$$\geq -\mathcal{A}_\varepsilon\overline{w} - 2L\overline{w} - L.$$

Thus, $\overline{w}$ is a bounded (with the bound depending on $\varepsilon > 0$) solution of the following:

(4.20)
$$\begin{cases} \overline{w}_s + \mathcal{A}_\varepsilon\overline{w} + 2L\overline{w} \geq -L, \qquad (s,x) \in [0,T) \times \mathbb{R}^n, \\ \overline{w}\big|_{s=T} \leq \frac{1}{2}\|g\|_\infty. \end{cases}$$

By Lemma 4.5 below, we obtain

(4.21)
$$\overline{w}(s,x) \leq C, \qquad \forall (s,x) \in [0,T] \times \mathbb{R}^n,$$

with the constant only depending on L and $\|g\|_\infty$ (and independent of $\varepsilon > 0$). Since w is nonnegative by definition (see (4.19)), (4.16) follows.   □

In the above, we have used the following lemma. In what follows, this lemma will be used again.

Lemma 4.5. *Let $\mathcal{A}_\varepsilon$ be given by (4.18) and w be a bounded solution of the following:*

$$(4.22) \qquad \begin{cases} w_s + \mathcal{A}_\varepsilon w + \lambda_0 w \geq -h_0, & (s, x) \in [0, T) \times \mathbb{R}^n, \\ w|_{s=T} \leq g_0, \end{cases}$$

for some constants $h_0, g_0 \geq 0$ and $\lambda_0 \in \mathbb{R}$, with the bound of w might depend on $\varepsilon > 0$, in general. Then, for any $\lambda > \lambda_0 \vee 0$,

$$(4.23) \qquad w(s, x) \leq e^{\lambda T}\left[g_0 \vee \frac{h_0}{\lambda - \lambda_0}\right], \quad \forall (s, x) \in [0, T] \times \mathbb{R}^n.$$

Proof. Fix any $\lambda > \lambda_0 \vee 0$. For any $\beta > 0$, we define

$$(4.24) \qquad \Phi(s, x) = e^{\lambda s} w(s, x) - \beta|x|^2, \qquad \forall (s, x) \in [0, T] \times \mathbb{R}^n.$$

Since $w(s, x)$ is bounded, we see that

$$(4.25) \qquad \lim_{|x| \to \infty} \Phi(s, x) = -\infty.$$

Thus, there exists a point $(\bar{s}, \bar{x}) \in [0, T] \times \mathbb{R}^n$ (depending on $\beta > 0$), such that

$$(4.26) \qquad \Phi(s, x) \leq \Phi(\bar{s}, \bar{x}), \qquad \forall (s, x) \in [0, T] \times \mathbb{R}^n.$$

In particular,

$$(4.27) \qquad e^{\lambda \bar{s}} w(\bar{s}, \bar{x}) - \beta|\bar{x}|^2 = \Phi(\bar{s}, \bar{x}) \geq \Phi(T, 0) = e^{\lambda T} w(T, 0),$$

which yields

$$(4.28) \qquad \beta|\bar{x}|^2 \leq e^{\lambda \bar{s}} w(\bar{s}, \bar{x}) - e^{\lambda T} w(T, 0) \leq C_\varepsilon.$$

We have two cases. First, if there exists a sequence $\beta \downarrow 0$, such that $\bar{s} = T$, then, for any $(s, x) \in [0, T] \times \mathbb{R}^n$, we have

$$(4.29) \qquad \begin{aligned} w(s, x) &\leq e^{-\lambda s}[\beta|x|^2 + \Phi(T, \bar{x})] \\ &\leq e^{-\lambda s}[\beta|x|^2 + e^{\lambda T} g_0 - \beta|\bar{x}|^2] \\ &\leq \beta|x|^2 + e^{\lambda T} g_0 \to e^{\lambda T} g_0, \quad \text{as } \beta \to 0. \end{aligned}$$

We now assume that for any $\beta > 0$, $\bar{s} < T$. In this case, we have

$$(4.30) \qquad \begin{aligned} 0 &\geq (\Phi_s + \mathcal{A}_\varepsilon \Phi)(\bar{s}, \bar{x}) \\ &= \lambda e^{\lambda \bar{s}} w + e^{\lambda \bar{s}}[w_s + \mathcal{A}_\varepsilon w] - \beta \mathcal{A}_\varepsilon(|x|^2)\big|_{x=\bar{x}} \\ &\geq (\lambda - \lambda_0) e^{\lambda \bar{s}} w - e^{\lambda \bar{s}} h_0 - \beta \mathcal{A}_\varepsilon(|x|^2)\big|_{x=\bar{x}}. \end{aligned}$$

Note that (see (4.28))

$$\mathcal{A}_\varepsilon(|x|^2)\big|_{x=\bar{x}} = 2n\varepsilon + |\sigma(\bar{s},\bar{x},\theta^\varepsilon(\bar{s},\bar{x}))|^2 + 2\langle b(\bar{s},\bar{x},\theta^\varepsilon(\bar{s},\bar{x})),\bar{x}\rangle$$
$$\leq 2n\varepsilon + C_\varepsilon + C_\varepsilon|\bar{x}| \leq C_\varepsilon + C_\varepsilon\beta^{-1/2}.$$

Hence, for any $(s,x) \in [0,T] \times \mathbb{R}^n$, we have

$$e^{\lambda s}w(s,x) - \beta|x|^2 = \Phi(s,x) \leq \Phi(\bar{s},\bar{x}) = e^{\lambda\bar{s}}w(\bar{s},\bar{x}) - \beta|\bar{x}|^2$$
$$\leq \frac{e^{\lambda\bar{s}}h_0}{\lambda-\lambda_0} + \frac{e^{\lambda\bar{s}}}{\lambda-\lambda_0}\beta\mathcal{A}_\varepsilon(|x|^2)\big|_{x=\bar{x}}$$
$$\leq \frac{e^{\lambda T}h_0}{\lambda-\lambda_0} + \frac{e^{\lambda T}}{\lambda-\lambda_0}(\beta C_\varepsilon + \sqrt{\beta}C_\varepsilon).$$

Sending $\beta \to 0$, we obtain

$$(4.31) \qquad w(s,x) \leq \frac{e^{\lambda T}h_0}{\lambda-\lambda_0}, \qquad \forall(s,x) \in [0,T] \times \mathbb{R}^n.$$

Combining (4.29) and (4.31), one obtains (4.23). $\square$

Proof of Theorem 4.3. We define (note (3.24))

$$w^{\delta,\varepsilon}(s,x) \stackrel{\Delta}{=} \widetilde{V}^{\delta,\varepsilon}(s,x,\theta^\varepsilon(s,x)) \geq 0, \quad \forall(s,x) \in [0,T] \times \mathbb{R}^n.$$

Then we obtain (using (3.25), (3.29) and (4.15))

$$0 = \widetilde{V}_s^{\delta,\varepsilon} + \varepsilon\Delta\widetilde{V}^{\delta,\varepsilon} + \frac{1}{2}\mathrm{tr}\,[\sigma\sigma^T\widetilde{V}_{xx}^{\delta,\varepsilon}] + \langle b, \widetilde{V}_x^{\delta,\varepsilon}\rangle + \langle h, \widetilde{V}_y^{\delta,\varepsilon}\rangle$$
$$+ \frac{1}{2}\inf_{|z|\leq 1/\delta}\mathrm{tr}\,[(\widetilde{V}_{xy}^{\delta,\varepsilon})^T\sigma z^T + \widetilde{V}_{xy}^{\delta,\varepsilon}z\sigma^T + \widetilde{V}_{yy}^{\delta,\varepsilon}zz^T]$$
$$(4.32)\quad = \left\{w_s^{\delta,\varepsilon} + \varepsilon\Delta w^{\delta,\varepsilon} + \frac{1}{2}\mathrm{tr}\,[\sigma\sigma^T w_{xx}^{\delta,\varepsilon}] + \langle b, w_x^{\delta,\varepsilon}\rangle\right\} + \varepsilon\Delta_y\widetilde{V}^{\delta,\varepsilon}$$
$$- \langle \widetilde{V}_y^{\delta,\varepsilon}, \theta_s^\varepsilon + \varepsilon\Delta\theta^\varepsilon + \frac{1}{2}\mathrm{tr}\,[\sigma\sigma^T\theta_{xx}^\varepsilon] + \theta_x^\varepsilon b - h\rangle$$
$$+ \frac{1}{2}\inf_{|z|\leq 1/\delta}\mathrm{tr}\,[2(z-\theta_x^\varepsilon\sigma)\sigma^T\widetilde{V}_{xy}^{\delta,\varepsilon} + (zz^T - \theta_x^\varepsilon\sigma\sigma^T(\theta_x^\varepsilon)^T)\widetilde{V}_{yy}^{\delta,\varepsilon}]$$
$$\leq \left\{w_s^{\delta,\varepsilon} + \varepsilon\Delta w^{\delta,\varepsilon} + \frac{1}{2}\mathrm{tr}\,[\sigma\sigma^T w_{xx}^{\delta,\varepsilon}] + \langle b, w_x^{\delta,\varepsilon}\rangle\right\} + \varepsilon C.$$

The above is true for all $\varepsilon,\delta > 0$ such that $|\theta_x^\varepsilon(s,x)\sigma(s,x,\theta^\varepsilon(s,x))| \leq \frac{1}{\delta}$, which is always possible for any fixed ε, and $\delta > 0$ sufficiently small. Then we obtain

$$\begin{cases} w_s^{\delta,\varepsilon} + \mathcal{A}_\varepsilon w^{\delta,\varepsilon} \geq -\varepsilon C, \quad \forall(s,x) \in [0,T] \times \mathbb{R}^n, \\ w^{\delta,\varepsilon}\big|_{s=T} = 0. \end{cases}$$

On the other hand, by (H1) and (H3), we see that corresponding to the control $Z_\delta(\cdot) = 0 \in \widetilde{\mathcal{Z}}_\delta[s,T]$, we have (by Gronwall's inequality) $|Y(T)| \leq$

$C(1 + |y|)$, almost surely. Thus, by the boundedness of g, we obtain (using Lemma 4.5)

$$0 \leq w^{\delta,\varepsilon}(s,x) \equiv \widetilde{V}^{\delta,\varepsilon}(s,x,\theta^\varepsilon(s,x))$$
$$\leq \widetilde{J}^{\delta,\varepsilon}(s,x,\theta^\varepsilon(s,x);0) \leq C(1 + |\theta^\varepsilon(s,x)|) \leq C.$$

Next, by Lemma 4.5 (with $\lambda_0 = g_0 = 0$, $\lambda = 1$ and $h_0 = \varepsilon C$), we must have $w^{\delta,\varepsilon}(s,x) \leq \varepsilon C e^T$, $\forall (s,x) \in [0,T] \times \mathbb{R}^n$. Thus, we obtain the following conclusion: There exists a constant $C_0 > 0$, such that for any $\varepsilon > 0$, one can find a $\delta = \delta(\varepsilon)$ with the property that

(4.33) $$0 \leq \widetilde{V}^{\delta,\varepsilon}(s,x,\theta^\varepsilon(s,x)) \leq \varepsilon C_0, \qquad \forall \delta \leq \delta(\varepsilon).$$

Then, by (3.28), (3.39) (with $\delta = 0$) and (4.33), we obtain

$$0 \leq V(0,x,\theta^\varepsilon(0,x)) \leq |\widetilde{V}^{0,0}(0,x,\theta^\varepsilon(0,x)) - \widetilde{V}^{\delta,\varepsilon}(0,x,\theta^\varepsilon(0,x))| + \varepsilon C_0$$
$$\leq \eta_0(\varepsilon + \delta, |x| + |\theta^\varepsilon(0,x)|) + \varepsilon C_0.$$

Now, we let $\delta \to 0$ and then $\varepsilon \to 0$ to get the right hand side of the above going to 0. This can be achieved due to (4.16). Finally, since $\theta^\varepsilon(s,x)$ is bounded, we can find a convergent subsequence. Thus, we obtain that $V(0,x,y) = 0$, for some $y \in \mathbb{R}^m$. This implies (1.13).   □

§5. Construction of Approximate Adapted Solutions

We have already noted that in order that the method of optimal control works completely, one has to actually find the optimal control of the *Problem (OC)*, with the initial state satisfying the constraint (1.13). But on the other hand, due to the non-compactness of the control set (i.e., there is no *a priori* bound for the process Z), the existence of the optimal control itself is a rather complicated issue. The conceivable routes are either to solve the problem by considering *relaxed* control, or to figure out an a priori compact set in which the process Z lives (it turns out that such a compact set can be found theoretically in some cases, as we will see in the next chapter). However, compared to the other methods that will be developed in the following chapters, the main advantage of the method of optimal control lies in that it provides a tractable way to construct the approximate solution for fairly large class of the FBSDEs, which we will focus on in this section.

To begin with, let us point out that in Corollary 3.9 we had a scheme of constructing the approximate solution, provided that one is able to start from the right initial position $(x,y) \in \mathcal{N}(\overline{V})$ (or equivalently, $V(0,x,y) = 0$). The draw back of that scheme is that one usually do not have a way to access the value function $\overline{V}$ directly, again due to the possible degeneracy of the forward diffusion coefficient σ and the non-compactness of the admissible control set $\mathcal{Z}[0,T]$. The scheme of the special case in §4 is also restrictive, because it involves some other subtleties such as, among others, the estimate (4.16).

To overcome these difficulties, we will first try to start from some initial state that is "close" to the nodal set $\mathcal{N}(\overline{V})$ in a certain sense. Note that the unique strong solution to the HJB equation (3.25), $\widetilde{V}^{\delta,\varepsilon}$, is the value function of a *regularized* control problem with the state equation (3.22), which is non-degenerate and with compact control set, thus many standard methods can be applied to study its analytical and numerical properties, on which our scheme will rely. For notational convenience, in this section we assume that all the processes involved are one dimensional (i.e., $n = m = d = 1$). However, one should be able to extend the scheme to general higher dimensional cases without substantial difficulties. Furthermore, throughout this section we assume that

(H4) $g \in C^2$; and there exists a constant $L > 0$, such that for all $(t, x, y, z) \in [0, T] \times \mathbb{R}^3$,

(5.1)
$$\begin{cases} |b(t,x,y,z)| + |\sigma(t,x,y,z)| + |h(t,x,y,z)| \le L(1+|x|); \\ |g'(x)| + |g''(x)| \le L. \end{cases}$$

We first give a lemma that will be useful in our discussion.

Lemma 5.1. *Let (H1) and (H4) hold. Then there exists a constant $C > 0$, depending only on L and T, such that for all $\delta, \varepsilon \ge 0$, and $(s, x, y) \in [0, T] \times \mathbb{R}^2$, it holds that*

(5.2)
$$\widetilde{V}^{\delta,\varepsilon}(s, x, y) \ge f(x, y) - C(1 + |x|^2),$$

where $f(x, y)$ is defined by (1.6).

Proof. First, it is not hard to check that the function f is twice continuously differentiable, such that for all $(x, y) \in \mathbb{R}^2$ the following hold:

(5.3)
$$\begin{cases} |f_x(x,y)| \le |g'(x)|, \quad |f_y(x,y)| \le 1, \\ f_{xx}(x,y) = \dfrac{(g(x) - y)g''(x)}{[1 + (y - g(x))^2]^{1/2}} + \dfrac{g'(x)^2}{[1 + (y - g))^2]^{3/2}}, \\ f_{yy}(x,y) = \dfrac{1}{[1 + (y - g(x))^2]^{\frac{3}{2}}} > 0, \quad f_{xy}(x,y) = -g'(x)f_{yy}(x,y). \end{cases}$$

Now for any $\delta, \varepsilon \ge 0$, $(s, x, y) \in [0, T] \times \mathbb{R}^2$ and $Z \in \widetilde{\mathcal{Z}}_\delta[s, T]$, let (X, Y) be the corresponding solution to the controlled system (3.22). Applying Itô's formula we have

(5.4)
$$\widetilde{J}^{\delta,\varepsilon}(s, x, y; Z) = Ef(X(T), Y(T))$$
$$= f(x, y) + E \int_s^T \Pi(t, X(t), Y(t), Z(t))dt,$$

where, denoting $(f_x = f_x(x,y), f_y = f_y(x,y),$ and so on),

(5.5)

$$\Pi(t,x,y,z) = f_x b(t,x,y,z) + f_y h(t,x,y,z)$$
$$+ \frac{1}{2}[f_{xx}\sigma^2(t,x,y,z) + 2f_{xy}\sigma(t,x,y,z)z + f_{yy}z^2]$$
$$\geq f_x b(t,x,y,z) + f_y h(t,x,y,z) + \frac{1}{2}\left[f_{xx} - \frac{f_{xy}^2}{f_{yy}}\right]\sigma^2(t,x,y,z)$$
$$\geq -C(1 + |x|^2),$$

where $C > 0$ depends only on the constant L in (H4), thanks to the estimates in (5.3). Note that (H4) also implies, by a standard arguments using Gronwall's inequality, that $E|X(t)|^2 \leq C(1 + |x|^2)$, $\forall t \in [0,T]$, uniformly in $Z(\cdot) \in \widetilde{\mathcal{Z}}_\delta[s,T]$, $\delta \geq 0$. Thus we derive from (5.4) and (5.5) that

$$\widetilde{V}^{\delta,\varepsilon}(s,x,y) = \inf_{Z \in \widetilde{\mathcal{Z}}_\delta[s,T]} \widetilde{J}^{\delta,\varepsilon}(s,x,y;Z)$$
$$= f(x,y) + \inf_{Z \in \widetilde{\mathcal{Z}}_\delta[s,T]} E \int_s^T \Pi(t,X(t),Y(t),Z(t))dt$$
$$\geq f(x,y) - C(1 + |x|^2),$$

proving the lemma. □

Next, for any $x \in \mathbb{R}$ and $r > 0$, we define

$$Q_x(r) \stackrel{\Delta}{=} \{y \in \mathbb{R} : f(x,y) \leq r + C(1 + |x|^2)\},$$

where $C > 0$ is the constant in (5.2). Since $\lim_{|y| \to \infty} f(x,y) = +\infty$, $Q_x(r)$ is a compact set for any $x \in \mathbb{R}$ and $r > 0$. Moreover, Lemma 5.1 shows that, for all $\delta, \varepsilon \geq 0$, one has

(5.6) $$\{y \in \mathbb{R} : \widetilde{V}^{\delta,\varepsilon}(0,x,y) \leq r\} \subseteq Q_x(r).$$

From now on we set $r = 1$. Recall that by Proposition 3.6 and Theorem 3.7, for any $\rho > 0$, and fixed $x \in \mathbb{R}$, we can first choose $\delta, \varepsilon > 0$ depending only on x and $Q_x(1)$, so that

(5.7) $$0 \leq \widetilde{V}^{\delta,\varepsilon}(0,x,y) < V(0,x,y) + \rho, \qquad \text{for all } y \in Q_x(1).$$

Now suppose that the FBSDE (1.1) is approximately solvable, we have from Proposition 1.4 that $\inf_{y \in \mathbb{R}} V(0,x,y) = 0$ (note that (H4) implies (H2)). By (5.6), we have

$$0 = \inf_{y \in \mathbb{R}} V(0,x,y) = \min_{y \in Q_x(1)} V(0,x,y).$$

Thus, by (5.7), we conclude the following

Lemma 5.2. Assume (H1) and (H4), and assume that the FBSDE (0.1) is approximately solvable. Then for any $\rho > 0$, there exist $\delta, \varepsilon > 0$ and depending only on ρ, x and $Q_x(1)$, such that

$$0 \leq \inf_{y \in \mathbb{R}} \widetilde{V}^{\delta,\varepsilon}(0,x,y) = \min_{y \in Q_x(1)} \widetilde{V}^{\delta,\varepsilon}(0,x,y) < \rho.$$

□

Our scheme of finding the approximate adapted solution of (0.1) starting from $X(0) = x$ can now be described as follows: for any integer k, we want to find $\{y^{(k)}\} \subset Q_x(1)$ and $\{Z^{(k)}\} \subset \mathcal{Z}[0,T]$ such that

(5.8) $$E f(X^{(k)}(T), Y^{(k)}(T)) \leq \frac{C_x}{k},$$

here and below $C_x > 0$ will denote generic constant depending only on L, T and x. To be more precise, we propose the following steps for each fixed k.

Step 1. Choose $0 < \delta < \frac{1}{k}$ and $0 < \varepsilon < \delta^4$, such that

$$\inf_{y \in \mathbb{R}} \widetilde{V}^{\delta,\varepsilon}(0, x, y) = \min_{y \in Q_x(1)} \widetilde{V}^{\delta,\varepsilon}(0, x, y) < \frac{1}{k}.$$

Step 2. For the given δ and ε, choose $y^{(k)} \in Q_x(1)$ such that

$$\widetilde{V}^{\delta,\varepsilon}(0, x, y^{(k)}) < \min_{y \in Q_x(1)} \widetilde{V}^{\delta,\varepsilon}(0, x, y) + \frac{1}{k}.$$

Step 3. For the given δ, ε, and $y^{(k)}$, find $Z^{(k)} \in \mathcal{Z}_\delta[0, T]$, such that

$$J(0, x, y^{(k)}; Z^{(k)}) = E f(X^{(k)}(T), Y^{(k)}(T)) \leq \widetilde{V}^{\delta,\varepsilon}(0, x, y^{(k)}) + \frac{C_x}{k},$$

where $(X^{(k)}, Y^{(k)})$ is the solution to (2.1) with $Y^{(k)}(0) = y^{(k)}$ and $Z = Z^{(k)}$; and C_x is a constant depending only on L, T and x.

It is obvious that a combination of the above three steps will serve our purpose (5.8). We would like to remark here that in the whole procedure we do not use the exact knowledge about the nodal set $\mathcal{N}(\overline{V})$, nor do we have to solve any degenerate parabolic PDEs, which are the two most formidable parts in this problem. Now that the Step 1 is a consequence of Lemma 5.2 and Step 2 is a standard (nonlinear) minimizing problem, we only briefly discuss Step 3. Note that $\widetilde{V}^{\delta,\varepsilon}$ is the value function of a regularized control problem, by standard methods of constructing ε-optimal strategies using information of value functions (e.g., Krylov [1, Ch.5]), we can find a Markov type control $\widehat{Z}^{(k)}(t) = \alpha^{(k)}(t, \widehat{X}^{(k)}(t), \widehat{Y}^{(k)}(t))$, where $\alpha^{(k)}$ is some smooth function satisfying $\sup_{t,x,y} |\alpha^{(k)}(t, x, y)| \leq \frac{1}{\delta}$ and $(\widehat{X}^{(k)}, \widehat{Y}^{(k)})$ is the corresponding solution of (4.8) with $\widehat{Y}^{(k)}(0) = y^{(k)}$, so that

(5.9) $$\widetilde{J}^{\delta,\varepsilon}(0, x, y^{(k)}; \widehat{Z}^{(k)}) < \widetilde{V}^{\delta,\varepsilon}(0, x, y^{(k)}) + \frac{1}{k}.$$

The last technical point is that (5.9) is only true if we use the state equation (3.22), which is different from (2.1), the original control problem that leads to the approximate solution that we need. However, if we denote $(X^{(k)}, Y^{(k)})$ to be the solutions to (2.1) with $Y^{(k)}(0) = y^{(k)}$ and the feedback control $Z^{(k)}(t) = \alpha^{(k)}(X^{(k)}(t), Y^{(k)}(t))$, then a simple calculation shows that

$$
\begin{aligned}
0 \le J(0, x, y^{(k)}; Z^{(k)}) &= Ef(X^{(k)}(T), Y^{(k)}(T)) \\
&< Ef(\widehat{X}^{(k)}(T), \widehat{Y}^{(k)}(T)) + C_\alpha \sqrt{2\varepsilon} \\
&< \widetilde{V}^{\delta,\varepsilon}(0, x, y^{(k)}) + \frac{1}{k} + C_\alpha \sqrt{2\varepsilon},
\end{aligned}
$$

(5.10)

thanks to (5.9), where C_α is some constant depending only on L, T and the Lipschitz constant of $\alpha^{(k)}$. But on the other hand, in light of Lemma 5.1 of Krylov [1], the Lipschitz constant of $\alpha^{(k)}$ can be shown to depend only on the bounds of the coefficients of the system (2.1) (i.e., b, h, σ, and $\widehat{\sigma}(z) \equiv z$) and their derivatives. Therefore using assumptions (H1) and (H4), and noting that $\sup_t |Z^{(k)}(t)| \le \sup |\alpha^{(k)}| \le \frac{1}{\delta}$, we see that, for fixed δ, C_α is no more than $C(1 + |x| + 1/\delta)$ where C is some constant depending only on L. Consequently, note the requirement we posed on ε and δ in Step 1, we have

$$
(5.11) \qquad C_\alpha \sqrt{2\varepsilon} < C(1 + |x| + \frac{1}{\delta})\sqrt{2\delta^4} \le 2\sqrt{2}C(1 + |x|)\delta \le \frac{C_x - 1}{k},
$$

where $C_x \overset{\Delta}{=} C(1 + |x|)2\sqrt{2} + 1$. Finally, we note that the process $Z^{(k)}(\cdot)$ obtain above is $\{\mathcal{F}_t\}_{t \ge 0}$-adapted and hence it is in $\mathcal{Z}_\delta[0, T]$ (instead of $\widetilde{\mathcal{Z}}_\delta[0, T]$). This, together with (5.10)–(5.11), fulfills Step 3.

Chapter 4

Four Step Scheme

In this chapter, we introduce a direct method for solving FBSDEs. Since this method contains four major steps, it has been called the *Four Step Scheme*.

§1. A Heuristic Derivation of Four Step Scheme

Let us consider the following FBSDE:

$$
(1.1) \quad \begin{cases} dX(t) = b(t, X(t), Y(t), Z(t))dt + \sigma(t, X(t), Y(t), Z(t))dW(t), \\ dY(t) = h(t, X(t), Y(t), Z(t))dt + Z(t)dW(t), \\ X(0) = x, \qquad Y(T) = g(X(T)). \end{cases}
$$

We assume throughout this section that the functions b, σ, h and g are deterministic. As we have seen in the previous chapter that for any given $x \in \mathbb{R}^n$, the solvability of (1.1) is essentially equivalent to the following:

$$
V(0, x, \theta(0, x)) = 0,
$$

where $\theta(s, x)$ is the "solution" of some parabolic system and $V(s, x, y)$ is the value function of the optimal control problem associated with the FBSDE (1.1). Assuming the Markov property (since coefficients are deterministic!) we suspect that $V(t, X(t), \theta(t, X(t))) = 0$, and $Y(t) = \theta(t, X(t))$ should hold for all t. In other words, we see a strong indication that there might some special relations among the components of an adapted solution (X, Y, Z), which we now explore.

Suppose that (X, Y, Z) is an adapted solution to (1.1). We assume that that Y and X are related by

$$
(1.2) \qquad Y(t) = \theta(t, X(t)), \qquad \forall t \in [0, T], \qquad \text{a.s. } \mathbf{P},
$$

where θ is some function to be determined. Let us assume that $\theta \in C^{1,2}([0, T] \times \mathbb{R}^n)$. Then by Itô's formula, we have for $1 \le k \le m$:

$$
\begin{aligned}
dY^k(t) &= d\theta^k(t, X(t)) \\
&= \Big\{ \theta_t^k(t, X(t)) + \langle \theta_x^k(t, X(t)), b(t, X(t), \theta(t, X(t)), Z(t)) \rangle \\
& \quad + \frac{1}{2}\text{tr}\, [\theta_{xx}^k(t, X(t))(\sigma\sigma^T)(t, X(t), \theta(t, X(t)), Z(t))] \Big\} dt \\
& \quad + \langle \theta_x^k(t, X(t)), \sigma(t, X(t), \theta(t, X(t)), Z(t))dW(t) \rangle.
\end{aligned}
$$

(1.3)

Comparing (1.3) and (1.1), we see that if θ is the right choice, it should be that, for $k = 1, \cdots, m$,

(1.4)
$$\begin{cases} h^k(t, X(t), \theta(t, X(t))) \\ \quad = \theta_t^k(t, X(t)) + \langle \theta_x^k(t, X(t)), b(t, X(t), \theta(t, X(t)), Z(t)) \rangle \\ \quad + \dfrac{1}{2}\mathrm{tr}\left[\theta_{xx}^k(t, X(t))(\sigma\sigma^T)(t, X(t), \theta(t, X(t)), Z(t))\right]; \\ \theta(T, X(T)) = g(X(T)), \end{cases}$$

and

(1.5) $\theta_x(t, X(t))\sigma(t, X(t), \theta(t, X(t)), Z(t)) = Z(t).$

The above heuristic arguments suggest the following *Four Step Scheme* for solving the FBSDE (1.1).

The Four Step Scheme:

  Step 1. Find a function $z(t, x, y, p)$ that satisfies the following:

(1.6)
$$\begin{aligned} &z(t, x, y, p) = p\sigma(t, x, y, z(t, x, y, p)), \\ &\forall (t, x, y, p) \in [0, T] \times \mathbb{R}^n \times \mathbb{R}^m \times \mathbb{R}^{m\times n}. \end{aligned}$$

  Step 2. Using the function z obtained in above to solve the following parabolic system for $\theta(t, x)$:

(1.7)
$$\begin{cases} \theta_t^k + \dfrac{1}{2}\mathrm{tr}\left[\theta_{xx}^k(\sigma\sigma^T)(t, x, \theta, z(t, x, \theta, \theta_x))\right] \\ \quad + \langle b(t, x, \theta, z(t, x, \theta, \theta_x)), \theta_x^k \rangle - h^k(t, x, \theta, z(t, x, \theta, \theta_x)) = 0, \\ \hspace{3.5cm} (t, x) \in [0, T) \times \mathbb{R}^n, \quad 1 \le k \le m, \\ \theta(T, x) = g(x), \hspace{1cm} x \in \mathbb{R}^n. \end{cases}$$

  Step 3. Using θ and z obtained in Steps 1–2 to solve the following forward SDE:

(1.8)
$$\begin{cases} dX(t) = \tilde{b}(t, X(t))dt + \tilde{\sigma}(t, X(t))dW(t), \quad t \in [0, T], \\ X(0) = x, \end{cases}$$

where

(1.9)
$$\begin{cases} \tilde{b}(t, x) = b(t, x, \theta(t, x), z(t, x, \theta(t, x), \theta_x(t, x))), \\ \tilde{\sigma}(t, x) = \sigma(t, x, \theta(t, x), z(t, x, \theta(t, x), \theta_x(t, x))). \end{cases}$$

  Step 4. Set

(1.10)
$$\begin{cases} Y(t) = \theta(t, X(t)), \\ Z(t) = z\big(t, X(t), \theta(t, X(t)), \theta_x(t, X(t))\big). \end{cases}$$

If the above scheme is realizable, (X, Y, Z) would give an adapted solution of (1.1). As a matter of fact, we have the following result.

Theorem 1.1. *Let (1.6) admit a unique solution $z(t,x,y,p)$ which is uniformly Lipschitz continuous in (x,y,p) with $z(t,0,0,0)$ being bounded. Let (1.7) admit a classical solution $\theta(t,x)$ with bounded θ_x and θ_{xx}. Let functions b and σ be uniformly Lipschitz continuous in (x,y,z) with $b(t,0,0,0)$ and $\sigma(t,0,0,0)$ being bounded. Then the process $(X(\cdot),Y(\cdot),Z(\cdot))$ determined by (1.8)–(1.10) is an adapted solution to (1.1). Moreover, if h is also uniformly Lipschitz continuous in (x,y,z), σ is bounded, and there exists a constant $\beta \in (0,1)$, such that*

$$
(1.11) \qquad
\begin{aligned}
&\left| [\sigma(s,x,y,z) - \sigma(s,x,y,\widetilde{z})]^T \theta_x^k(s,x) \right| \le \beta|z - \widetilde{z}|, \\
&\forall (s,x,y) \in [0,T] \times \mathbb{R}^n \times \mathbb{R}^m, \ z,\widetilde{z} \in \mathbb{R}^{m \times d},
\end{aligned}
$$

then the adapted solution is unique, which is determined by (1.8)–(1.10).

Proof. Under our conditions both $\widetilde{b}(t,x)$ and $\widetilde{\sigma}(t,x)$ (see (1.9)) are uniformly Lipschitz continuous in x. Thus, for any $x \in \mathbb{R}^n$, (1.8) has a unique strong solution. Then, by defining $Y(t)$ and $Z(t)$ via (1.10) and applying Itô's formula, we can easily check that (1.1) is satisfied. Hence, (X,Y,Z) is a solution of (1.1).

It remains to show the uniqueness. We claim that any adapted solution (X,Y,Z) of (1.1) must be of the form we constructed using the Four Step Scheme. To show this, let (X,Y,Z) be any solution of (1.1). We define

$$
(1.12) \qquad \widetilde{Y}(t) = \theta(t,X(t)), \quad \widetilde{Z}(t) = z(t,X(t),\theta(t,X(t)),\theta_x(t,X(t))).
$$

By our assumption, (1.6) admits a unique solution. Thus, (1.12) implies

$$
(1.13) \qquad \widetilde{Z}(t) = \theta_x(t,X(t))\sigma(t,X(t),\widetilde{Y}(t),\widetilde{Z}(t)), \qquad \text{a.s. } t \in [0,T].
$$

Now, applying Itô's formula to $\theta(t,X(t))$, noting (1.7) and (1.10), we have the following (for notational simplicity, we suppress t in $X(t)$, etc.):

$$
\begin{aligned}
d\widetilde{Y}^k(t) &= d\theta^k(t,X(t)) \\
&= \Big\{ \theta_t^k(t,X) + \langle \theta_x^k(t,X), b(t,X,Y,Z) \rangle + \tfrac{1}{2}\mathrm{tr}\,[\theta_{xx}^k(t,X)(\sigma\sigma^T)(t,X,Y,Z)] \Big\} dt \\
&\quad + \langle \theta_x^k(t,X), \sigma(t,X,Y,Z)dW(t) \rangle \\
&= \Big\{ \langle \theta_x^k(t,X), b(t,X,Y,Z) - b(t,X,\widetilde{Y},\widetilde{Z}) \rangle \\
&\quad + \tfrac{1}{2}\mathrm{tr}\,\Big[\theta_{xx}^k(t,X)\{(\sigma\sigma^T)(t,X,Y,Z) - (\sigma\sigma^T)(t,X,\widetilde{Y},\widetilde{Z})\} \Big] \\
&\quad + h^k(t,X,\widetilde{Y},\widetilde{Z}) \Big\} dt + \langle \theta_x^k(t,X), \sigma(t,X,Y,Z)dW(t) \rangle.
\end{aligned}
$$

Then, it follows from (1.1) and (1.13) that

$$
E|\tilde{Y}(t) - Y(t)|^2 = -E \int_t^T \sum_{k=1}^n \Big\{ 2(\tilde{Y}^k - Y^k) \cdot
$$

(1.14)
$$
\Big[\langle \theta_x^k(s, X), b(s, X, Y, Z) - b(s, X, \tilde{Y}, \tilde{Z}) \rangle
$$
$$
+ \frac{1}{2} \mathrm{tr} \big\{ \theta_{xx}^k(s, X) [(\sigma\sigma^T)(s, X, Y, Z) - (\sigma\sigma^T)(s, X, \tilde{Y}, \tilde{Z})] \big\}
$$
$$
+ h^k(s, X, \tilde{Y}, \tilde{Z}) - h^k(s, X, Y, Z) \Big]
$$
$$
+ \Big| \{\sigma(s, X, Y, Z) - \sigma(s, X, \tilde{Y}, \tilde{Z})\}^T \theta_x^k(s, X) + \tilde{Z} - Z \Big|^2 \Big\} ds.
$$

Since, by (1.11), the boundedness of θ_x, and the uniform Lipschitz continuity of σ, we have

$$
\Big| \{\sigma(s, X, Y, Z) - \sigma(s, X, \tilde{Y}, \tilde{Z})\}^T \theta_x^k(s, X) + \tilde{Z} - Z \Big|^2
$$
$$
\geq |\tilde{Z} - Z|^2 - |\sigma(s, X, Y, Z) - \sigma(s, X, \tilde{Y}, \tilde{Z})(\theta_x^k)^T(s, X)|^2
$$
$$
\geq (1 - \beta)|\tilde{Z} - Z|^2 - C|\tilde{Y} - Y|^2,
$$

here and in the sequel $C > 0$ is again a generic constant which may vary from line to line. Thus (1.14) leads to that

$$
E|\tilde{Y}(t) - Y(t)|^2 + (1 - \beta) \int_t^T E|Z(s) - \tilde{Z}(s)|^2 ds
$$

(1.15)
$$
\leq C \int_t^T E\{|\tilde{Y}(s) - Y(s)|^2 + |\tilde{Y}(s) - Y(s)||\tilde{Z}(s) - Z(s)|\} ds
$$
$$
\leq C_\varepsilon \int_t^T E|\tilde{Y}(s) - Y(s)|^2 ds + \varepsilon \int_t^T |\tilde{Z}(s) - Z(s)|^2 ds,
$$

where $\varepsilon > 0$ is arbitrary and C_ε depends on ε. Since $\beta < 1$, choosing $\varepsilon < 1 - \beta$ and applying Gronwall's inequality, we conclude that

(1.16) $Y(t) = \tilde{Y}(t), \qquad Z(t) = \tilde{Z}(t), \qquad$ a.s.,   a.e. $t \in [0, T]$

Thus any solution of (1.1) must have the form that we have constructed, proving our claim.

Finally, let (X, Y, Z) and $(\tilde{X}, \tilde{Y}, \tilde{Z})$ be any two solutions of (1.1). By the previous argument we have

(1.17)
$$
\begin{cases}
Y(t) = \theta(t, X(t)), & Z(t) = z(t, X(t), \theta(t, X(t)), \theta_x(t, X(t))), \\
\tilde{Y}(t) = \theta(t, \tilde{X}(t)), & \tilde{Z}(t) = z(t, \tilde{X}(t), \theta(t, \tilde{X}(t)), \theta_x(t, \tilde{X}(t))).
\end{cases}
$$

Hence $X(t)$ and $\tilde{X}(t)$ satisfy exactly the same forward SDE (1.8) with the same initial state x. Thus we must have

$$
X(t) = \tilde{X}(t), \qquad \forall t \in [0, T], \text{ a.s. } \mathbf{P},
$$

which in turn shows that (by (1.17))

$$Y(t) = \widetilde{Y}(t), \ Z(t) = \widetilde{Z}(t), \qquad \forall t \in [0, T], \text{ a.s. } \mathbf{P}.$$

The proof is now complete. $\qquad\qquad\qquad\qquad\qquad\qquad\qquad\qquad\qquad$ □

Remark 1.2. We note that the uniqueness of FBSDE (1.1) requires the condition (1.11), which is very hard to be verified in general and therefore looks *ad hoc*. However, we should note this condition is trivially true if σ is independent of z! Since the dependence of σ on variable z also causes difficulty in solving (1.6), the first step of the Four Step Scheme, in what follows to simplify discussion we often assume that $\sigma = \sigma(t, x, y)$ when the generality is not the main issue.

§2. Non-Degenerate Case — Several Solvable Classes

From the previous subsection, we see that to solve FBSDE (1.1), one needs only to look when the Four Step Scheme can be realized. In this subsection, we are going to find several such classes of FBSDEs.

§2.1. A general case

Let us make the following assumptions.

(A1) $d = n$; and the functions b, σ, h and g are smooth functions taking values in $\mathbb{R}^n$, $\mathbb{R}^m$, $\mathbb{R}^{n\times n}$, $\mathbb{R}^{m\times n}$ and $\mathbb{R}^m$, respectively, and with first order derivatives in x, y, z being bounded by some constant $L > 0$.

(A2) The function σ is independent of z and there exists a positive continuous function $\nu(\cdot)$ and a constant $\mu > 0$, such that for all $(t, x, y, z) \in [0, T] \times \mathbb{R}^n \times \mathbb{R}^m \times \mathbb{R}^{n\times m}$

$$(2.1) \qquad\qquad \nu(|y|)I \leq \sigma(t, x, y)\sigma(t, x, y)^T \leq \mu I,$$

$$(2.2) \qquad\qquad |b(t, x, 0, 0)| + |h(t, x, 0, z)| \leq \mu.$$

(A3) There exists a constant $\alpha \in (0, 1)$, such that g is bounded in $C^{2+\alpha}(\mathbb{R}^n)$.

Throughout this section, by "smooth" we mean that the involved functions possess partial derivatives of all necessary orders. We prefer not to indicate the exact order of smoothness for the sake of simplicity of presentation.

Since σ is independent of z, equation (1.6) is (trivially) uniquely solvable for z. In the present case, FBSDEs (1.1) reads as follows:

$$(2.3) \quad \begin{cases} dX(t) = b(t, X(t), Y(t), Z(t))dt + \sigma(t, X(t), Y(t))dW(t), \\ dY(t) = h(t, X(t), Y(t), Z(t))dt + Z(t)dW(t), \quad t \in [0, T], \\ X(0) = x, \qquad Y(T) = g(X(T)), \end{cases}$$

and (1.7) takes the following form:

$$(2.4) \quad \begin{cases} \theta_t^k + \dfrac{1}{2}\mathrm{tr}\left[\theta_{xx}^k(\sigma\sigma^T)(t,x,\theta)\right] + \langle\, b(t,x,\theta,z(t,x,\theta,\theta_x)),\theta_x^k\,\rangle \\ \qquad\qquad - h^k(t,x,\theta,z(t,x,\theta,\theta_x)) = 0, \\ \qquad\qquad (t,x) \in (0,T) \times \mathbb{R}^n, \quad 1 \le k \le m, \\ \theta(T,x) = g(x), \qquad x \in \mathbb{R}^n. \end{cases}$$

Let us first try to apply the result of Ladyzenskaja et al [1]. Consider the following initial boundary value problem:

$$(2.5) \quad \begin{cases} \theta_t^k + \displaystyle\sum_{i,j=1}^n a_{ij}(t,x,\theta)\theta_{x_ix_j} + \sum_{i=1}^n b_i(t,x,\theta,z(t,x,\theta,\theta_x))\theta_{x_i}^k \\ \qquad\qquad - h^k(t,x,\theta,z(t,x,\theta,\theta_x)) = 0, \\ \qquad\qquad (t,x) \in [0,T] \times B_R, \ 1 \le k \le m, \\ \theta\,\big|_{\partial B_R} = g(x), \qquad |x| = R, \\ \theta(T,x) = g(x), \qquad x \in B_R, \end{cases}$$

where B_R is the ball centered at the origin with radius $R > 0$ and

$$\begin{cases} \left(a_{ij}(t,x,y)\right) = \dfrac{1}{2}\sigma(t,x,y)\sigma(t,x,y)^T, \\ (b_1(t,x,y,z),\cdots,b_n(t,x,y,z))^T = b(t,x,y,z), \\ (h^1(t,x,y,z),\cdots,h^m(t,x,y,z))^T = h(t,x,y,z). \end{cases}$$

Clearly, under the present situation, the function $z(t,x,y,p)$ determined by (1.6) is smooth. We now give a lemma, which is an analogue of Ladyzenskaja et al [1, Chapter VII, Theorem 7.1].

Lemma 2.1. *Suppose that all the functions a_{ij}, b_i, h^k and g are smooth. Suppose also that for all $(t,x,y) \in [0,T] \times \mathbb{R}^n \times \mathbb{R}^m$ and $p \in \mathbb{R}^{m \times n}$, it holds that*

$$(2.6) \qquad \nu(|y|)I \le \left(a_{ij}(t,x,y)\right) \le \mu(|y|)I,$$

$$(2.7) \qquad |b(t,x,y,z(t,x,y,p))| \le \mu(|y|)(1 + |p|),$$

$$(2.8) \qquad \left|\dfrac{\partial}{\partial x_\ell}a_{ij}(t,x,y)\right| + \left|\dfrac{\partial}{\partial y^k}a_{ij}(t,x,y)\right| \le \mu(|y|),$$

for some continuous functions $\mu(\cdot)$ and $\nu(\cdot)$, with $\nu(r) > 0$;

$$(2.9) \qquad |h(t,x,y,z(t,x,y,p))| \le [\varepsilon(|y|) + P(|p|,|y|)](1 + |p|^2),$$

where $P(|p|,|y|) \to 0$, as $|p| \to \infty$ and $\varepsilon(|y|)$ is small enough;

$$(2.10) \qquad \sum_{k=1}^m h^k(t,x,y,z(t,x,y,p))y^k \ge -L(1 + |y|^2),$$

for some constant $L > 0$. Finally, suppose that g is bounded in $C^{2+\alpha}(\mathbb{R}^n)$ for some $\alpha \in (0,1)$. Then (2.5) admits a unique classical solution.   □

In the case g is bounded in $C^{2+\alpha}(\mathbb{R}^n)$, the solution of (2.5) and its partial derivatives $\theta(t,x)$, $\theta_t(t,x)$, $\theta_x(t,x)$ and $\theta_{xx}(t,x)$ are all bounded uniformly in $R > 0$ since only the interior type Schauder estimate is used. Using Lemma 2.1, we can now prove the solvability of (2.4) under our assumptions.

Theorem 2.2. *Let (A1)–(A3) hold. Then (2.4) admits a unique classical solution $\theta(t,x)$ which is bounded and $\theta_t(t,x)$, $\theta_x(t,x)$ and $\theta_{xx}(t,x)$ are all bounded as well. Consequently, FBSDE (2.3) is uniquely solvable.*

Proof. We first check that all the required conditions in Lemma 2.1 are satisfied. Since σ is independent of z, we see that the function $z(t,x,y,p)$ determined by (1.6) satisfies

$$(2.11) \quad |z(t,x,y,p)| \leq C|p|, \quad \forall (t,x,y,p) \in [0,T] \times \mathbb{R}^n \times \mathbb{R}^m \times \mathbb{R}^{m \times n}.$$

Now, we see that (2.6) and (2.8) follow from (A1) and (A2); (2.7) follows from (A1), (2.2) and (2.11); and (2.9)–(2.10) follow from (A1) and (2.2). Therefore, by Lemma 2.1 there exists a unique bounded solution $\theta(t,x;R)$ of (2.5) for which $\theta_t(t,x;R)$, $\theta_x(t,x;R)$ and $\theta_{xx}(t,x;R)$ together with $\theta(t,x)$ are bounded uniformly in $R > 0$. Using a diagonalization argument one further shows that there exists a subsequence $\theta(t,x,R)$ which converges uniformly to $\theta(t,x)$ as $R \to \infty$. Thus $\theta(t,x)$ is a classical solution of (2.4), and $\theta_t(t,x)$, $\theta_x(t,x)$ and $\theta_{xx}(t,x)$, as well as $\theta(t,x)$ itself, are all bounded.

Noting that all the functions together with the possible solutions are smooth with required bounded partial derivatives, the uniqueness follows from a standard argument using Gronwall's inequality.

Finally, by Theorem 1.1, FBSDE (2.3) is uniquely solvable.   □

§2.2. The case when h has linear growth in z

Although Theorem 2.2 gives a general solvability result of the FBSDE (2.3), condition (2.2) in (A2) is rather restrictive; for instance, the case that the coefficient $h(t,x,y,z)$ is linearly growing in z is excluded. This case, however, is very important for applications in optimal stochastic control theory. For example in the Pontryagin maximum principle for optimal stochastic control, the adjoint equation is of the form that the corresponding h is affine in z. Thus we would like to discuss this case separately.

In order to relax the condition (2.2), we compensate by considering the following special FBSDE:

$$(2.12) \quad \begin{cases} dX(t) = b(t, X(t), Y(t), Z(t))dt + \sigma(t, X(t))dW(t), \\ dY(t) = h(t, X(t), Y(t), Z(t))dt + Z(t)dW(t), \\ X(0) = x, \quad Y(T) = g(X(T)). \end{cases}$$

We assume that σ is independent of y and z, but we allow h to have a linear growth in z. In this case, the parabolic system looks like the following (compare with (2.4)):

(2.13)
$$\begin{cases} \theta_t^k + \dfrac{1}{2}\text{tr}\,(\theta_{xx}^k \sigma(t,x)\sigma(t,x)^T) + \langle\, b(t,x,\theta,z(t,x,\theta,\theta_x)),\theta_x^k\,\rangle \\ \qquad\qquad - h^k(t,x,\theta,z(t,x,\theta,\theta_x)) = 0, \\ \qquad\qquad\qquad (t,x) \in [0,T] \times \mathbb{R}^n, \quad 1 \le k \le m, \\ \theta(T,x) = g(x), \qquad x \in \mathbb{R}^n. \end{cases}$$

Since now h has linear growth in z, the result of Ladyzenskaja et al [1] does not apply. We use the result of Wiegner [1] instead. To this end, let us rewrite the above parabolic system in divergence form:

(2.14)
$$\begin{cases} \theta_t^k + \displaystyle\sum_{i,j=1}^n (a_{ij}(t,x)\theta_{x_i})_{x_j} = f^k(t,x,\theta,\theta_x), \\ \qquad\qquad\qquad (t,x) \in [0,T] \times \mathbb{R}^m, \quad 1 \le k \le m, \\ \theta(T,x) = g(x), \qquad x \in \mathbb{R}^n, \end{cases}$$

where

(2.15)
$$\begin{cases} \big(a_{ij}(t,x)\big) = \dfrac{1}{2}\sigma(t,x)\sigma(t,x)^T, \\ f^k(t,x,y,p) = \displaystyle\sum_{i,j=1}^n a_{ijx_j}(t,x)p_i^k - \sum_{i=1}^n b_i(t,x,y,z(t,x,y,p))p_i^k \\ \qquad\qquad + h^k(t,x,y,z(t,x,y,p)). \end{cases}$$

By Wiegner [1], we know that for any $T > 0$, (2.14) has a unique classical solution, global in time, provided the following conditions hold:

(2.16) $\qquad \nu I \le \big(a_{ij}(t,x)\big) \le \mu I, \qquad \forall (t,x) \in [0,T] \times \mathbb{R}^n,$

(2.17)
$$\sum_{k=1}^m y^k f^k(t,x,y,p) \le \varepsilon_0 |p|^2 + C(1+|y|^2),$$
$$\forall (t,x,y,p) \in [0,T] \times \mathbb{R}^n \times \mathbb{R}^m \times \mathbb{R}^{n \times m},$$

where $\nu, \mu, C, \varepsilon_0$ are constants with ε_0 being small enough. (To fit the framework of Wiegner [1], we have taken $H = |y|^2$, $c^k \equiv 0$ and $r^k \equiv 0$, $k = 1, \cdots, m$. See Wiegner [1] for details). Therefore, we need the following assumption:

(A2)′ There exist positive constants ν, μ, such that

(2.18) $\qquad \nu I \le \sigma(t,x)\sigma(t,x)^T \le \mu I, \qquad \forall (t,x) \in [0,T] \times \mathbb{R}^n,$

(2.19)
$$|b(t,x,y,z)|, \; |h(t,x,0,0)| \le \mu,$$
$$\forall (t,x,y,z) \in [0,T] \times \mathbb{R}^n \times \mathbb{R}^m \times \mathbb{R}^{m \times n}.$$

Theorem 2.3. *Suppose that (A1), (A2)' and (A3) hold. Then (2.12) admits a unique adapted solution (X, Y, Z).*

Proof. In the present case, for the function $z(t, x, y, p)$ determined by (1.6), we still have (2.11). Also, conditions (2.16) and (2.17) hold, which will lead to the existence and uniqueness of classical solutions of (2.14) or (2.13). Next, applying Theorem 1.1, we can show that there exists a unique adapted solution (X, Y, Z) of (2.12). $\qquad\square$

Since $h(t, x, y, z)$ is only assumed to be uniformly Lipschitz continuous in (y, z) (see (A1)), we have

$$(2.20) \qquad \begin{aligned} |h(t, x, y, z)| &\leq C(1 + |y| + |p|), \\ &\forall (t, x, y, z) \in [0, T] \times \mathbb{R}^n \times \mathbb{R}^m \times \mathbb{R}^{m \times n}. \end{aligned}$$

In other words, the function h is allowed to have a linear growth in (y, z).

§2.3. The case when $m = 1$

Unlike the previous cases, this is the case in which the existence of adapted solutions can be derived from a more general system than (2.4) and (2.13). The main reason is that in this case, function $\theta(t, x)$ is scalar valued, and the theory of quasilinear parabolic equations is much more satisfactory than that for parabolic systems. Consequently, the corresponding results for the FBSDEs will allow more complicated nonlinearities. Remember that in the present case, the backward component is one dimensional, but the forward part is still n dimensional.

We can now consider (1.1) with $m = 1$. Here W is an n-dimensional standard Brownian motion, b, σ, h and g take values in $\mathbb{R}^n$, $\mathbb{R}^{n \times n}$, $\mathbb{R}$ and $\mathbb{R}$, respectively. Also, X, Y and Z take values in $\mathbb{R}^n$, $\mathbb{R}$ and $\mathbb{R}^n$, respectively. In what follows we will try to use our Four Step Scheme to solve (1.1). To this end, we first need to solve (1.6) for z. In the present case, using the convention that all the vector are column vectors, we should rewrite (1.6) as follows:

$$(2.21) \qquad z = \sigma(t, x, y, z)^T p.$$

Let us introduce the following assumption.

(A2)'' There exist a positive continuous function $\nu(\cdot)$ and constants $C, \beta > 0$, such that for all $(t, x, y, z) \in [0, t] \times \mathbb{R}^n \times \mathbb{R} \times \mathbb{R}^n$,

$$(2.22) \qquad \nu(|y|)I \leq \sigma(t, x, y, z)\sigma(t, x, y, z)^T \leq CI,$$

$$(2.23) \qquad \begin{aligned} \langle [\sigma(t, x, y, z)^T]^{-1} z &- [\sigma(t, x, y, \widehat{z})^T]^{-1}\widehat{z}, z - \widehat{z} \rangle \\ &\geq \beta|z - \widehat{z}|^2, \end{aligned}$$

$$(2.24) \qquad |b(t, x, 0, 0)| + |h(t, x, 0, 0)| \leq C.$$

We note that condition (2.23) amounts to saying that the map $z \mapsto [\sigma(t, x, y, z)^T]^{-1} z$ is uniformly monotone. This is a sufficient condition for (2.21) to be uniquely solvable for z. Some other conditions are also possible, for example, the map $z \mapsto -[\sigma(t, x, y, z)^T]^{-1} z$ is uniformly monotone.

We have the following result for the unique solvability of FBSDE (1.1) with $m = 1$.

Theorem 2.4. *Let (A1) with $m = 1$, (A2)$''$ hold. Then there exists a unique smooth function $z(t, x, y, p)$ that solves (2.21) and satisfies (2.11). In addition, if (A3) also holds, then FBSDE (1.1) (with $m = 1$) admits an adapted solution determined by the Four Step Scheme.*

The proof is omitted here.

We should note that the well-posedness of (1.7) in the present case ($m = 1$) follows from Ladyzenskaja et al. [1, Chapter V, Theorem 8.1]. We see that the condition (2.24) together with (A1) means that the functions b and h are allowed to have linear growth in y and z. Also, note that we do not claim the uniqueness of adapted solutions since a condition similar to (1.11) is not easy to be made explicit.

§3. Infinite Horizon Case

In this section, we are concerned with the following FBSDE:

$$(3.1) \quad \begin{cases} dX(t) = b(X(t), Y(t))dt + \sigma(X(t), Y(t))dW(t), & t \in [0, \infty), \\ dY(t) = [h(X(t))Y(t) - 1]dt - \langle Z(t), dW(t) \rangle, & t \in [0, \infty), \\ X(0) = x, \\ Y(t) \quad \text{is bounded a.s.}, \text{ uniformly in } t \in [0, \infty). \end{cases}$$

Note that the time duration here is $[0, \infty)$. Thus, (3.1) is an FBSDE in an infinite time duration. In this section, we only consider the case $m = 1$, i.e., $Y(\cdot)$ is a scalar-valued process. Hence, $Z(\cdot)$ is valued in $\mathbb{R}^d$. Note that $X(\cdot)$ is still taking values in $\mathbb{R}^n$.

§3.1. The nodal solution

First of all, let us introduce the following notion.

Definition 3.1. A process $\{(X(t), Y(t), Z(t))\}_{t \geq 0}$ is called an *adapted solution* of (3.1) if for any $T > 0$, $(X, Y, Z)|_{[0,T]} \in \mathcal{M}[0, T]$, and

$$(3.2) \quad \begin{cases} X(t) = x + \int_0^t b(X(s), Y(s))ds + \int_0^t \sigma(X(s), Y(s))dW(s), \\ Y(t) = Y(T) - \int_t^T [h(X(s))Y(s) - 1]ds + \int_t^T \langle Z(s), dW(s) \rangle, \\ \qquad\qquad 0 \leq t \leq T < \infty, \end{cases}$$

such that $\exists M > 0$, $|Y(t)| \leq M$, $\forall t$, **P**-a.s. Moreover, if an adapted solution (X, Y, Z) is such that for some $\theta \in C^2(\mathbb{R}^n) \cap C_b^1(\mathbb{R}^n)$, the following relations hold:

$$(3.3) \quad \begin{cases} Y(t) = \theta(X(t)), \\ Z(t) = \sigma(X(t), \theta(X(t)))^\top \theta_x(X(t)), \end{cases} \qquad t \in [0, \infty),$$

then we call (X, Y, Z) a *nodal solution* of (3.1), with the *representing function* θ.

Let us now make some assumptions.

(H1) The functions σ, b, h are C^1 with bounded partial derivatives and there exist constants $\lambda, \mu > 0$, and some continuous increasing function $\nu : [0, \infty) \to [0, \infty)$, such that

$$(3.4) \quad \lambda I \leq \sigma(x, y)\sigma(x, y)^\top \leq \mu I, \qquad (x, y) \in \mathbb{R}^n \times \mathbb{R},$$

$$(3.5) \quad |b(x, y)| \leq \nu(|y|), \qquad (x, y) \in \mathbb{R}^n \times \mathbb{R},$$

$$(3.6) \quad \inf_{x \in \mathbb{R}^n} h(x) \equiv \delta > 0, \qquad \sup_{x \in \mathbb{R}^n} h(x) \equiv \gamma < \infty.$$

The following result plays an important role below.

Lemma 3.2. *Let (H1) hold. Then the following equation admits a classical solution* $\theta \in C^{2+\alpha}(\mathbb{R}^n)$:

$$(3.7) \quad \frac{1}{2}\mathrm{tr}\left(\theta_{xx}\sigma(x, \theta)\sigma^\top(x, \theta)\right) + \langle b(x, \theta), \theta_x \rangle - h(x)\theta + 1 = 0, \ x \in \mathbb{R}^n.$$

such that

$$(3.8) \quad \frac{1}{\gamma} \leq \theta(x) \leq \frac{1}{\delta}, \qquad x \in \mathbb{R}^n.$$

Sketch of the proof. Let $B_R(0)$ be the ball of radius $R > 0$ centered at the origin. We consider the equation (3.7) in $B_R(0)$ with the homogeneous Dirichlet boundary condition. By [Gilbarg-Trudinger, Theorem 14.10], there exists a solution $\theta^R \in C^{2+\alpha}(B_R(0))$ for some $\alpha > 0$. By the maximum principle, we have

$$(3.9) \quad 0 \leq \theta^R(x) \leq \frac{1}{\delta}, \qquad x \in B_R(0).$$

Next, for any fixed $x_0 \in \mathbb{R}^n$, and $R > |x_0| + 2$, by Gilbarg-Trudinger [1, Theorem 14.6], we have

$$(3.10) \quad |\theta_x^R(x)| \leq C, \qquad x \in B_1(x_0),$$

where the constant C is independent of $R > |x_0| + 2$. This, together with the boundedness of σ and the first partial derivatives of σ, b, h, implies that as a linear equation in θ (regarding $\sigma(x, \theta(x))$ and $b(x, \theta(x))$ as known functions), the coefficients are bounded in C^1. Hence, by Schauder's interior estimates, we obtain that

$$(3.11) \qquad \|\theta^R\|_{C^{2+\alpha}(B_1(x_0))} \le C, \qquad \forall R > |x_0| + 2.$$

Then, we can let $R \to \infty$ along some sequence to get a limit function $\theta(x)$. By the standard diagonalization argument, we may assume that θ is defined in the whole of $\mathbb{R}^n$. Clearly, $\theta \in C^{2+\alpha}(\mathbb{R}^n)$ and is a classical solution of (3.7). Finally, by the maximum principle again, we obtain (3.8). □

Now, we come up with the following existence of nodal solutions to (3.1). This result is essentially the infinite horizon version of the Four Step Scheme presented in the previous sections.

Theorem 3.3. *Let (H1) hold. Then there exists at least one nodal solution (X, Y, Z) of (3.1), with the representing function θ being the solution of (3.7). Conversely, if (X, Y, Z) is a nodal solution of (3.1) with the representing function θ. Then θ is a solution of (3.7).*

Proof. By Lemma 3.2, we can find a classical solution $\theta \in C^{2+\alpha}(\mathbb{R}^n)$ of (3.7). Now, we consider the following (forward) SDE:

$$(3.12) \qquad \begin{cases} dX(t) = b(X(t), \theta(X(t)))dt + \sigma(X(t), \theta(X(t)))dW(t), \ t > 0, \\ X(0) = x. \end{cases}$$

Since θ_x is bounded and b and σ are uniformly Lipschitz, (3.12) admits a unique strong solution $X(t)$, $t \in [0, \infty)$. Next, we define $Y(\cdot)$ and $Z(\cdot)$ by (3.3). Then, by Itô's formula, we see immediately that (X, Y, Z) is an adapted solution of (3.1). By Definition 3.1, it is a nodal solution of (3.1).

Conversely, let (X, Y, Z) be a nodal solution of (3.1) with the representing function θ. Since θ is C^2, we can apply Itô's formula to $Y(t) = \theta(X(t))$. This leads to that

$$(3.13) \qquad \begin{aligned} dY(t) = & \Big[\langle b(X(t), \theta(X(t))), \theta_x(X(t)) \rangle \\ & + \frac{1}{2}\mathrm{tr}\left(\theta_{xx}(X(t))\sigma\sigma^\top(X(t), \theta(X(t)))\right) \Big] dt \\ & + \langle \theta_x(X(t)), \sigma(X(t), \theta(X(t)))dW(t) \rangle . \end{aligned}$$

Comparing (3.13) with (3.1) and noting that $Y(t) = \theta(X(t))$, we obtain that
$$(3.14)$$
$$\langle b(X(t), \theta(X(t))), \theta_x(X(t)) \rangle + \frac{1}{2}\mathrm{tr}\left[\theta_{xx}(X(t))\sigma\sigma^\top(X(t), \theta(X(t)))\right]$$
$$= h(X(t))\theta(X(t)) - 1, \qquad \forall t \ge 0, \ \mathbf{P}\text{-a.s.}$$

Define a continuous function $F : \mathbb{R}^n \to \mathbb{R}$ by

$$(3.15) \qquad \begin{aligned} F(x) \overset{\triangle}{=} & \langle b(x, \theta(x)), \theta_x(x) \rangle + \frac{1}{2}\mathrm{tr}\left[\theta_{xx}(x)\sigma\sigma^\top(x, \theta(x))\right] \\ & - h(x)\theta(x) + 1. \end{aligned}$$

We shall prove that $F \equiv 0$. In fact, process X actually satisfies the following FSDE

$$(3.16) \qquad \begin{cases} dX(t) = \widetilde{b}(X(t))dt + \widetilde{\sigma}(X(t))dW(t), & t \geq 0; \\ X_0 = x, \end{cases}$$

where $\widetilde{b}(x) \overset{\Delta}{=} b(x, \theta(x))$ and $\widetilde{\sigma}(x) \overset{\Delta}{=} \sigma(x, \theta(x))$. Therefore, X is a time-homogeneous Markov process with some transition probability density $p(t, x, y)$. Since both $\widetilde{b}$ and $\widetilde{\sigma}$ are bounded and satisfy a Lipschitz condition; and since $\widetilde{a} \overset{\Delta}{=} \sigma \sigma^\top$ is uniformly positive definite, it is well known (see, for example, Friedman [1,2]) that for each $y \in \mathbb{R}^n$, $p(\cdot, \cdot, y)$ is the fundamental solution of the following parabolic PDE:

$$(3.17) \qquad \frac{1}{2} \sum_{i,j=1}^{n} \widetilde{a}^{ij}(x) \frac{\partial^2 p}{\partial x_i \partial x_j} + \sum_{i=1}^{n} \widetilde{b}^i(x) \frac{\partial p}{\partial x_i} - \frac{\partial p}{\partial t} = 0,$$

and it is positive everywhere. Now by (3.14), we have that $F(X(t)) = 0$ for all $t \geq 0$, **P**-a.s. , whence

$$(3.18) \qquad 0 = E_{0,x} \left[F(X(t))^2 \right] = \int_{\mathbb{R}^n} p(t, x, y) F(y)^2 dy, \quad \forall t > 0.$$

By the positivity of $p(t, x, y)$, we have $F(y) = 0$ almost everywhere under the Lebesgue measure in $\mathbb{R}^n$. The result then follows from the continuity of F.   □

Theorem 3.3 tells us that if (3.7) has multiple solutions, we have the non-uniqueness of the nodal solutions (and hence the non-uniqueness of the adapted solutions) to (3.1); and the number of the nodal solutions will be exactly the same as that of the solutions to (3.7). However, if the solution of (3.7) is unique, then the nodal solution of (3.1) will be unique as well. Note that we are not claiming the uniqueness of adapted solutions to (3.1).

§3.2. Uniqueness of nodal solutions

In this subsection we study the uniqueness of the nodal solutions to (3.1). We first consider the one dimensional case, that is, when X and Y are both one-dimensional processes. However, the Brownian motion $W(t)$ is still d-dimensional ($d \geq 1$). For simplicity, we denote

$$(3.19) \qquad a(x, y) = \frac{1}{2} |\sigma(x, y)|^2, \qquad (x, y) \in \mathbb{R}^2.$$

Let us make the some further assumptions:

(H2) Let $m = n = 1$ and the functions a, b, h satisfy the following:

$$(3.20) \qquad h(x) \text{ is strictly increasing in } x \in \mathbb{R}.$$

$$
(3.21) \quad
\begin{cases}
a(x,y)h(x) - (h(x)y - 1)\displaystyle\int_0^1 a_y(x, y + \beta(\widehat{y} - y))d\beta \geq \eta > 0, \\[4mm]
\displaystyle\int_0^1 \big[a(x,y)b_y(x, y + \beta(\widehat{y} - y)) \\[2mm]
\qquad - a_y(x, y + \beta(\widehat{y} - y))b(x,y)\big]d\beta \geq 0, \quad y, \widehat{y} \in [\tfrac{1}{\gamma}, \tfrac{1}{\delta}], x \in \mathbb{R}.
\end{cases}
$$

Condition (3.21) essentially says that the coefficients b, σ and h should be somewhat "compatible." Although a little complicated, (3.21) is still quite explicit and not hard to verify. For example, a sufficient conditions for (3.21) is

$$
(3.22) \quad
\begin{cases}
a(x,y)h(x) - (h(x)y - 1)a_y(x, w) \geq \eta > 0, \\[2mm]
a(x,y)b_y(x, w) - a_y(x, w)b(x, y) \geq 0, \\[2mm]
\qquad\qquad y, w \in [\tfrac{1}{\gamma}, \tfrac{1}{\delta}], \ x \in \mathbb{R}.
\end{cases}
$$

It is readily seen that the following will guarantee (3.22) (if (H1) is assumed):

$$
(3.23) \qquad a_y(x, y) = 0, \quad b_y(x, y) \geq 0, \qquad (x, y) \in \mathbb{R} \times [\tfrac{1}{\gamma}, \tfrac{1}{\delta}].
$$

In particular, if both a and b are independent of y, then (3.21) holds automatically.

Our main result of this subsection is the following uniqueness theorem.

Theorem 3.4. *Let (H1)–(H2) hold. Then (3.1) has a unique adapted solution. Moreover, this solution is nodal.*

To prove the above result, we need several lemmas.

Lemma 3.5. *Let h be strictly increasing and θ solves*

$$
(3.24) \qquad a(x, \theta)\theta_{xx} + b(x, \theta)\theta_x - h(x)\theta + 1 = 0, \qquad x \in \mathbb{R}.
$$

Suppose x_M is a local maximum of θ and x_m is a local minimum of θ with $\theta(x_m) \leq \theta(x_M)$. Then $x_m > x_M$.

Proof. Since h is strictly increasing, from (3.24) we see that θ is not identically constant in any interval. Therefore $x_m \neq x_M$. Now, let us look at x_M. It is clear that $\theta_x(x_M) = 0$ and $\theta_{xx}(x_M) \leq 0$. Thus, from (3.24) we obtain that

$$
(3.25) \qquad\qquad\qquad \theta(x_M) \leq \frac{1}{h(x_M)}.
$$

Similarly, we have

$$
(3.26) \qquad\qquad\qquad \theta(x_m) \geq \frac{1}{h(x_m)}.
$$

Since $\theta(x_m) \leq \theta(x_M)$, we have

$$
(3.27) \qquad\qquad\qquad \frac{1}{h(x_m)} \leq \frac{1}{h(x_M)},
$$

whence $x_M < x_m$ because $h(x)$ is strictly increasing. $\qquad\qquad\square$

Lemma 3.6. *Let (H1)–(H2) hold. Then (3.24) admits a unique solution.*

Proof. By Lemma 3.2 we know that (3.24) admits at lease one classical solution θ. We first show that θ is monotone decreasing. Suppose not, assume that it has a local minimum at x_m. Since $\theta \in C^1$ and θ is not constant on any interval as we pointed out before, $\theta_x > 0$ near x_m. Using Lemma 3.5, one further concludes that $\theta_x \geq 0$ over (x_m, ∞). In other words, θ is monotone increasing on (x_m, ∞). The boundedness of θ then leads to that $\lim_{x\to\infty} \theta(x)$ exists.

Next we show that

$$(3.28) \qquad \lim_{x\to\infty} \theta_x(x) = \lim_{x\to\infty} \theta_{xx}(x) = 0,$$

To see this, we first apply Taylor's formula and use the boundedness of θ_{xx} to conclude that there exists $M > 0$ such that for any $x \in (x_m, \infty)$ and $h > 0$

$$\theta(x+h) - \theta(x) - Mh^2 \leq \theta_x(x)h \leq \theta(x+h) - \theta(x) + Mh^2.$$

Since $\lim_{x\to\infty} \theta(x+h) - \theta(x) = 0$, we have

$$-Mh^2 \leq \varliminf_{x\to\infty} \theta_x(x)h \leq \varlimsup_{x\to\infty} \theta_x(x)h \leq Mh^2.$$

Dividing h and letting $h \to 0$ we derive $\lim_{x\to\infty} \theta_x(x) = 0$. Further, note that

$$\theta_{xx} = \frac{1}{a(x,\theta)}[-b(x,\theta)\theta_x + h(x)\theta - 1].$$

The boundedness of θ, θ_x, and θ_{xx} and the assumption (H1) then show that θ_{xxx} exists and is continuous and bounded as well. Thus apply Taylor's expansion to the third order and repeat the discussion above one shows further that $\lim_{x\to\infty} \theta_{xx}(x) = 0$ as well, proving (3.28). Consequently, by (3.24) we have

$$(3.29) \qquad \lim_{x\to\infty} \theta(x) = \frac{1}{h(+\infty)}.$$

On the other hand, by (3.20), we see that

$$(3.30) \qquad \lim_{x\to\infty} \theta(x) > \theta(x_m) \geq \frac{1}{h(x_m)} > \frac{1}{h(+\infty)},$$

which contradicts (3.29). This means that θ has no local minimum. Similarly one shows that θ can not have any local maximum either, hence it must be monotone on $\mathbb{R}$. Finally, since

$$(3.31) \qquad \theta(-\infty) = \frac{1}{h(-\infty)} > \frac{1}{h(+\infty)} = \theta(+\infty),$$

it is necessary that θ is monotone decreasing.

Next, let θ and $\widehat{\theta}$ be two solutions of (3.24). Then, $w \equiv \widehat{\theta} - \theta$ satisfies

$$
\begin{aligned}
0 = {} & a(x, \widehat{\theta})w_{xx} + b(x, \widehat{\theta})w_x \\
& - \left(h(x) - \int_0^1 \left[a_y(x, \theta + \beta w)\theta_{xx} + b_y(x, \theta + \beta w)\theta_x \right] d\beta \right) w \\
= {} & a(x, \widehat{\theta})w_{xx} + b(x, \widehat{\theta})w_x - c(x)w,
\end{aligned}
$$

(3.32)

where

$$
\begin{aligned}
c(x) = {} & h(x) - \int_0^1 \left[a_y(x, \theta + \beta w) \frac{h(x)\theta - 1 - b(x, \theta)\theta_x}{a(x, \theta)} \right. \\
& \left. + b_y(x, \theta + \beta w)\theta_x \right] d\beta \\
= {} & \frac{a(x, \theta)h(x) - (h(x)\theta - 1)\int_0^1 a_y(x, \theta + \beta(\widehat{\theta} - \theta))d\beta}{a(x, \theta)} \\
& + |\theta_x| \int_0^1 \left[a(x, \theta)b_y(x, \theta + \beta(\widehat{\theta} - \theta)) \right. \\
& \left. - a_y(x, \beta\theta + \beta(\widehat{\theta} - \theta))b(x, \theta) \right] d\beta \geq \frac{\eta}{\mu}.
\end{aligned}
$$

(3.33)

Here, we have used the fact that $\theta_x(x) = -|\theta_x(x)|$ (since θ is decreasing in x) and (3.21) as well as (3.7). From (H1), we also see that $a(x, \widehat{\theta}) \geq 0$ and $b(x, \widehat{\theta})$ are bounded. Thus, by the lemma that will be proved below, we obtain $w = 0$, proving the uniqueness. $\qquad\square$

Lemma 3.7. *Let w be a bounded classical solution of the following equation:*

$$
(3.34) \qquad \tilde{a}(x)w_{xx} + \tilde{b}(x)w_x - c(x)w = 0, \qquad x \in \mathbb{R},
$$

with $c(x) \geq c_0 > 0$, $\tilde{a}(x) \geq 0$, $x \in \mathbb{R}^n$, and with $\tilde{a}$ and $\tilde{b}$ bounded. Then $w(x) \equiv 0$.

Proof. For any $\alpha > 0$, let us consider $\Phi_\alpha(x) = w(x) - \alpha|x|^2$. Since w is bounded, there exists some x_α at which Φ_α attains its global maximum. Thus, $\Phi_\alpha'(x_\alpha) = 0$ and $\Phi_\alpha''(x_\alpha) \leq 0$, which means that

$$
(3.35) \qquad w_x(x_\alpha) = 2\alpha x_\alpha, \qquad w_{xx}(x_\alpha) \leq 2\alpha.
$$

Now, by (3.34),

$$
\begin{aligned}
(3.36) \qquad c(x_\alpha)w(x_\alpha) &= \tilde{a}(x_\alpha)w_{xx}(x_\alpha) + \tilde{b}(x_\alpha)w_x(x_\alpha) \\
&\leq 2\alpha\big(\tilde{a}(x_\alpha) + \tilde{b}(x_\alpha)x_\alpha\big).
\end{aligned}
$$

For any $x \in \mathbb{R}$, by the definition of x_α, we have (note the boundedness of $\tilde{a}$ and $\tilde{b}$)

$$
\begin{aligned}
(3.37) \qquad w(x) - \alpha|x|^2 &\leq w(x_\alpha) - \alpha|x_\alpha|^2 \\
&\leq \frac{\alpha}{c_0}\big(2\tilde{a}(x_\alpha) + 2\tilde{b}(x_\alpha)x_\alpha - |x_\alpha|^2\big) \leq C\alpha.
\end{aligned}
$$

Sending $\alpha \to 0$, we obtain $w(x) \le 0$. Similarly, we can show that $w(x) \ge 0$. Thus $w(x) \equiv 0$. $\qquad\blacksquare$

Proof of Theorem 3.4. Let (X, Y, Z) be any adapted solution of (3.1). Under (H1)–(H2), by Lemma 3.6, equation (3.24) admits a unique classical solution θ with $\theta_x \le 0$. We set

(3.38)
$$\begin{cases} \widetilde{Y}(t) = \theta(X(t)), \\ \widetilde{Z}(t) = \sigma\big(X(t), \theta(X(t))\big)^{\top} \theta_x(X(t)), \end{cases} \qquad t \in [0, \infty).$$

By Itô's formula, we have (note (3.19))

(3.39)
$$\begin{aligned} d\widetilde{Y}(t) = &\Big[\theta_x(X(t))b(X(t), Y(t)) + \theta_{xx}(X(t))a(X(t), Y(t))\Big] dt \\ &+ \big\langle \sigma(X(t), Y(t))^{\top} \theta_x(X(t)), dW(t) \big\rangle. \end{aligned}$$

Hence, with (3.1), we obtain (note (3.24)) that for any $0 \le r < t < \infty$,

(3.40)
$$\begin{aligned} &E[\widetilde{Y}(r) - Y(r)]^2 - E[\widetilde{Y}(t) - Y(t)]^2 \\ &= -E \int_r^t \Big\{ 2[\widetilde{Y} - Y]\big[\theta_x(X)b(X, Y) + \theta_{xx}(X)a(X, Y) \\ &\qquad\qquad - h(X)Y + 1\big] + |\sigma(X, Y)\theta_x(X) - Z|^2 \Big\} ds \\ &\le -2E \int_r^t [\widetilde{Y} - Y]\Big[\theta_x(X)\big(b(X, Y) - b(X, \widetilde{Y})\big) \\ &\qquad\qquad + \theta_{xx}(X)\big(a(X, Y) - a(X, \widetilde{Y})\big) - h(X)(Y - \widetilde{Y})\Big] ds \\ &= -2E \int_r^t \Big\{ [\widetilde{Y} - Y]^2 \Big[|\theta_x(X)| \int_0^1 b_y(X, \widetilde{Y} + \beta(Y - \widetilde{Y}))d\beta \\ &\qquad\qquad - \theta_{xx}(X) \int_0^1 a_y(X, \widetilde{Y} + \beta(Y - \widetilde{Y}))d\beta + h(X)\Big] \Big\} ds \\ &\equiv -2E \int_r^t c(s) |\widetilde{Y}(s) - Y(s)|^2 ds, \end{aligned}$$

where (note the equation (3.24))

$$c(s) = h(X) + |\theta_x(X)| \int_0^1 b_y(X, \tilde{Y} + \beta(Y - \tilde{Y}))d\beta$$

$$+ \frac{b(X, \tilde{Y})\theta_x(X) - h(X)\tilde{Y} + 1}{a(X, \tilde{Y})}$$

$$\cdot \int_0^1 a_y(X, \tilde{Y} + \beta(Y - \tilde{Y}))d\beta$$

(3.41)
$$= \frac{1}{a(X, Y)}\Big\{a(X, Y)h(X)$$

$$- [h(X)\tilde{Y} - 1] \int_0^1 a_y(X, \tilde{Y} + \beta(Y - \tilde{Y}))d\beta$$

$$+ |\theta_x(X)| \int_0^1 [a(X, \tilde{Y})b_y(X, \tilde{Y} + \beta(Y - \tilde{Y}))$$

$$- b(X, \tilde{Y})a_y(X, \tilde{Y} + \beta(Y - \tilde{Y}))]d\beta \geq \frac{\eta}{\mu}.$$

Denote $\varphi(t) = E[\tilde{Y}(t) - Y(t)]^2$ and $\alpha = \frac{2\eta}{\mu} > 0$. Then (3.40) can be written as

(3.42) $$\varphi(r) \leq \varphi(t) - \alpha \int_r^t \varphi(s)ds, \qquad 0 \leq r < t < \infty.$$

Thus,

(3.43)
$$\left(e^{-\alpha t} \int_r^t \varphi(s)ds\right)' = e^{-\alpha t}\left(\varphi(t) - \alpha \int_r^t \varphi(s)ds\right)$$

$$\geq e^{-\alpha t}\varphi(r), \qquad t \in [r, \infty).$$

Integrating it over $[r, T]$, we obtain (note Y and $\tilde{Y}$ are bounded, and so is φ)

(3.44) $$\frac{e^{-\alpha r} - e^{-\alpha T}}{\alpha}\varphi(r) \leq e^{-\alpha T} \int_r^T \varphi(s)ds \leq CTe^{-\alpha T}, \qquad T > 0.$$

Therefore, sending $T \to \infty$, we see that $\varphi(r) = 0$. This implies that

(3.45) $$Y(r) = \tilde{Y}(r) \equiv \theta(X(r)), \qquad r \in [0, \infty), \text{ a.s. } \omega \in \Omega.$$

Consequently, from the second equality in (3.40), one has

(3.46) $$Z(s) = \tilde{Z}(s) = \sigma(X(s), \theta(X(s)))^T \theta_x(X(s)), \qquad \forall s \in [0, \infty).$$

Hence, (X, Y, Z) is a nodal solution. Finally, suppose (X, Y, Z) and $(\hat{X}, \hat{Y}, \hat{Z})$ are any adapted solutions of (3.1). Then, by the above proof, we must have

(3.47) $$Y(t) = \theta(X(t)), \quad \hat{Y}(t) = \theta(\hat{X}(t)), \qquad t \in [0, \infty).$$

Thus, by (3.1), we see that $X(\cdot)$ and $\widehat{X}(\cdot)$ satisfy the same forward SDE with the same initial condition (see (3.12)). By the uniqueness of the strong solution to such an SDE, $X = \widehat{X}$. Consequently, $Y = \widehat{Y}$ and $Z = \widehat{Z}$. This proves the theorem.   □

Let us indicate an obvious extension of Theorem 3.4 to higher dimensions.

Theorem 3.8. *Let (H1) hold and suppose there exists a solution θ to (3.7) satisfying*

(3.48)
$$
\begin{aligned}
h(x) - \int_0^1 \Big[\sum_{i,j=1}^n a_y^{ij}\big(x, (1-\beta)\theta(x) + \beta\widehat{\theta}\big)\theta_{x_i x_j}(x) \\
- \sum_{i=1}^n b_y^i\big(x, (1-\beta)\theta(x) + \beta\widehat{\theta}\big)\theta_{x_i}(x) \Big] d\beta \geq \eta > 0,
\end{aligned}
$$
$$
x \in \mathbb{R}^n, \widehat{\theta} \in [\tfrac{1}{\gamma}, \tfrac{1}{\delta}].
$$

Then (3.1) has a unique adapted solution. Moreover, this solution is nodal with θ being the representing function.

Sketch of the proof. First of all, by an equality similar to (3.32), we can prove that (3.7) has no other solution except $\theta(x)$. Then, by a proof similar to that of Theorem 3.4, we obtain the conclusion here.   □

Corollary 3.9. *Let (H1) hold and both a and b be independent of y. Then (3.1) has a unique adapted solution and it is nodal.*

Proof. In the present case, condition (3.48) trivially holds. Thus, Theorem 3.8 applies.   □

§3.3. The limit of finite duration problems

In this subsection, we will prove the following result, which gives a relationship between the FBSDEs in finite and infinite time durations.

Theorem 3.10. *Let (H1)–(H2) hold and let θ be a solution of (3.7) with the property (3.48). Let (X, Y, Z) be the nodal solution of (3.1) with the representing function θ, and $(X^K, Y^K, Z^K) \in \mathcal{M}[0, K]$ be the adapted solution of (3.1) with $[0, \infty)$ replaced by $[0, K]$, and $Y^K(K) = g(X(K))$ for some bounded smooth function g. Then*

(3.49) $\lim\limits_{K \to \infty} E\big\{ |X^K(t) - X(t)|^2 + |Y^K(t) - Y(t)|^2 + E|Z^K(t) - Z(t)|^2 \big\} = 0,$

uniformly in t on any compact sets.

To prove the above result, we need the following lemma.

Lemma 3.11. *Suppose that*

$$(3.50) \quad \begin{cases} \lambda I \leq (a^{ij}(t,x)) \leq \mu I, \\ |b^i(t,x)| \leq C, \quad 1 \leq i \leq n, \\ c(t,x) \geq \eta > 0, \\ |w_0(x)| \leq M, \end{cases} \quad (t,x) \in [0,\infty) \times \mathbb{R}^n,$$

with some positive constants λ, μ, η, C and M. Let w be the classical solution of the following equation:

$$(3.51) \quad \begin{cases} w_t - \displaystyle\sum_{ij=1}^{n} a^{ij}(t,x)w_{x_i x_j} - \sum_{i=1}^{n} b^i(t,x)w_{x_i} + c(t,x)w = 0, \\ \qquad\qquad\qquad\qquad (t,x) \in [0,\infty) \times \mathbb{R}^n, \\ w\big|_{t=0} = w_0(x). \end{cases}$$

Then

$$(3.52) \quad |w(t,x)| \leq Me^{-\eta t}, \qquad (t,x) \in [0,\infty) \times \mathbb{R}^n.$$

Proof. First, let $R > 0$ and consider the following initial-boundary value problem:

$$(3.53) \quad \begin{cases} w_t^R - \displaystyle\sum_{i,j=1}^{n} a^{ij}(t,x)w_{x_i x_j}^R - \sum_{i=1}^{n} b^i(t,x)w_{x_i}^R + c(t,x)w^R = 0, \\ \qquad\qquad\qquad\qquad (t,x)[0,\infty) \times \in B_R, \\ w^R\big|_{\partial B_R} = 0, \\ w^R\big|_{t=0} = w_0(x)\chi^R(x), \end{cases}$$

where B_R is the ball of radius $R > 0$ centered at 0 and χ^R is some "cut-off" function. Then we know that (3.53) admits a unique classical solution $w^R \in C^{2+\alpha,1+\alpha/2}(B_R \times [0,\infty))$ for some $\alpha > 0$, where $C^{2+\alpha,1+\alpha/2}$ is the space of all functions $v(x,t)$ which are C^2 in x and C^1 in t with Hölder continuous $v_{x_i x_j}$ and v_t of exponent α and $\alpha/2$, respectively. Moreover, we have

$$(3.54) \quad |w^R(t,x)| \leq M, \qquad (t,x) \in [0,\infty) \times B_R,$$

and for any $x_0 \in \mathbb{R}^n$ and $T > 0$, $(0 < \alpha' < \alpha)$

$$(3.55) \quad w^R \xrightarrow{s} w, \qquad \text{in } C^{2+\alpha',1+\alpha'/2}([0,T] \times B_1(x_0)), \qquad \text{as } R \to \infty,$$

where w is the solution of (3.51). Now, we let $\psi(t,x) = Me^{-(\eta-\varepsilon)t}$ $(\varepsilon > 0)$.

Then

(3.56)
$$
\begin{cases}
\psi_t - \displaystyle\sum_{i,j=1}^{n} a^{ij}(t,x)\psi_{x_i x_j} - \sum_{i=1}^{n} b^i(t,x)\psi_{x_i} + c(t,x)\psi \\
\qquad = \bigl(c(t,x) - \eta + \varepsilon\bigr)M^{-(\eta-\varepsilon)t} \ge \varepsilon M^{-(\eta-\varepsilon)t} > 0, \\
\psi|_{\partial B_R} > 0 = w^R|_{\partial B_R}, \\
\psi|_{t=0} = M \ge w_0(x) = w^R|_{t=0}.
\end{cases}
$$

Thus, by Friedman [1, Chapter 2, Theorem 16], we have

(3.57) $w^R(t,x) \le \psi(t,x) = M^{-(\eta-\varepsilon)t}, \qquad (t,x) \in [0,\infty) \times B_R.$

Similarly, we can prove that

(3.58) $w^R(t,x) \ge -M e^{-(\eta-\varepsilon)t}, \qquad (t,x) \in [0,\infty) \times B_R.$

Since the right hand sides of (3.57)–(3.58) are independent of R, we see that

(3.59) $|w(t,x)| \le M e^{-(\eta-\varepsilon)t}, \qquad (t,x) \in [0,\infty) \times \mathbb{R}^n.$

Hence, (3.52) follows by sending $\varepsilon \to 0$.   □

  Proof of Theorem 3.10.   By the result from §2, we know that (X^K, Y^K, Z^K) satisfies

(3.60)
$$
\begin{cases}
Y^K(t) = \theta^K(t, X^K(t)), \\
Z^K(t) = \sigma(X^K(t), \theta^K(t, X^K(t)))^T \theta_x^K(t, X^K(t)), \\
\qquad t \in [0,K], \text{ a.s. } \omega \in \Omega,
\end{cases}
$$

where θ^K is the solution of the parabolic equation:

(3.61)
$$
\begin{cases}
\theta_t^K + \displaystyle\sum_{i,j=1}^{n} a^{ij}(x,\theta^K)\theta_{x_i x_j}^K + \sum_{i=1}^{n} b^i(x,\theta^K)\theta_{x_i}^K - h(x)\theta^K + 1 = 0, \\
\qquad\qquad\qquad\qquad (x,t) \in \mathbb{R}^n \times [0,T), \\
\theta^K|_{t=T} = g(x),
\end{cases}
$$

with $a = \frac{1}{2}\sigma\sigma^T$. Next, we define φ to be the solution of

(3.62)
$$
\begin{cases}
\varphi_t - \displaystyle\sum_{i,j=1}^{n} a^{ij}(x,\varphi)\varphi_{x_i x_j} - \sum_{i=1}^{n} b^i(x,\varphi)\varphi_{x_i} + h(x)\varphi - 1 = 0, \\
\qquad\qquad\qquad\qquad (t,x) \in [0,\infty) \times \mathbb{R}^n, \\
\varphi|_{t=0} = g(x).
\end{cases}
$$

Clearly, we have

(3.63) $\theta^K(t,x) = \varphi(K-t,x), \qquad (t,x) \in [0,K] \times \mathbb{R}^n.$

Now, we let $w(t, x) = \varphi(t, x) - \theta(x)$. Then

(3.64)
$$
\begin{cases}
\begin{aligned}
w_t &- \sum_{i,j=1}^n a^{ij}(x, \varphi) w_{x_i x_j} - \sum_{i=1}^n b(x, \varphi) w_{x_i} \\
&- \Big[h(x) - \int_0^1 \Big(\sum_{i,j=1}^n a_y^{ij}(x, \theta + \beta w) \theta_{x_i x_j} \\
&\qquad + \sum_{i=1}^n b_y^i(x, \theta + \beta w) \theta_{x_i} \Big) \, d\beta \Big] w = 0,
\end{aligned} \\
w\Big|_{t=0} = g(x) - \theta(x).
\end{cases}
$$

We note that both $\varphi(x, t)$ and $\theta(x)$ lie in $[\frac{1}{\gamma}, \frac{1}{\delta}]$. Thus, by condition (3.48) and Lemma 3.11, we see that

(3.65)
$$
|\theta^K(t, x) - \theta(x)| = |\varphi(K - t, x) - \theta(x)| \leq \frac{1}{\delta} e^{-\eta(K-t)},
$$
$$
(t, x) \in [0, K] \times \mathbb{R}^n \times [0, K], \quad K > 0.
$$

Now, we look at the following forward SDEs:

(3.66)
$$
\begin{cases}
\begin{aligned}
dX^K(t) &= b(X(t)^K, \theta^K(t, X(t)^K)) dt \\
&\quad + \sigma(X(t)^K, \theta^K(t, X(t)^K)) dW(t),
\end{aligned} \\
X^K(0) = x.
\end{cases}
$$

(3.67)
$$
\begin{cases}
dX(t) = b(X(t), \theta(X(t))) dt + \sigma(X(t), \theta(X(t))) dW(t), \\
X(0) = x.
\end{cases}
$$

By Itô's formula, we have

(3.68)
$$
\begin{aligned}
&E|X^K(t) - X(t)|^2 \\
&= E \int_0^t \Big[2 \langle X^K - X, b(X^K, \theta^K(s, X^K)) - b(X, \theta(X)) \rangle \\
&\qquad + \operatorname{tr} \Big([\sigma(X^K, \theta^K(s, X^K)) - \sigma(X, \theta(X))] \cdot \\
&\qquad\qquad [\sigma(X^K, \theta^K(s, X^K)) - \sigma(X, \theta(X))]^{\top} \Big) \Big] ds \\
&\leq CE \int_0^t \Big[|X^K - X|(|X^K - X| + |\theta^K(s, X^K) - \theta(X^K)|) \\
&\qquad + (|X^K - X| + |\theta^K(s, X^K) - \theta(X^K)|)^2 \Big] ds \\
&\leq C \int_0^t \Big[E|X^K(s) - X(s)|^2 + e^{-2\eta(K-s)} \Big] ds \\
&\leq C \int_0^t |X^K(s) - X(s)|^2 ds + C e^{-2\eta(K-t)}.
\end{aligned}
$$

Applying Gronwall's inequality, we obtain that

$$(3.69) \qquad E\left(|X^K(t) - X(t)|^2\right) \le Ce^{-2\eta(K-t)}, \qquad t \in [0, K], \ K > 0.$$

Furthermore,

$$
\begin{aligned}
(3.70) \qquad E\left(|Y^K(t) - Y(t)|^2\right) &= E\left(|\theta^K(t, X^K(t)) - \theta(X(t))|^2\right) \\
&\le 2E\left(|\theta^K(t, X^K(t)) - \theta(X^K(t))|^2\right) \\
&\quad + 2E\left(|\theta(X^K(t)) - \theta(X(t))|^2\right) \\
&\le Ce^{-2\eta(K-t)} + CE\left(|X^K(t) - X(t)|^2\right) \\
&\le Ce^{-2\eta(K-t)}, \quad t \in [0, K], \ K > 0.
\end{aligned}
$$

Similarly, we have

$$(3.71) \qquad E\left(|Z^K(t) - Z(t)|^2\right) \le Ce^{-2\eta(K-t)}, \quad t \in [0, K], \ K > 0.$$

Finally, letting $K \to \infty$, the conclusion follows.   □

Chapter 5

Linear, Degenerate Backward Stochastic Partial Differential Equations

§1. Formulation of the Problem

We note that in the previous chapter, all the coefficients b, σ, h and g are deterministic, i.e., they are all independent of $\omega \in \Omega$. If one tries to apply the Four Step Scheme to FBSDEs with random coefficients, i.e., b, σ, h and g are possibly depending on $\omega \in \Omega$ explicitly, then it will lead to the study of general degenerate nonlinear backward partial differential equations (BSPDEs, for short). In this chapter, we restrict ourselves to the study of the following linear BSPDE:

$$(1.1) \quad \begin{cases} du = \big\{ -\dfrac{1}{2}\nabla\cdot(ADu) - \langle\, a, Du\,\rangle - cu - \nabla\cdot(Bq) - \langle\, b, q\,\rangle - f \big\}dt \\ \qquad\quad + \langle\, q, dW(t)\,\rangle, \qquad (t, x) \in [0, T] \times \mathbb{R}^n, \\ u\big|_{t=T} = g, \end{cases}$$

where Du is the gradient of u,

$$\begin{cases} \nabla\cdot\xi = \displaystyle\sum_{i=1}^{n} \partial_{x_i}\xi_i, \quad \forall \xi = (\xi_1, \cdots, \xi_n) \in C^1(\mathbb{R}^n; \mathbb{R}^n), \\ \nabla\cdot\Phi = (\nabla\cdot\Phi_1, \cdots, \nabla\cdot\Phi_m)^T, \quad \forall \Phi = (\Phi_1, \cdots, \Phi_m) \in C^1(\mathbb{R}^n; \mathbb{R}^{n\times m}), \end{cases}$$

and

$$(1.2) \quad \begin{cases} A : [0, T] \times \mathbb{R}^n \times \Omega \to S^n, \\ B : [0, T] \times \mathbb{R}^n \times \Omega \to \mathbb{R}^{n\times d}, \\ a : [0, T] \times \mathbb{R}^n \times \Omega \to \mathbb{R}^n, \\ b : [0, T] \times \mathbb{R}^n \times \Omega \to \mathbb{R}^d, \\ c, \; f : [0, T] \times \mathbb{R}^n \times \Omega \to \mathbb{R}, \\ g : \mathbb{R}^n \times \Omega \to \mathbb{R}, \end{cases}$$

are random fields (S^n is the set of all ($n \times n$) symmetric matrices). We assume that $W = \{W(t) : t \in [0, T]\}$ is a d-dimensional Brownian motion defined on some complete filtered probability space $(\Omega, \mathcal{F}, \{\mathcal{F}_t\}_{t\geq 0}, \mathbf{P})$, with $\{\mathcal{F}_t\}_{t\geq 0}$ being the natural filtration generated by W, augmented by all the $\mathbf{P}$-null sets in $\mathcal{F}$.

In our discussions, we will always assume that A and B are differentiable in x. In such a case, (1.1) is equivalent to an equation of a general form. To see this, we note that

$$(1.3) \quad \begin{cases} \mathrm{tr}\,[AD^2u] = \nabla\cdot(ADu) - \langle\, \nabla\cdot A, Du\,\rangle; \\ \mathrm{tr}\,[B^T Dq] = \nabla\cdot(Bq) - \langle\, \nabla\cdot B, q\,\rangle, \end{cases}$$

where D^2u is the Hessian of u and

$$Dq \stackrel{\Delta}{=} (Dq_1, \cdots, Dq_d) \stackrel{\Delta}{=} \begin{pmatrix} \partial_{x_1}q_1 & \cdots & \partial_{x_1}q_d \\ \vdots & \cdots & \vdots \\ \partial_{x_n}q_1 & \cdots & \partial_{x_n}q_d \end{pmatrix}.$$

Therefore, if we define

(1.4) $\tilde{a} = a + \dfrac{1}{2}\nabla \cdot A; \qquad \tilde{b} = b + \nabla \cdot B,$

then (1.1) is the same as

(1.5)
$$\begin{cases} du = \left\{ -\dfrac{1}{2}\text{tr}\,[AD^2u] - \langle \tilde{a}, Du \rangle - cu - \text{tr}\,[B^T Dq] \right. \\ \qquad\qquad \left. - \langle \tilde{b}, q \rangle - f \right\} dt + \langle q, dW(t) \rangle, \quad (t,x) \in [0,T] \times \mathbb{R}^n, \\ u\big|_{t=T} = g, \end{cases}$$

Since (1.1) and (1.5) are equivalent, all the results for (1.1) can be automatically carried over to (1.5) and vice versa. For notational convenience, we will concentrate on (1.1) for well-posedness (§§2–5) and on (1.5) for comparison theorems (§6).

Next, we introduce the following definition.

Definition 1.1. If A and B satisfy the following:

(1.6) $A(t,x) - B(t,x)B(t,x)^T \geq 0, \quad \text{a.e. } (t,x) \in [0,T] \times \mathbb{R}^n, \text{ a.s.},$

we say that equation (1.1) is *parabolic*; if there exists a constant $\delta > 0$, such that

(1.7) $A(t,x) - B(t,x)B(t,x)^T \geq \delta I, \quad \text{a.e. } (t,x) \in [0,T] \times \mathbb{R}^n, \text{ a.s.}$

we say that (1.1) is *super-parabolic*; whereas, if (1.6) holds and there exists a set $G \subseteq [0,T] \times \mathbb{R}^n$ of positive Lebesgue measure, such that

(1.8) $\det\left[A(t,x) - B(t,x)B(t,x)^T\right] = 0, \quad \forall (t,x) \in G, \text{ a.s.}$

we say that (1.1) is *degenerate parabolic*.

We see that in the above definition, only A and B are involved. Thus, the above three notions are adopted to equation (1.5) as well.

Note that if (1.1) is super-parabolic, it is necessary that $A(t,x)$ is uniformly positive definite, i.e.,

(1.9) $A(t,x) \geq \delta I > 0, \quad \text{a.e. } (t,x) \in [0,T] \times \mathbb{R}^n, \text{ a.s.}$

However, if $A(t,x)$ is uniformly positive definite and (1.6) holds, we do not necessarily have the super-parabolicity of (1.1). As a matter of fact, if $A(t,x)$ satisfies (1.9) and

(1.10) $A(t,x) = B(t,x)B(t,x)^T, \quad \text{a.e. } (t,x) \in [0,T] \times \mathbb{R}^n, \text{ a.s.}$

then (1.1) is degenerate parabolic. This is the case if we have the BSPDE from the Four Step Scheme for FBSDEs with random coefficients (see Chapter 4, §6).

Now, we introduce the notion of solutions to (1.1). In what follows, we denote $B_R = \{x \in \mathbb{R}^n \mid |x| < R\}$ for any $R > 0$.

Definition 1.2. Let $\{(u(t, x; \omega), q(t, x; \omega)), (t, x, \omega) \in [0, T] \times \mathbb{R}^n \times \Omega\}$ be a pair of random fields.

(i) (u, q) is called an *adapted classical solution* of (1.1) if

(1.11)
$$
\begin{cases}
u \in C_{\mathcal{F}}([0, T]; L^2(\Omega; C^2(\overline{B}_R))), \\
q \in L^2_{\mathcal{F}}(0, T; C^1(\overline{B}_R; \mathbb{R}^d)),
\end{cases} \quad \forall R > 0,
$$

such that almost surely the following holds for all $(t, x) \in [0, T] \times \mathbb{R}^n$:

(1.12)
$$
\begin{aligned}
u(t, x) = g(x) + \int_t^T \Big\{ & \frac{1}{2} \nabla \cdot [A(s, x) Du(s, x)] + \langle a(s, x), Du(s, x) \rangle \\
& + c(s, x) u(s, x) + \nabla \cdot [B(s, x) q(s, x)] \\
& + \langle b(s, x), q(s, x) \rangle + f(s, x) \Big\} ds \\
& - \int_t^T \langle q(s, x), dW(s) \rangle.
\end{aligned}
$$

(ii) (u, q) is called an *adapted strong solution* of (1.1) if

(1.13)
$$
\begin{cases}
u \in C_{\mathcal{F}}([0, T]; L^2(\Omega; H^2(B_R))), \\
q \in L^2_{\mathcal{F}}(0, T; H^1(B_R; \mathbb{R}^d)),
\end{cases} \quad \forall R > 0,
$$

such that almost surely (1.12) holds for all $t \in [0, T]$, a.e. $x \in \mathbb{R}^n$.

(iii) (u, q) is called an *adapted weak solution* of (1.1) if

(1.14)
$$
\begin{cases}
u \in C_{\mathcal{F}}([0, T]; L^2(\Omega; H^1(B_R))), \\
q \in L^2_{\mathcal{F}}(0, T; L^2(B_R; \mathbb{R}^d)),
\end{cases} \quad \forall R > 0,
$$

such that almost surely for all $\varphi \in C_0^\infty(\mathbb{R}^n)$ and all $t \in [0, T]$,

(1.15)
$$
\begin{aligned}
& \int_{\mathbb{R}^n} u(t, x) \varphi(x) dx - \int_{\mathbb{R}^n} g(x) \varphi(x) dx \\
= & \int_t^T \int_{\mathbb{R}^n} \Big\{ -\frac{1}{2} \langle A(s, x) Du(s, x), D\varphi(x) \rangle \\
& + \langle a(s, x), Du(s, x) \rangle \varphi(x) + c(s, x) u(s, x) \varphi(s, x) \\
& - \langle B(s, x) q(s, x), D\varphi(x) \rangle + \langle b(s, x), q(s, x) \rangle \varphi(x) \\
& + f(s, x) \varphi(x) \Big\} dx ds - \int_t^T \langle \int_{\mathbb{R}^n} q(s, x) \varphi(x) dx, dW(s) \rangle.
\end{aligned}
$$

We note that in the definition of adapted classical solution, we need A and B to be C^1 in x; in the definition of adapted strong solution, we need A and B to be differentiable in x almost everywhere; and in the definition of adapted weak solution, we need only the coefficients $\{A, B, a, b, c\}$ to be bounded and f and g to be locally square integrable.

It is clear that for (1.1), if (u, q) is an adapted classical solution, it is an adapted strong solution; if (u, q) is an adapted strong solution, it is an adapted weak solution. The following result tells the other way around, which will be useful later. In the following proposition, by "the coefficients are regular enough" we mean that all the coefficients have the required differentiability, continuity and integrability.

Proposition 1.3. *Let the coefficients of (1.1) be regular enough. Let* (u, q) *be an adapted weak solution of (1.1). If in addition, (1.13) holds, then* (u, q) *is an adapted strong solution of (1.1). Further, if (1.11) holds, then* (u, q) *is an adapted classical solution of (1.1).*

Proof. Let (u, q) be an adapted weak solution of (1.1) such that (1.13) holds. Then, from (1.15), by integration by parts, we have

$$
\int_{\mathbb{R}^n} \{u(t, x) - g(x)\} \varphi(x) dx
$$

$$
(1.16) \qquad = \int_{\mathbb{R}^n} \Big\{ \int_t^T \big\{ \tfrac{1}{2} \nabla \cdot [A(s, x) Du(s, x)] + \langle a(s, x), Du(s, x) \rangle
$$

$$
+ c(s, x) u(s, x) + \nabla \cdot [B(s, x) q(s, x)] + \langle b(s, x), q(s, x) \rangle
$$

$$
+ f(s, x) \big\} ds - \int_t^T \langle q(s, x), dW(s) \rangle \Big\} \varphi(x) dx.
$$

The above is true for all $\varphi \in C_0^\infty(\mathbb{R}^n)$. Then, (1.12) follows, proving that (u, q) is an adapted strong solution. The other assertion is obvious. $\qquad \square$

Although the parabolicity condition (1.6) is not necessary in Definition 1.2 and Proposition 1.3, we will see later that such a condition is very crucial for our studying the well-posedness of BSPDE (1.1).

§2. Well-posedness of Linear BSPDEs

In this section, we state the results of well-posedness for BSPDE (1.1). The proofs of them will be carried out in later sections.

To begin with, let us introduce the following assumption concerning the coefficients of equation (1.1). Let $m \geq 1$.

$(\mathrm{H})_m$ Functions $\{A, B, a, b, c\}$ satisfy the following:

$$
(2.1) \qquad \begin{cases} A \in L_{\mathcal{F}}^\infty(0, T; C_b^{m+1}(\mathbb{R}^n; S^n)), \\ B \in L_{\mathcal{F}}^\infty(0, T; C_b^{m+1}(\mathbb{R}^n; \mathbb{R}^{n \times d})), \\ a \in L_{\mathcal{F}}^\infty(0, T; C_b^m(\mathbb{R}^n; \mathbb{R}^n)), \\ b \in L_{\mathcal{F}}^\infty(0, T; C_b^m(\mathbb{R}^n; \mathbb{R}^d)), \\ c \in L_{\mathcal{F}}^\infty(0, T; C_b^m(\mathbb{R}^n)). \end{cases}
$$

We note that $(H)_m$ implies that the partial derivatives of A and B in x up to order $(m+1)$, and those of a, b and c up to order m are bounded uniformly in (t, x, ω) by a constant $K_m > 0$. This constant will be referred in the statements of Theorems 2.1, 2.2 and 2.3.

In what follows, we let

$$
\begin{cases}
\alpha \overset{\Delta}{=} (\alpha_1, \cdots, \alpha_n), & \alpha_i\text{'s are nonnegative integers,} \\
|\alpha| \overset{\Delta}{=} \displaystyle\sum_{i=1}^{n} \alpha_i, \quad \partial^\alpha \overset{\Delta}{=} \partial_{x_1}^{\alpha_1} \cdots \partial_{x_n}^{\alpha_n}, \quad x^\alpha \overset{\Delta}{=} x_1^{\alpha_1} \cdots x_n^{\alpha_n}.
\end{cases}
$$

Any α of the above form is called a *multi-index*. If $\beta = (\beta_1, \cdots, \beta_n)$ is another multi-index, by $\beta \leq \alpha$, we mean that $\beta_i \leq \alpha_i$ for each $i = 1, \cdots, n$, and by $\beta < \alpha$, we mean $\beta \leq \alpha$ and at least for one i, one has $\beta_i < \alpha_i$.

Now, we state the following result concerning the well-posedness of BSPDE (1.1).

Theorem 2.1. *Suppose that the parabolicity condition (1.6) holds and $(H)_m$ holds for some $m \geq 1$. Suppose further that the coefficient $B(t, x)$ satisfies the following "symmetry condition":*

(2.2)
$$
\left[B(\partial_{x_i} B^T) \right]^T = B(\partial_{x_i} B^T),
$$
$$
\text{a.e. } (t, x) \in [0, T] \times \mathbb{R}^n, \text{ a.s. }, \quad 1 \leq i \leq n.
$$

Then for any random fields f and g satisfying

(2.3)
$$
\begin{cases}
f \in L^2_{\mathcal{F}}(0, T; H^m(\mathbb{R}^n)), \\
g \in L^2_{\mathcal{F}_T}(\Omega; H^m(\mathbb{R}^n)),
\end{cases}
$$

BSPDE (1.1) admits a unique adapted weak solution (u, q), such that the following estimate holds:

(2.4)
$$
\begin{aligned}
\max_{t \in [0,T]} & E\|u(t, \cdot)\|_{H^m}^2 + E \int_0^T \|q(t, \cdot)\|_{H^{m-1}}^2 dt \\
& + \sum_{|\alpha| \leq m} E \int_0^T \int_{\mathbb{R}^n} \Big\{ \langle (A - BB^T) D(\partial^\alpha u), D(\partial^\alpha u) \rangle \\
& \qquad\qquad\qquad + \big| B^T [D(\partial^\alpha u)] + \partial^\alpha q \big|^2 \Big\} dx\, dt \\
\leq CE & \Big\{ \int_0^T \|f(t, \cdot)\|_{H^m}^2 dt + \|g\|_{H^m}^2 \Big\},
\end{aligned}
$$

where the constant $C > 0$ only depends on m, T and K_m.

Furthermore, if $m \geq 2$, the weak solution (u, q) becomes the unique adapted strong solution of (1.1); and if $m > 2 + n/2$, then (u, q) is the unique adapted classical solution of (1.1).

The symmetry condition (2.2) is technical. It will play a very important role in proving the existence of adapted solutions. However, we point out that such a condition is not needed for the uniqueness of adapted weak (and hence strong and classical) solutions. See §3.1 for details. Several examples satisfying such a condition are listed below:

$$(2.5) \quad \begin{cases} d = n = 1; B \text{ is a scalar;} \\ B \quad \text{is independent of } x; \\ B(t, x) = \varphi(t, x) B_0(t), \text{ where } \varphi \text{ is a scalar-valued random field.} \end{cases}$$

The following result tells us that the symmetry condition (2.2) can be removed if the parabolicity condition (1.6) is strengthened.

Theorem 2.2. *Suppose (1.6) holds and $(H)_m$ with $m \geq 1$ is in force. Suppose further that for some $\varepsilon_0 > 0$, either*

$$(2.6) \qquad A - BB^T \geq \varepsilon_0 BB^T \geq 0, \quad \text{a.e. } (t, x) \in [0, T] \times \mathbb{R}^n, \text{ a.s.},$$

or

$$(2.7) \qquad A - BB^T \geq \varepsilon_0 \sum_{|\alpha|=1} (\partial^\alpha B)(\partial^\alpha B^T) \geq 0,$$

$$\text{a.e. } (t, x) \in [0, T] \times \mathbb{R}^n, \text{ a.s.}$$

Then the conclusion of Theorem 2.1 remains true and the estimate (2.4) is improved to the following:

$$(2.8) \qquad \begin{aligned} \max_{t \in [0,T]} &E\|u(t, \cdot)\|_{H^m}^2 + E \int_0^T \|q(t, \cdot)\|_{H^m}^2 dt \\ &+ \sum_{|\alpha| \leq m} E \int_0^T \int_{\mathbb{R}^n} \langle AD(\partial^\alpha u), D(\partial^\alpha u) \rangle \, dx dt \\ &\leq CE \Big\{ \int_0^T \|f(t, \cdot)\|_{H^m}^2 dt + \|g\|_{H^m}^2 \Big\}, \end{aligned}$$

where the constant $C > 0$ only depends on m, T, K_m and ε_0.

In addition, if A is uniformly positive definite, i.e., (1.9) holds for some $\delta > 0$ (this is the case if (1.1) is super-parabolic, i.e., (1.7) holds), then (2.8) can further be improved to the following:

$$(2.9) \qquad \begin{aligned} \max_{t \in [0,T]} &E\|u(t, \cdot)\|_{H^m}^2 + E \int_0^T \{\|u(t, \cdot)\|_{H^{m+1}}^2 + \|q(t, \cdot)\|_{H^m}^2\} dt \\ &\leq CE \Big\{ \int_0^T \|f(t, \cdot)\|_{H^{m-1}}^2 dt + \|g\|_{H^m}^2 \Big\}. \end{aligned}$$

We note here that conditions (2.6) and (2.7) together with (1.9) are still weaker than the super-parabolicity condition (1.7). For example, if $n > d$ and B is an $(n \times d)$ matrix, then BB^T is always degenerate. We can easily find an A such that (2.6), (2.7) and (1.9) hold but (1.7) fails. Let us also note that if (2.6) or (2.7) holds, we have

(2.10) $|B^T \xi|^2 \leq \langle A\xi, \xi \rangle, \quad \forall \xi \in \mathbb{R}^n$, a.e. $(t, x) \in [0, T] \times \mathbb{R}^n$, a.s.

Thus, (2.8) follows from (2.4) easily.

In the above theorems, we have assumed that f and g are square integrable in $x \in \mathbb{R}^n$ globally. This excludes the case that f and g approach infinity as $|x|$ goes to infinity. In some important applications, such a case happens very often. Thus, in the rest of this section, we would like to extend the above theorems a little further so that f and g are allowed to have certain growth as $|x| \to \infty$. To this end, let us make an observation. Suppose (u, q) is an adapted classical solution of (1.1). Let $\lambda > 0$ and denote $\langle x \rangle \overset{\Delta}{=} \sqrt{|x|^2 + 1}$. Set

(2.11) $\begin{cases} v(t, x) = e^{-\langle \lambda \rangle} u(t, x), \\ p(t, x) = e^{-\langle \lambda \rangle} q(t, x), \end{cases} \quad (t, x) \in [0, T] \times \mathbb{R}^n.$

Then, by a direct computation, we see that (v, p) satisfies the following BSPDE: (compare with (1.1))

(2.12) $\begin{cases} dv = \{ -\dfrac{1}{2} \nabla \cdot (AD v) - \langle \bar{a}, Dv \rangle - \bar{c}v - \nabla \cdot (Bp) - \langle \bar{b}, p \rangle - \bar{f} \} dt \\ \qquad + \langle p, dW(t) \rangle, \qquad (t, x) \in [0, T] \times \mathbb{R}^n, \\ v\big|_{t=T} = \bar{g}, \end{cases}$

with

(2.13) $\begin{cases} \bar{a} = a + \lambda A \left[\frac{x}{\langle x \rangle} \right], \\ \bar{c} = c + \dfrac{\lambda^2}{2} \langle A \left[\frac{x}{\langle x \rangle} \right], \left[\frac{x}{\langle x \rangle} \right] \rangle + \dfrac{\lambda}{2} \nabla \cdot \left(A \left[\frac{x}{\langle x \rangle} \right] \right) + \lambda \langle a, \left[\frac{x}{\langle x \rangle} \right] \rangle, \\ \bar{b} = b + \lambda B^T \left[\frac{x}{\langle x \rangle} \right], \\ \bar{f} = e^{-\langle \lambda \rangle} f(t, x), \quad \bar{g} = e^{-\langle \lambda \rangle} g(t, x). \end{cases}$

Conversely, if (v, p) is an adapted classical solution of (2.12), then (u, q), which is determined through (2.11), is an adapted classical solution of (1.1). Clearly, the same equivalence between (1.1) and (2.12) holds for adapted strong and weak solutions, respectively.

On the other hand, from (2.13) we see easily that the group $\{A, B, a, b, c\}$ satisfies $(H)_m$ if and only if $\{A, B, \bar{a}, \bar{b}, \bar{c}\}$ satisfies $(H)_m$. This is due to the fact that for any multi-index α, it holds that

$$|\partial^\alpha \langle x \rangle| \leq C, \qquad \forall x \in \mathbb{R}^n,$$

with the constant $C > 0$ only depending on $|\alpha|$. Hence, from Theorems 2.1 and 2.2, we can derive the following result.

Theorem 2.3. *Let $m \geq 1$ and $(H)_m$ hold for $\{A, B, a, b, c\}$. Let (1.6) and (2.2) hold. Let $\lambda > 0$ such that*

(2.14)
$$
\begin{cases}
e^{-\lambda\langle\cdot\rangle} f \in L^2_{\mathcal{F}}(0, T; H^m(\mathbb{R}^n)), \\
e^{-\lambda\langle\cdot\rangle} g \in L^2_{\mathcal{F}_T}(\Omega; H^m(\mathbb{R}^n)).
\end{cases}
$$

Then BSPDE (1.1) admits a unique adapted weak solution (u, q), such that the following estimate holds:

(2.15)
$$
\begin{aligned}
&\max_{t\in[0,T]} E\|e^{-\lambda\langle\cdot\rangle}u(t,\cdot)\|^2_{H^m} + E\int_0^T \|e^{-\lambda\langle\cdot\rangle}q(t,\cdot)\|^2_{H^{m-1}}dt \\
&+ \sum_{|\alpha|\leq m} E\int_0^T \int_{\mathbb{R}^n} \Big\{\langle (A - BB^T)D[\partial^\alpha(e^{-\lambda\langle\cdot\rangle}u)], D[\partial^\alpha(e^{-\lambda\langle\cdot\rangle}u)]\rangle \\
&\qquad\qquad\qquad + \Big|B^T\{D[\partial^\alpha(e^{-\lambda\langle\cdot\rangle}u)]\} + \partial^\alpha(e^{-\lambda\langle\cdot\rangle}q)\Big|^2\Big\}dxdt \\
&\leq CE\Big\{\int_0^T \|e^{-\lambda\langle\cdot\rangle}f(t,\cdot)\|^2_{H^m}dt + \|e^{-\lambda\langle\cdot\rangle}g\|^2_{H^m}\Big\},
\end{aligned}
$$

where the constant $C > 0$ only depends on m, T and K_m.

Furthermore, if $m \geq 2$, the weak solution (u, q) becomes the unique adapted strong solution of (1.1); and if $m > 2 + n/2$, then (u, q) is the unique adapted classical solution of (1.1).

In the case that (2.2) is replaced by (2.6) or (2.7), the above conclusion remains true and the estimate (2.15) can be improved to the following:

(2.16)
$$
\begin{aligned}
&\max_{t\in[0,T]} E\|e^{-\lambda\langle\cdot\rangle}u(t,\cdot)\|^2_{H^m} + E\int_0^T \|e^{-\lambda\langle\cdot\rangle}q(t,\cdot)\|^2_{H^m}dt \\
&+ \sum_{|\alpha|\leq m} E\int_0^T \int_{\mathbb{R}^n} \langle AD[\partial^\alpha(e^{-\lambda\langle\cdot\rangle}u)], D[\partial^\alpha(e^{-\lambda\langle\cdot\rangle}u)]\rangle \, dxdt \\
&\leq CE\Big\{\int_0^T \|e^{-\lambda\langle\cdot\rangle}f(t,\cdot)\|^2_{H^m}dt + \|e^{-\lambda\langle\cdot\rangle}g\|^2_{H^m}\Big\},
\end{aligned}
$$

Finally, if in addition, (1.9) holds for some $\delta > 0$, then (2.16) can further be improved to the following:

(2.17)
$$
\begin{aligned}
&\max_{t\in[0,T]} E\|e^{-\lambda\langle\cdot\rangle}u(t,\cdot)\|^2_{H^m} \\
&+ E\int_0^T \Big\{\|e^{-\lambda\langle\cdot\rangle}u(t,\cdot)\|^2_{H^{m+1}} + \|e^{-\lambda\langle\cdot\rangle}q(t,\cdot)\|^2_{H^m}\Big\}dt \\
&\leq CE\Big\{\int_0^T \|e^{-\lambda\langle\cdot\rangle}f(t,\cdot)\|^2_{H^{m-1}}dt + \|e^{-\lambda\langle\cdot\rangle}g\|^2_{H^m}\Big\}.
\end{aligned}
$$

Clearly, (2.14) means that f and g can have an exponential growth as $|x| \to \infty$. This is good enough for many applications.

We note that $\{A, B, a, b, c\}$ satisfies (H)$_m$ if and only if $\{A, B, \tilde{a}, \tilde{b}, c\}$ satisfies (H)$_m$, where $\tilde{a}$ and $\tilde{b}$ are given by (1.4). Thus, we have the exact statements as Theorems 2.1, 2.2 and 2.3 for BSPDE (1.5) with a and b replaced by $\tilde{a}$ and $\tilde{b}$.

§3. Uniqueness of Adapted Solutions

In this section, we are going to establish the uniqueness of adapted weak, strong and classical solutions to our BSPDEs. From the discussion right before Proposition 1.3, we see that it suffices for us to prove the uniqueness of adapted weak solutions.

§3.1. Uniqueness of adapted weak solutions

For convenience, we denote

$$(3.1) \quad \begin{cases} \mathcal{L}u \stackrel{\Delta}{=} \dfrac{1}{2}\nabla\cdot[ADu] + \langle a, Du\rangle + cu, \\ \mathcal{M}q \stackrel{\Delta}{=} \nabla\cdot[Bq] + \langle b, q\rangle. \end{cases}$$

Then, equation (1.1) is the same as the following:

$$(3.2) \quad \begin{cases} du = -\{\mathcal{L}u + \mathcal{M}q + f\}dt + \langle q, dW(t)\rangle, \quad (t,x) \in [0,T] \times \mathbb{R}^n, \\ u\big|_{t=T} = g. \end{cases}$$

In this section, we are going to prove the following result.

Theorem 3.1. *Let (2.3) hold and the following hold:*

$$(3.3) \quad \begin{cases} A \in L_{\mathcal{F}}^\infty(0, T; L^\infty(\mathbb{R}^n; S^n)), \\ B \in L_{\mathcal{F}}^\infty(0, T; L^\infty(\mathbb{R}^n; \mathbb{R}^{n \times d})), \\ a \in L_{\mathcal{F}}^\infty(0, T; L^\infty(\mathbb{R}^n; \mathbb{R}^n)), \\ b \in L_{\mathcal{F}}^\infty(0, T; L^\infty(\mathbb{R}^n; \mathbb{R}^d)), \\ c \in L_{\mathcal{F}}^\infty(0, T; L^\infty(\mathbb{R}^n)). \end{cases}$$

Then, the adapted weak solution (u, q) of (3.2) is unique in the class

$$(3.4) \quad \begin{cases} u \in C_{\mathcal{F}}([0, T]; L^2(\Omega; H^1(\mathbb{R}^n))), \\ q \in L_{\mathcal{F}}^2(0, T; L^2(\mathbb{R}^n; \mathbb{R}^d)). \end{cases}$$

To prove the above uniqueness theorem, we need some preliminaries. First of all, let us recall the *Gelfand triple* $H^1(\mathbb{R}^n) \hookrightarrow L^2(\mathbb{R}^n) \hookrightarrow H^{-1}(\mathbb{R}^n)$. Here, $H^{-1}(\mathbb{R}^n)$ is the dual space of $H^1(\mathbb{R}^n)$, and the embeddings are dense and continuous. We denote the duality paring between $H^1(\mathbb{R}^n)$ and $H^{-1}(\mathbb{R}^n)$ by $\langle\cdot,\cdot\rangle_0$, and the inner product and the norm in $L^2(\mathbb{R}^n)$ by

$(\cdot,\cdot)_0$ and $|\cdot|_0$, respectively. Then, by identifying $L^2(\mathbb{R}^n)$ with its dual $L^2(\mathbb{R}^n)^*$ (using Riesz representation theorem), we have the following:

$$
\begin{aligned}
\langle \psi, \varphi \rangle_0 &= (\psi, \varphi)_0 \\
(3.5) \qquad &= \int_{\mathbb{R}^n} \psi(x)\varphi(x)dx, \quad \forall \psi \in L^2(\mathbb{R}^n), \ \varphi \in H^1(\mathbb{R}^n),
\end{aligned}
$$

and

$$
(3.6) \qquad
\begin{cases}
\sum_{i=1}^{n} \partial_i \psi_i \in H^{-1}(\mathbb{R}^n), \quad \forall \psi_i \in L^2(\mathbb{R}^n), \ 1 \le i \le n, \\
\langle \sum_{i=1}^{n} \partial_i \psi_i, \varphi \rangle_0 \overset{\Delta}{=} - \int_{\mathbb{R}^n} \psi_i(x)\partial_i \varphi(x)dx, \quad \forall \varphi \in H^1(\mathbb{R}^n).
\end{cases}
$$

Next, let (u, q) be an adapted weak solution of (3.2) satisfying (3.4). Note that in (3.4), the integrability of (u, q) in x is required to be global. By (3.5)–(3.6), we see that

$$
(3.7) \qquad \mathcal{L}u + \mathcal{M}q \in L^2_{\mathcal{F}}(0, T; H^{-1}(\mathbb{R}^n)).
$$

In the present case, from (1.15), for any $\varphi \in H^1(\mathbb{R}^n)$ (not just $C_0^\infty(\mathbb{R}^n)$), we have

$$
(3.8) \qquad
\begin{cases}
d(u, \varphi)_0 = - \langle \mathcal{L}u + \mathcal{M}q + f, \varphi \rangle_0 + \langle (q, \varphi)_0, dW(t) \rangle, \quad t \in [0, T], \\
(u, \varphi)_0 \big|_{t=T} = (g, \varphi)_0.
\end{cases}
$$

Here, $(q, \varphi)_0 \overset{\Delta}{=} ((q_1, \varphi)_0, \cdots, (q_d, \varphi)_0)$ and $q = (q_1, \cdots, q_d)$. Sometimes, we say that (3.2) holds in $H^{-1}(\mathbb{R}^n)$ if (3.8) holds for all $\varphi \in H^1(\mathbb{R}^n)$.

In proving the uniqueness of the adapted weak solutions, the following special type of Itô's formula is very crucial.

Lemma 3.2. Let $\xi \in L^2_{\mathcal{F}}(0, T; H^{-1}(\mathbb{R}^n))$ and (u, q) satisfy (3.4), such that

$$
(3.9) \qquad du = \xi dt + \langle q, dW(t) \rangle, \quad t \in [0, T].
$$

Then

$$
(3.10) \qquad
\begin{aligned}
|u(t)|_0^2 = |u(0)|_0^2 &+ \int_0^t \big\{ 2 \langle \xi(s), u(s) \rangle_0 + |q(s)|_0^2 \big\} ds \\
&+ 2 \int_0^t \langle (q(s), u(s))_0, dW(s) \rangle, \quad t \in [0, T].
\end{aligned}
$$

Although the above seems to be a very special form of general Itô's formula, it is enough for our purpose. We note that the processes u, q and ξ take values in different spaces $H^1(\mathbb{R}^n)$, $L^2(\mathbb{R}^n)$ and $H^{-1}(\mathbb{R}^n)$, respectively. This makes the proof of (3.10) a little nontrivial. We postpone the proof of Lemma 3.2 to the next subsection.

Proof of Theorem 3.1. Let (u, q) be any adapted weak solution of (3.2) with f and g being zero, such that (3.4) holds. We need to show that $(u, q) = 0$, which gives the uniqueness of adapted weak solution. Applying Lemma 3.2, we have (note (3.7))

(3.11)
$$
\begin{aligned}
E|u(t)|_0^2 &= E \int_t^T \left\{ 2 \langle \mathcal{L}u(s) + \mathcal{M}q(s), u(s) \rangle_0 - |q(s)|_0^2 \right\} ds \\
&= E \int_t^T \int_{\mathbb{R}^n} \left\{ - \langle ADu, Du \rangle + \langle a, D(u^2) \rangle + 2cu^2 \right. \\
&\qquad\qquad \left. - 2 \langle q, B^T Du \rangle + 2 \langle bu, q \rangle - |q|^2 \right\} dx ds \\
&= E \int_t^T \int_{\mathbb{R}^n} \left\{ - \langle (A - BB^T)Du, Du \rangle \right. \\
&\qquad\qquad - |q + B^T Du - bu|^2 \\
&\qquad\qquad \left. + [b^2 + 2c - \nabla \cdot (a + Bb)]u^2 \right\} ds \\
&\le C \int_t^T E|u(s)|_0^2 ds, \qquad t \in [0, T].
\end{aligned}
$$

By Gronwall's inequality, we obtain

$$
E|(t)|_0^2 = 0, \quad t \in [0, T].
$$

Hence, $u = 0$. By (3.11) again, we must also have $q = 0$. This proves the uniqueness of adapted weak solutions to (3.2). □

§3.2. An Itô formula

In this subsection, we are going to present a special type of Itô's formula in abstract spaces for which Lemma 3.2 is a special case.

Let V and H be two separable Hilbert spaces such that the embedding $V \hookrightarrow H$ is dense and continuous. We identify H with its dual H' (by Riesz representation theorem). The dual of V is denoted by V'. Then we have the *Gelfand triple* $V \hookrightarrow H = H' \hookrightarrow V'$. We denote the inner product and the induced norm of H by $(\cdot, \cdot)_0$ and $|\cdot|_0$, respectively. The duality paring between V and V' is denoted by $\langle \cdot, \cdot \rangle_0$, and the norms of V and V' are denoted by $\| \cdot \|$ and $\| \cdot \|_*$, respectively. We know that the following holds:

(3.12)
$$
\langle u, v \rangle_0 = (u, v)_0, \quad \forall u \in H, \; v \in V.
$$

Due to this reason, H is usually called the *pivot space*. It is also known (see [Lions]) that in the present setting, there exists a symmetric linear operator $\mathcal{A} \in \mathcal{L}(V, V')$, such that

(3.13)
$$
\langle \mathcal{A}v, v \rangle_0 \le -\|v\|^2, \quad \forall v \in V.
$$

Now, let us state the following result which is more general than Lemma 3.2.

Lemma 3.3. *Let*

(3.14)
$$
\begin{cases}
u \in C_{\mathcal{F}}([0,T];V), \\
q \in L_{\mathcal{F}}^2(0,T;H)^d, \\
\xi \in L_{\mathcal{F}}^2(0,T;V'),
\end{cases}
$$

satisfying

(3.15)
$$
du = \xi dt + \langle q, dW(t) \rangle, \quad t \in [0,T].
$$

Then

(3.16)
$$
|u(t)|_0^2 = |u(0)|_0^2 + \int_0^t \big\{ 2 \langle \xi(s), u(s) \rangle_0 + |q(s)|_0^2 \big\} ds
$$
$$
+ 2 \int_0^t \langle (q(s), u(s))_0, dW(s) \rangle, \quad t \in [0,T].
$$

In the above, $q \in L_{\mathcal{F}}^2(0,T;H)^d$ means that $q = (q_1, \cdots, q_d)$ with $q_i \in L_{\mathcal{F}}^2(0,T;H)$. In what follows, we will see the expression $q \in L_{\mathcal{F}}^2(0,T;V)^d$ whose meaning is similar. Before giving a rigorous proof of the above result, let us try to prove it in an obvious (naive) way. From (3.16), we see that the trouble mainly comes from ξ since it takes values in V'. Thus, it is pretty natural that we should find a sequence $\xi_k \in L_{\mathcal{F}}^2(0,T;H)$, such that

(3.17)
$$
\xi_k \to \xi, \quad \text{in } L_{\mathcal{F}}^2(0,T;V'), \quad (k \to \infty),
$$

and let u_k be defined by

(3.18)
$$
u_k(t) = u(0) + \int_0^t \xi_k(s) ds + \int_0^t \langle q(s), dW(s) \rangle, \quad t \in [0,T].
$$

Since the processes u_k, ξ_k and q are all taking values in H, we have

(3.19)
$$
|u_k(t)|_0^2 = |u(0)|_0^2 + \int_0^t \big\{ 2 \langle \xi_k(s), u_k(s) \rangle_0 + |q(s)|_0^2 \big\} ds
$$
$$
+ 2 \int_0^t \langle (q(s), u_k(s))_0, dW(s) \rangle, \quad t \in [0,T].
$$

This can be proved by projecting (3.18) to finite dimensional spaces, using usual Itô's formula, then pass to the limit. Having (3.19), one then hopes to pass to the limit to obtain (3.16). This can be done provided one has the following convergence:

$$
u_k \to u, \quad \text{in } L_{\mathcal{F}}^2(0,T;V).
$$

However, (3.17)–(3.18) only guarantees

$$
u_k \to u, \quad \text{in } L_{\mathcal{F}}^2(0,T;V').
$$

Thus, the convergence of u_k to u is not *strong* enough and such an approach does not work! In what follows, we will see that to prove (3.16), much more has to be involved.

Let us now state two standard lemmas for deterministic evolution equations whose proofs are omitted here (see Lions [1]).

Lemma 3.4. Let $v : [0, T] \to V'$ be absolutely continuous, such that

$$
(3.20) \qquad \begin{cases} v \in L^2(0, T; V), \\ \dot{v} \in L^2(0, T; V'). \end{cases}
$$

Then $v \in C([0, T]; H)$ and

$$
(3.21) \qquad \frac{d}{dt} |v(t)|_0^2 = 2 \langle \dot{v}(t), v(t) \rangle_0, \quad \text{a.e. } t \in [0, T].
$$

Lemma 3.5. Let $\mathcal{A} \in \mathcal{L}(V, V')$ be symmetric satisfying (3.13). Then for any $v_0 \in H$ and $f \in L^2(0, T; V')$, the following problem

$$
(3.22) \qquad \begin{cases} \dot{v} = \mathcal{A}v + f, \quad t \in [0, T], \\ v(0) = v_0, \end{cases}
$$

admits a unique solution v satisfying (3.20) and

$$
(3.23) \qquad |v(t)|_0^2 + \int_0^t \|v(s)\|^2 ds \le |v_0|_0^2 + \int_0^t \|f(s)\|_*^2 ds, \quad t \in [0, T].
$$

Moreover, it holds

$$
(3.24) \qquad |v(t)|_0^2 = |v_0|_0^2 + 2 \int_0^t \langle \mathcal{A}v(s) + f(s), v(s) \rangle_0 ds, \quad t \in [0, T].
$$

Now, we consider stochastic evolution equations. We first have the following result.

Lemma 3.6. Let v be an $\{\mathcal{F}_t\}_{t \ge 0}$-adapted V'-valued processes which is absolutely continuous almost surely and q be an $\{\mathcal{F}_t\}_{t \ge 0}$-adapted H-valued process such that the following holds:

$$
(3.25) \qquad \begin{cases} v \in L^2_{\mathcal{F}}(0, T; V), \\ \dot{v} \in L^2_{\mathcal{F}}(0, T; V'), \\ q \in L^2_{\mathcal{F}}(0, T; V)^d. \end{cases}
$$

Let

$$
(3.26) \qquad M(t) = \int_0^t \langle q(s), dW(s) \rangle, \quad t \in [0, T].
$$

Then, $M \in C_{\mathcal{F}}([0, T]; V)$ and

$$
(3.27) \qquad |M(t)|_0^2 = 2 \int_0^t \langle (M(s), q(s))_0, dW(s) \rangle + \int_0^t |q(s)|_0^2 ds,
$$
$$
t \in [0, T], \text{ a.s.}
$$

$$(3.28) \quad d(v(t), M(t))_0 = \langle \dot{v}(t), M(t) \rangle_0 dt + \langle (v(t), q(t))_0, dW(t) \rangle,$$

$$\text{a.e. } t \in [0, T], \text{ a.s.}$$

Proof. First of all, it is clear that $M \in C_{\mathcal{F}}([0, T]; V)$ and (3.27) holds since we may regard both M and q as H-valued processes. We now prove (3.28). Take a sequence of absolutely continuous processes v_k with the following properties:

$$(3.29) \quad \begin{cases} v_k \in L_{\mathcal{F}}^2(0, T; V), \\ \dot{v}_k \in L_{\mathcal{F}}^2(0, T; H), \\ v_k \to v, \quad \text{in } L_{\mathcal{F}}^2(0, T; V), \\ \dot{v}_k \to \dot{v}, \quad \text{in } L_{\mathcal{F}}^2(0, T; V'). \end{cases}$$

Now, in H, we have (note (3.12))

$$(3.30) \quad \begin{aligned} d(v_k(t), M(t))_0 &= (\dot{v}_k(t), M(t))_0 dt + \langle (v_k(t), q(t))_0, dW(t) \rangle \\ &= \langle \dot{v}_k(t), M(t) \rangle_0 dt + \langle (v_k(t), q(t))_0, dW(t) \rangle. \end{aligned}$$

Pass to the limit in the above, using (3.29), we obtain (3.28). $\qquad\square$

Lemma 3.7. Let $\mathcal{A} \in \mathcal{L}(V, V')$ be symmetric satisfying (3.13). Then, for any f, q, u_0 satisfying

$$(3.31) \quad \begin{cases} f \in L_{\mathcal{F}}^2(0, T; V'), \\ q \in L_{\mathcal{F}}^2(0, T; H)^d, \\ u_0 \in H, \end{cases}$$

the following problem

$$(3.32) \quad \begin{cases} du = (\mathcal{A}u + f)dt + \langle q, dW(t) \rangle, \quad t \in [0, T], \\ u(0) = u_0, \end{cases}$$

admits a unique solution $u \in L_{\mathcal{F}}^2(0, T; V) \cap C_{\mathcal{F}}([0, T]; H)$, such that

$$(3.33) \quad \begin{aligned} |u(t)|_0^2 &= |u_0|_0^2 + \int_0^t \{2\langle \mathcal{A}u(s) + f(s), u(s) \rangle_0 + |q(s)|_0^2\} ds \\ &\quad + 2\int_0^t \langle (q(s), u(s))_0, dW(s) \rangle, \quad \forall t \in [0, T], \text{ a.s.} \end{aligned}$$

Proof. We first let $q \in L_{\mathcal{F}}^2(0, T; V)^d$ and define $M(t)$ by (3.26). Consider the following problem:

$$(3.34) \quad \begin{cases} \dot{v} = \mathcal{A}v + f + \mathcal{A}M, \quad t \in [0, T], \\ v(0) = u_0. \end{cases}$$

By Lemma 3.5, for almost all $\omega \in \Omega$, (3.34) admits a unique solution v. Obviously (by the variation of constants formula, if necessary), v is $\{\mathcal{F}_t\}_{t\geq0}$-adapted. Thus, we have

$$\begin{cases} v \in L^2_{\mathcal{F}}(0,T;V), \\ \dot{v} \in L^2_{\mathcal{F}}(0,T,V'), \end{cases}$$

which implies (by Lemma 3.4) $v \in C_{\mathcal{F}}([0,T];H)$ and (by (3.24))

$$(3.35) \qquad |v(t)|^2_0 = |u_0|^2_0 + 2 \int_0^t \langle \mathcal{A}v(s) + f(s) + \mathcal{A}M(s), v(s) \rangle_0 ds,$$

$$\forall t \in [0,T], \text{ a.s.}$$

Set $u(t) = v(t) + M(t)$. Then, we see that $u \in L^2_{\mathcal{F}}(0,T;V) \cap C_{\mathcal{F}}([0,T];H)$ is a solution of (3.32). We now combining (3.27)–(3.28) and (3.34)–(3.35) to obtain the following:

$$
\begin{aligned}
|u(t)|^2_0 &= |v(t)|^2_0 + |M(t)|^2_0 + 2(v(t), M(t))_0 \\
&= |u_0|^2_0 + 2 \int_0^t \langle \mathcal{A}v(s) + f(s) + \mathcal{A}M(s), v(s) \rangle_0 ds \\
&\quad + 2 \int_0^t \langle (M(s), q(s))_0, dW(s) \rangle + \int_0^t |q(s)|^2_0 ds \\
&\quad + 2 \int_0^t \langle \dot{v}(s), M(s) \rangle_0 ds + 2 \int_0^t \langle (v(s), q(s))_0, dW(s) \rangle \\
(3.36) \quad &= |u_0|^2_0 + 2 \int_0^t \langle \mathcal{A}u(s) + f(s), v(s) \rangle_0 ds \\
&\quad + 2 \int_0^t \langle (u(s), q(s))_0, dW(s) \rangle + \int_0^t |q(s)|^2_0 ds \\
&\quad + 2 \int_0^t \langle \mathcal{A}u(s) + f(s), M(s) \rangle_0 ds \\
&= |u_0|^2_0 + \int_0^t \{ 2 \langle \mathcal{A}u(s) + f(s), u(s) \rangle_0 + |q(s)|^2_0 \} ds \\
&\quad + 2 \int_0^t \langle (u(s), q(s))_0, dW(s) \rangle.
\end{aligned}
$$

Next, we claim that solution to (3.32) is unique (for any f, q and u_0 satisfying (3.31)). As a matter of fact, if $\widehat{u}$ is another solution to (3.32), then $u - \widehat{u}$ is a solution of (3.32) with f, q and u_0 all being zero. Applying (3.36) to $u - \widehat{u}$, we obtain (see (3.13))

$$|u(t) - \widehat{u}(t)|^2_0 = 2 \int_0^t \langle A[u(s) - \widehat{u}(s)], u(s) - \widehat{u}(s) \rangle_0 ds \leq 0,$$

which results in $u = \widehat{u}$. Thus, we have proved our lemma for the case $q \in L^2_{\mathcal{F}}(0,T;V)^d$. Now, for general case, i.e., $q \in L^2_{\mathcal{F}}(0,T;H)^d$, we take a sequence $q_k \in L^2_{\mathcal{F}}(0, T; V)^d$ with

$$q_k \to q, \quad \text{in } L^2_{\mathcal{F}}(0, T; H)^d.$$

Let u_k be the solution of (3.32) with q being replaced by q_k. Then applying (3.36) to $u_k - u_\ell$, we have (note (3.13))

(3.37)
$$E|u_k(t) - u_\ell(t)|^2_0 + 2E \int_0^t \|u_k(s) - u_\ell(s)\|^2 ds$$

$$\leq E \int_0^T |q_k(s) - q_\ell(s)|^2_0 ds \to 0, \qquad k, \ell \to \infty.$$

This means that the sequence $\{u_k\}$ is Cauchy in $L^2_{\mathcal{F}}(0, T; V) \cap C_{\mathcal{F}}([0, T]; H)$. Hence, there exists a limit u of $\{u_\}k$ in this space. Clearly, u is a solution of (3.32). Also, we have a similar equality (3.33) for each u_k. Pass to the limit, we obtain the equality (3.33) for u (with general $q \in L^2_{\mathcal{F}}(0, T; H)^d$). $\qquad \square$

Now, we are ready to prove Lemma 3.3.

Proof of Lemma 3.3. Set

$$\begin{cases} u_0 = u(0) \in H, \\ f \overset{\Delta}{=} \xi - \mathcal{A}u \in L^2_{\mathcal{F}}(0, T; V'). \end{cases}$$

Then u is a solution of (3.32) with (3.31) holds. Hence, (3.33) holds, which yields (3.16). $\qquad \square$

Now, by taking $V = H^1(\mathbb{R}^n)$, $H = L^2(\mathbb{R}^n)$ and $V' = H^{-1}(\mathbb{R}^n)$, we see that Lemma 3.2 follows immediately from Lemma 3.3.

§4. Existence of Adapted Solutions

The proofs of existence of adapted solutions is based on the following fundamental lemma.

Lemma 4.1. *Let the parabolicity condition (1.6) and the symmetry condition (2.2) hold. Let $(H)_m$ hold for some $m \geq 1$. Then there exists a constant $C > 0$, such that for any $u \in C_0^\infty(\mathbb{R}^n)$ and $q \in C_0^\infty(\mathbb{R}^n; \mathbb{R}^d)$, it holds*

(4.1)
$$\int_{\mathbb{R}^n} \Big\{ \sum_{|\alpha| \leq m} \{ \langle (A - BB^T)D(\partial^\alpha u), D(\partial^\alpha u) \rangle$$

$$+ |B^T D(\partial^\alpha u) + \partial^\alpha q|^2 \} + \sum_{|\alpha| \leq m-1} |\partial^\alpha q|^2 \Big\} dx$$

$$\leq C \int_{\mathbb{R}^n} \sum_{|\alpha| \leq m} \big\{ -2(\partial^\alpha u)\partial^\alpha(\mathcal{L}u + \mathcal{M}q) + |\partial^\alpha q|^2 + |\partial^\alpha u|^2 \big\} dx,$$

$$\text{a.e.} \, t \in [0, T], \text{ a.s.}$$

If (2.6) or (2.7) holds instead of (2.2), the above can be replaced by the following:

(4.2)
$$\int_{\mathbf{R}^n} \Big\{ \sum_{|\alpha| \le m} \langle AD(\partial^\alpha u), D(\partial^\alpha u) \rangle + \sum_{|\alpha| \le m} |\partial^\alpha q|^2 \Big\} dx$$
$$\le C \int_{\mathbf{R}^n} \sum_{|\alpha| \le m} \Big\{ -2(\partial^\alpha u)\partial^\alpha (\mathcal{L}u + \mathcal{M}q) + |\partial^\alpha q|^2 + |\partial^\alpha u|^2 \Big\} dx,$$

a.e. $t \in [0, T]$, a.s.

Furthermore, if (2.6) or (2.7) holds and $A(t, x)$ is uniformly positive definite, then (4.2) can be improved to the following:

(4.3)
$$\int_{\mathbf{R}^n} \Big\{ \sum_{|\alpha| \le m+1} |\partial^\alpha u|^2 + \sum_{|\alpha| \le m} |\partial^\alpha q|^2 \Big\} dx$$
$$\le C \int_{\mathbf{R}^n} \sum_{|\alpha| \le m} \Big\{ -2(\partial^\alpha u)\partial^\alpha (\mathcal{L}u + \mathcal{M}q) + |\partial^\alpha q|^2 + |\partial^\alpha u|^2 \Big\} dx,$$

a.e. $t \in [0, T]$, a.s.

We note that the square root of the left hand side of (4.1) is a norm in the space $C_0^\infty(\mathbf{R}^n) \times C_0^\infty(\mathbf{R}^n; \mathbf{R}^d)$. Thus, if we denote the completion of the space $C_0^\infty(\mathbf{R}^n) \times C_0^\infty(\mathbf{R}^n; \mathbf{R}^d)$ under this norm by $\mathcal{H}^m(t, \omega)$ (note that it depends on $(t, \omega) \in [0, T] \times \Omega$), then we have the following inclusions:

$$C_0^\infty(\mathbf{R}^n) \times C_0^\infty(\mathbf{R}^n; \mathbf{R}^d) \subset \mathcal{H}^m(t, \omega) \subseteq H^m(\mathbf{R}^n) \times H^{m-1}(\mathbf{R}^n; \mathbf{R}^d).$$

It is clear that estimate (4.1) also holds for any $(u, q) \in \mathcal{H}^m(t, \omega)$. A similar argument holds for (4.2) and (4.3).

Since the proof of the above lemma is rather technical and lengthy, we postpone its proof to the next section.

Before going further, let us recall the following fact concerning the differentiability of stochastic integrals with respect to the parameter. Let $h \in L_{\mathcal{F}}^2(0, T; C_b^m(\mathbf{R}^n; \mathbf{R}^d))$. Then it can be shown that the stochastic integral with parameter: $\int_0^t \langle h(s, x, \cdot), dW(s) \rangle$ has a modification that belongs to $L_{\mathcal{F}}^2(0, T; C_b^{m-1}(\mathbf{R}^n; \mathbf{R}^m))$ and it satisfies

(4.4)
$$\partial^\alpha \int_0^t \langle h(s, x, \cdot), dW(s) \rangle = \int_0^t \langle \partial^\alpha h(s, x, \cdot), dW(s) \rangle,$$

for $|\alpha| = 1, 2, \cdots, m - 1$.

Consequently, if $h \in L_{\mathcal{F}}^2(0, T; C_b^\infty)$, then

$$\int_0^\cdot \langle h(s, \cdot, \cdot), dW(s) \rangle \in L_{\mathcal{F}}^2(0, T; C_b^\infty),$$

and (4.4) holds for all multi-index α.

In the rest of this section, we prove the existence of adapted weak solutions to (1.1) (or equivalently (3.2)) under conditions of Theorems 2.1 or 2.2.

We first assume that conditions of Theorem 2.1 hold.

Let us take an orthonormal basis $\{\varphi_k\}_{k \geq 1} \subseteq C_0^\infty(\mathbb{R}^n)$ for the Hilbert space $H^m \equiv H^m(\mathbb{R}^n)$, whose inner product is denoted by

$$(\varphi, \psi)_m \equiv \int_{\mathbb{R}^n} \sum_{|\alpha| \leq m} (\partial^\alpha \varphi)(\partial^\alpha \psi) dx, \quad \forall \varphi, \psi \in H^m.$$

The induced norm is denoted by $|\cdot|_m$. When $q = (q_1, \cdots, q_d)$, $p = (p_1, \cdots, p_d) \in (H^m)^d$, we denote

$$(q, p)_m = \sum_{i=1}^{d} (q_i, p_i)_m,$$

which should not be misunderstood from the context. As a usual convention, $H^0 \equiv L^2(\mathbb{R}^n)$. Let $k \geq 1$ be fixed. Consider the following linear BSDE (not BSPDE):

$$(4.5) \quad \begin{cases} du^{kj}(t) = \Big\{ -\sum_{i=1}^{k} \big[(\mathcal{L}\varphi_i, \varphi_j)_m u^{ki}(t) - \langle (\mathcal{M}\varphi_i, \varphi_j)_m, q^{ki}(t) \rangle \big] \\ \qquad\qquad\quad - (f, \varphi_j)_m \Big\} dt + \langle q^{kj}(t), dW(t) \rangle, \\ u^{kj}(T) = (g, \varphi_j)_m, \qquad 1 \leq j \leq k. \end{cases}$$

By the result of Chapter 1, we know that there exists a unique adapted solution

$$(4.6) \quad \begin{cases} u^{kj}(\cdot) \in C_{\mathcal{F}}([0,T]; \mathbb{R}), \\ q^{kj}(\cdot) \in L_{\mathcal{F}}^2(0,T; \mathbb{R}^d), \end{cases} \quad 1 \leq j \leq k.$$

We define

$$(4.7) \quad \begin{cases} u^k(t,x,\omega) = \sum_{j=1}^{k} u^{kj}(t,\omega)\varphi_j(x), \\ q^k(t,x,\omega) = \sum_{j=1}^{k} q^{kj}(t,\omega)\varphi_j(x), \end{cases} \quad (t,x,\omega) \in [0,T] \times \mathbb{R}^n \times \Omega.$$

Then we see that for any fixed $(t,\omega) \in [0,T] \times \Omega$,

$$u^k(t,\cdot,\omega) \in C_0^\infty(\mathbb{R}^n), \quad q^k(t,\cdot,\omega) \in C_0^\infty(\mathbb{R}^n; \mathbb{R}^d).$$

Also, the following holds:

$$(4.8) \quad \begin{cases} du^k = \big\{ -P_k[\mathcal{L}u^k + \mathcal{M}q^k] - f^k \big\} dt + \langle q^k, dW(t) \rangle, \\ u^k \big|_{t=T} = g^k, \end{cases}$$

where

$$P_k : H^m \to \text{span}\{\varphi_1, \cdots, \varphi_k\} \triangleq H_k^m,$$

is the orthogonal projection (in H^m), and

$$f^k = P_k f, \qquad g^k = P_k g.$$

Note that, as processes, u^k and $q_1^k, \cdots, q_d^k$ are taking values in H_k^m, where $q^k = (q_1^k, \cdots, q_d^k)$. Thus, in particular,

(4.9) $$P_k u^k = u^k, \qquad k \geq 1.$$

Next, we want to derive a proper estimate for (u^k, q^k). By Lemma 4.1, we have the following:

(4.10)
$$\int_{\mathbf{R}^n} \Big\{ \sum_{|\alpha| \leq m} \big\{ \langle (A - BB^T) D(\partial^\alpha u^k), D(\partial^\alpha u^k) \rangle$$
$$+ |B^T D(\partial^\alpha u^k) + \partial^\alpha q^k|^2 \big\} + \sum_{|\alpha| \leq m-1} |\partial^\alpha q^k|^2 \Big\} dx$$
$$\leq C \int_{\mathbf{R}^n} \sum_{|\alpha| \leq m} \big\{ -2(\partial^\alpha u^k)\partial^\alpha (\mathcal{L} u^k + \mathcal{M} q^k) + |\partial^\alpha q^k|^2$$
$$+ |\partial^\alpha u^k|^2 \big\} dx.$$

On the other hand, applying Itô-Ventzel's formula to $|\partial^\alpha u^k|^2$, we have from (4.8) that

(4.11)
$$E \int_{\mathbf{R}^n} \sum_{|\alpha| \leq m} \big\{ |\partial^\alpha g^k(x)|^2 - |\partial^\alpha u^k(t, x)|^2 \big\} dx$$
$$= E \int_t^T \int_{\mathbf{R}^n} \sum_{|\alpha| \leq m} \big\{ 2(\partial^\alpha u^k)\partial^\alpha [P_k(-\mathcal{L} u^k - \mathcal{M} q^k) - f^k]$$
$$+ |\partial^\alpha q^k|^2 \big\} dx ds$$
$$= E \int_t^T \big\{ -2(u^k, P_k(\mathcal{L} u^k + \mathcal{M} q^k) + f^k)_m + |q^k|_m^2 \big\} ds$$
$$= E \int_t^T \big\{ -2(u^k, \mathcal{L} u^k + \mathcal{M} q^k + f^k)_m + |q^k|_m^2 \big\} ds$$
$$= E \int_t^T \int_{\mathbf{R}^n} \sum_{|\alpha| \leq m} \big\{ -2(\partial^\alpha u^k)\partial^\alpha (\mathcal{L} u^k + \mathcal{M} q^k)$$
$$+ |\partial^\alpha q^k|^2 - 2(\partial^\alpha u^k)(\partial^\alpha f^k) \big\} dx ds.$$

The third equality in above is due to (4.9). Then combining (4.10)–(4.11), we obtain

$$\int_{\mathbf{R}^n} \Big\{ \sum_{|\alpha| \le m} \big\{ \langle (A - BB^T)D(\partial^\alpha u^k), D(\partial^\alpha u^k) \rangle$$

$$+ |B^T D(\partial^\alpha u^k) + \partial^\alpha q^k|^2 \big\} + \sum_{|\alpha| \le m-1} |\partial^\alpha q^k|^2 \Big\} dx$$

$$(4.12) \quad \le C \Big\{ E|g^k|_m^2 - E|u^k(t)|_m^2$$

$$+ E \sum_{|\alpha| \le m} \int_t^T \int_{\mathbf{R}^n} [2(\partial^\alpha u^k)(\partial^\alpha f^k) + |\partial^\alpha u^k|^2] \, dx ds \Big\}$$

$$\le C \Big\{ E|g^k|_m^2 - E|u^k(t)|_m^2 + \int_t^T E[|u^k(s)|_m^2 + |f^k(s)|_m^2] \, ds \Big\}.$$

By Gronwall's inequality, we obtain

$$\max_{t \in [0,T]} E|u^k(t)|_m^2 + E \int_0^T |q^k(t)|_{m-1}^2 \, dt$$

$$+ \sum_{|\alpha| \le m} E \int_0^T \int_{\mathbf{R}^n} \Big\{ \langle (A - BB^T)D(\partial^\alpha u^k), D(\partial^\alpha u^k) \rangle$$

$$(4.13)$$

$$+ \Big| B^T[D(\partial^\alpha u^k)] + \partial^\alpha q^k \Big|^2 \Big\} dx dt$$

$$\le CE \Big\{ \int_0^T |f^k(t)|_m^2 + |g^k|_m^2 \Big\}.$$

Note that the constant $C > 0$ in (4.13) only depends on T, m and K_m. From (4.13), we may assume that

$$(4.14) \quad \begin{cases} u^k \to u, & \text{weak}^* \text{ in } L_{\mathcal{F}}^\infty(0, T; L^2(\Omega; H^\ell)), \quad 0 \le \ell \le m, \\ q^k \to q, & \text{weakly in } L_{\mathcal{F}}^2(0, T; H^\ell)^d, \quad 0 \le \ell \le m-1, \end{cases}$$

and for any $|\alpha| \le m$,

$$(4.15) \quad \begin{cases} (A - BB^T)^{1/2} D(\partial^\alpha u^k) \to (A - BB^T)^{1/2} D(\partial^\alpha u), \\ B^T[D(\partial^\alpha u^k)] + \partial^\alpha q^k \to B^T[D(\partial^\alpha u)] + \partial^\alpha q, \\ \qquad\qquad\qquad\qquad\qquad \text{weakly in } L_{\mathcal{F}}^2(0, T; H^0). \end{cases}$$

By taking limits in (4.13), we see that (u, q) satisfies the estimate (2.4) with the constant $C > 0$ only depending on T, m and K_m. We are going to prove that (u, q) is a weak solution of (3.2). To this end, let us take $\rho \in H^1(0, T)$ such that

$$(4.16) \quad \begin{cases} \rho(0) = 0, \quad \rho(T) = 1, \\ 0 \le \rho(t) \le 1, \quad t \in [0, T]. \end{cases}$$

Let $\ell > 0$ be fixed and $k \geq \ell$. For any $\varphi \in H_\ell^m \subset C_0^\infty(\mathbb{R}^n)$, from (4.8) and the fact $P_k\varphi = \varphi$, we have

$$
\begin{aligned}
(g^k, \varphi)_m = \int_0^T &\Big\{ \dot\rho(t)(u^k(t), \varphi)_m \\
&- \rho(t)(\mathcal{L}u^k(t) + \mathcal{M}q^k(t) + f^k(t), \varphi)_m \Big\} dt \\
+ \int_0^T &\rho(t) \langle (q^k(t), \varphi)_m, dW(t) \rangle .
\end{aligned}
$$
(4.17)

By the definition of $\mathcal{L}$ and $\mathcal{M}$, using integration by parts, we obtain

$$
\begin{aligned}
(g^k, \varphi)_m = \int_0^T &\Big\{ \dot\rho(t)(u^k(t), \varphi)_m \\
&- \rho(t)\Big[-\frac{1}{2}(A(t)Du^k(t) + B(t)q^k(t), D\varphi)_m \\
&+ (\langle a(t), Du^k(t) \rangle +c(t)u^k(t) + \langle b(t), q^k(t) \rangle +f^k(t), \varphi)_m] \Big\} dt \\
+ \int_0^T &\rho(t) \langle (q^k(t), \varphi)_m, dW(t) \rangle, \qquad \text{a.s.}
\end{aligned}
$$
(4.18)

If we denote

$$
\begin{cases}
F(x,\omega) = g - \displaystyle\int_0^T \Big\{ \dot\rho(t)u^k(t) - \rho(t)[\langle a(t), Du^k(t) \rangle +c(t)u^k(t) \\
\qquad\qquad + \langle b(t), q^k(t) \rangle +f^k(t)] \Big\} dt - \displaystyle\int_0^T \langle \rho(t)q^k(t), dW(t) \rangle, \\
G(x,\omega) = \displaystyle\int_0^T \rho(t)[\frac{1}{2}A(t)Du^k(t) + B(t)q^k(t)]dt,
\end{cases}
$$
(4.19)

then (4.18) reads

(4.20) $\qquad (F, \varphi)_m = (G, D\varphi)_m, \qquad \forall \varphi \in C_0^\infty(\mathbb{R}^n), \text{ a.s.}$

By the lemma below, we must have

(4.21) $\qquad (F, \varphi)_0 = (G, D\varphi)_0, \qquad \forall \varphi \in C_0^\infty(\mathbb{R}^n), \text{ a.s.}$

This tells us that

$$
\begin{aligned}
(g^k, \varphi)_0 = \int_0^T &\Big\{ \dot\rho(t)(u^k(t), \varphi)_0 \\
&- \rho(t) \langle \mathcal{L}u^k(t) + \mathcal{M}q^k(t) + f^k(t), \varphi \rangle_0 \Big\} dt \\
+ \int_0^T &\rho(t) \langle (q^k(t), \varphi)_0, dW(t) \rangle, \qquad \text{a.s.}
\end{aligned}
$$
(4.22)

We want to pass to the limit in (4.22) to obtain a similar equality for (u, q). By (4.14) with $\ell = 1$ for u^k and $\ell = 0$ for q^k, together with the convergence of (f^k, g^k) to (f, g), we can pass to the limit in (4.22) weakly in $L^2(\Omega)$ for all terms except the last term which is involving the Itô integral. To treat this last term, we define $K : L^2_{\mathcal{F}}(0, T; H^0)^d \to L^2(\Omega)$ by

$$(4.23) \qquad Kp = \int_0^T \rho(t) \langle (p(t), \varphi)_0, dW(t) \rangle, \quad \forall p \in L^2_{\mathcal{F}}(0, T; H^0)^d.$$

Then

$$
\begin{aligned}
(4.24) \qquad E|Kp|^2 &= E \int_0^T |\rho(t)(p(t), \varphi)_0|^2 dt \\
&\leq |\varphi|_0^2 E \int_0^T |p(t)|_0^2 dt, \quad \forall p \in L^2_{\mathcal{F}}(0, T; H^0)^d.
\end{aligned}
$$

This means that K is a bounded linear operator. Thus, for any $\eta \in L^2(\Omega)$, one has

$$
\begin{aligned}
(4.25) \qquad &E\left(\eta \int_0^T \rho(t) \langle (q^k(t) - q(t), \varphi)_0, dW(t) \rangle \right) \\
&= (\eta, K(q^k - q))_{L^2(\Omega)} = (K^* \eta, q^k - q)_{L^2_{\mathcal{F}}(0, T; H^0)^d} \to 0.
\end{aligned}
$$

Thus, we obtain

$$
\begin{aligned}
(4.26) \qquad (g, \varphi)_0 = &\int_0^T \Big\{ \dot{\rho}(t)(u(t), \varphi)_0 \\
&- \rho(t) \langle \mathcal{L}u(t) + \mathcal{M}q(t) + f(t), \varphi \rangle_0 \Big\} dt \\
&+ \int_0^T \rho(t) \langle (q(t), \varphi)_0, dW(t) \rangle, \qquad \text{a.s.}
\end{aligned}
$$

Now, fixed any $t \in (0, T)$. For any $\varepsilon > 0$, we let

$$(4.27) \qquad \rho_\varepsilon(s) = \begin{cases} 0, & s \leq t - \varepsilon/2, \\ \frac{1}{2} + \frac{s-t}{\varepsilon}, & t - \varepsilon/2 < s < t + \varepsilon/2, \\ 1, & s \geq t + \varepsilon/2. \end{cases}$$

Choosing $\rho = \rho_\varepsilon$ in (4.26) and letting $\varepsilon \to 0$, we obtain

$$
\begin{aligned}
(4.28) \qquad (g, \varphi)_0 = &(u(t), \varphi)_0 - \int_t^T \langle \mathcal{L}u(t) + \mathcal{M}q(t) + f(t), \varphi \rangle_0 dt \\
&+ \int_t^T \langle (q(t), \varphi)_0, dW(t) \rangle, \quad \forall \varphi \in C_0^\infty(\mathbb{R}^n), \text{ a.s.}
\end{aligned}
$$

This means that (u, q) is an adapted weak solution of (3.2). By Theorem 3.2, it is unique.

In the case that $m \geq 2$, from (2.4), we see that (1.13) holds and thus, by Proposition 1.3, (u, q) is an adapted strong solution (3.2). In the case $m > 2 + n/2$, by Sobolev's embedding theorem, (1.11) holds and therefore, (u, q) is an adapted classical solution of (3.2) by Proposition 1.3 again.

Finally, let us look at the case when conditions of Theorem 2.2 hold. Suppose (2.6) or (2.7) holds instead of (2.2). By Lemma 4.1, we still have (4.1), which is now equivalent to (4.2). Then all the proof that we have presented above remains true. Moreover, we have estimate (2.8). In the case that A is uniformly positive definite, a little more careful estimate leads to (2.9). In fact, for the present case, in (4.12), we can use integration by parts to get

(4.29)
$$E \sum_{|\alpha| \leq m} \int_t^T \int_{\mathbb{R}^n} 2(\partial^\alpha u^k)(\partial^\alpha f^k) dx ds$$
$$\leq CE \int_t^T \{|u^k(s)|_{m+1}^2 + |f^k(s)|_{m-1}^2\} ds.$$

We leave the details of the proof to the interested readers. $\qquad\square$

We now prove the following lemma which has been used in the above proof.

Lemma 4.2. *Let $F \in H^m(\mathbb{R}^n)$ and $G \in H^m(\mathbb{R}^n)^n$, such that*

(4.30)
$$(F, \varphi)_m = (G, D\varphi)_m, \qquad \forall \varphi \in C_0^\infty(\mathbb{R}^n).$$

Then

(4.31)
$$(F, \varphi)_0 = (G, D\varphi)_0, \qquad \forall \varphi \in C_0^\infty(\mathbb{R}^n).$$

Proof. Let $\mathcal{S} \triangleq \mathcal{S}(\mathbb{R}^n)$ be the set of all $\varphi \in C^\infty(\mathbb{R}^n)$, such that

(4.32)
$$\Phi_{\alpha,\beta}(\varphi) \triangleq \sup_{x \in \mathbb{R}^n} |x^\alpha \partial^\beta \varphi(x)| < \infty, \qquad \forall \alpha, \beta.$$

Under the family of semi-norms $\Phi_{\alpha,\beta}$, $\mathcal{S}$ is a Fréchet space. Also, $C_0^\infty(\mathbb{R}^n)$ is a dense subset of $\mathcal{S}$. Thus, (4.30) holds for all $\varphi \in \mathcal{S}$. Next, by Hörmander [1, p.161], Fourier transformation $\varphi \mapsto \widehat{\varphi}$ is an isomorphism of $\mathcal{S}$ onto itself. Applying Parseval's formula to (4.30), we obtain

(4.33) $\displaystyle \int_{\mathbb{R}^n} \left[\widehat{F}(\xi) - \langle \widehat{G}(\xi), \xi \rangle\right] \left(\sum_{|\alpha| \leq m} |\xi^\alpha|^2 \right) \widehat{\varphi}(\xi) d\xi = 0, \qquad \forall \varphi \in C_0^\infty(\mathbb{R}^n).$

Now, for any $\psi \in C_0^\infty(\mathbb{R}^n)$, we have $\widehat{\psi}(\xi) \left(\sum_{|\alpha| \leq m} |\xi^\alpha|^2 \right)^{-1} \in \mathcal{S}$. Thus, there exists a $\varphi \in \mathcal{S}$, such that

(4.34)
$$\widehat{\varphi}(\xi) = \widehat{\psi}(\xi) \left(\sum_{|\alpha| \leq m} |\xi^\alpha|^2 \right)^{-1}.$$

Combining (4.33) and (4.34), using Parseval's formula again, we obtain

$$(F, \psi)_m = (G, D\psi)_m, \qquad \forall \psi \in C_0^\infty(\mathbb{R}^n).$$

This proves our lemma. $\qquad\square$

§5. A Proof of the Fundamental Lemma

In this section, we are going to prove Lemma 4.1, which has played an essential role in the proof of well-posedness theorems for (1.1).

Proof of Lemma 4.1. Let $\ell \overset{\triangle}{=} |\alpha| \leq m$. For any $u \in C_0^\infty(\mathbb{R}^n)$ and $q \in C_0^\infty(\mathbb{R}^n; \mathbb{R}^d)$, by definition of $\mathcal{L}$ and $\mathcal{M}$, and differentiation, we have

$$
\begin{aligned}
\mathcal{I}^\alpha \overset{\triangle}{=} & \int_{\mathbb{R}^n} \left\{ -2(\partial^\alpha u)\partial^\alpha(\mathcal{L}u + \mathcal{M}q) + |\partial^\alpha q|^2 \right\} dx \\
= & \int_{\mathbb{R}^n} \left\{ -2(\partial^\alpha u)\partial^\alpha \left[\frac{1}{2}\nabla\cdot(ADu) + \langle a, Du \rangle + cu \right. \right. \\
& \left. + \nabla\cdot[Bq] + \langle b, q \rangle \right] + |\partial^\alpha q|^2 \Big\} dx \\
= & \int_{\mathbb{R}^n} \left\{ -2(\partial^\alpha u)\left[\frac{1}{2}\nabla\cdot[AD(\partial^\alpha u)] + \langle a, D(\partial^\alpha u) \rangle + c(\partial^\alpha u) \right. \right. \\
& + \nabla\cdot[B(\partial^\alpha q)] + \langle b, \partial^\alpha q \rangle \Big] + |\partial^\alpha q|^2 \\
& - 2(\partial^\alpha u) \sum_{0 \leq \beta < \alpha} C_{\alpha\beta}\left[\frac{1}{2}\nabla\cdot[(\partial^{\alpha-\beta}A)D(\partial^\beta u)] \right. \\
& + \langle \partial^{\alpha-\beta}a, D(\partial^\beta u) \rangle + (\partial^{\alpha-\beta}c)(\partial^\beta u) \\
& \left. \left. + \nabla\cdot[(\partial^{\alpha-\beta}B)(\partial^\beta q)] + \langle \partial^{\alpha-\beta}b, \partial^\beta q \rangle \right] \right\} dx \\
\equiv & \ \mathcal{I}_0^\alpha + \mathcal{I}_1^\alpha + \mathcal{I}_2^\alpha + \mathcal{I}_3^\alpha,
\end{aligned}
$$

(5.1)

where $C_{\alpha\beta}$ is a positive integer depending on α and β, and

$$
\begin{aligned}
\mathcal{I}_0^\alpha = & \int_{\mathbb{R}^n} \left\{ -2(\partial^\alpha u)\left[\frac{1}{2}\nabla\cdot[AD(\partial^\alpha u)] + \langle a, D(\partial^\alpha u) \rangle + c(\partial^\alpha u) \right. \right. \\
& \left. \left. + \nabla\cdot[B(\partial^\alpha q)] + \langle b, \partial^\alpha q \rangle \right] + |\partial^\alpha q|^2 \right\} dx,
\end{aligned}
$$

(5.2)

$$
\begin{aligned}
\mathcal{I}_1^\alpha = & -2\int_{\mathbb{R}^n} \sum_{0 \leq \beta < \alpha} C_{\alpha\beta}(\partial^\alpha u)\left[\frac{1}{2}\nabla\cdot[(\partial^{\alpha-\beta}A)D(\partial^\beta u)] \right. \\
& + \langle \partial^{\alpha-\beta}a, D(\partial^\beta u) \rangle + (\partial^{\alpha-\beta}c)(\partial^\beta u) + \langle \partial^{\alpha-\beta}b, \partial^\beta q \rangle \Big] dx,
\end{aligned}
$$

(5.3)

$$
\mathcal{I}_2^\alpha = -2\int_{\mathbb{R}^n} \sum_{\substack{0 \leq \beta < \alpha \\ |\beta| < |\alpha|-1}} C_{\alpha\beta}(\partial^\alpha u)\,\nabla\cdot[(\partial^{\alpha-\beta}B)(\partial^\beta q)]\,dx,
$$

(5.4)

$$
\mathcal{I}_3^\alpha = -2\int_{\mathbb{R}^n} \sum_{\substack{0 \leq \beta < \alpha \\ |\beta| = |\alpha|-1}} C_{\alpha\beta}(\partial^\alpha u)\,\nabla\cdot[(\partial^{\alpha-\beta}B)(\partial^\beta q)]\,dx.
$$

(5.5)

We note that in the case $\ell = 0$, $\mathcal{I}_1^\alpha$, $\mathcal{I}_2^\alpha$ and $\mathcal{I}_3^\alpha$ are all absent. We now treat $\mathcal{I}_0^\alpha$, $\mathcal{I}_1^\alpha$, $\mathcal{I}_2^\alpha$ and $\mathcal{I}_3^\alpha$, separately.

Since A and B are C_b^{m+1} in x, we see immediately that

$$(5.6) \qquad |\mathcal{I}_1^\alpha| + |\mathcal{I}_2^\alpha| \leq C\big(|u|_\ell^2 + |q|_{\ell-1}^2\big).$$

Now, let us look at $\mathcal{I}_0^\alpha$ and $\mathcal{I}_3^\alpha$. Using integration by parts, we have

$$
\begin{aligned}
\mathcal{I}_0^\alpha &= \int_{\mathbf{R}^n} \Big\{ \langle AD(\partial^\alpha u), D(\partial^\alpha u) \rangle + 2\langle \partial^\alpha q, B^T D(\partial^\alpha u) \rangle + |\partial^\alpha q|^2 \\
&\qquad - \langle a, D[(\partial^\alpha u)^2] \rangle - 2c(\partial^\alpha u)^2 - 2\langle b(\partial^\alpha u), \partial^\alpha q \rangle \Big\} dx \\
(5.7) \quad &= \int_{\mathbf{R}^n} \Big\{ \langle (A - BB^T)D(\partial^\alpha u), D(\partial^\alpha u) \rangle + |B^T D(\partial^\alpha u)|^2 + |\partial^\alpha q|^2 \\
&\qquad + 2\langle \partial^\alpha q, B^T D(\partial^\alpha u) \rangle - 2\langle b(\partial^\alpha u), \partial^\alpha q \rangle \\
&\qquad - 2\langle B^T D(\partial^\alpha u), b(\partial^\alpha u) \rangle + [\nabla\cdot(a - Bb) - 2c](\partial^\alpha u)^2 \Big\} dx.
\end{aligned}
$$

In the meantime, let us look at each term in $\mathcal{I}_3^\alpha$. For $\beta < \alpha$ with $|\beta| = |\alpha|-1$, using integration by parts, we have

$$
\begin{aligned}
(5.8) \quad & -\int_{\mathbf{R}^n} (\partial^\alpha u)\, \nabla\cdot[(\partial^{\alpha-\beta}B)(\partial^\beta q)]\,dx \\
&= \int_{\mathbf{R}^n} (\partial^\beta u)\, \nabla\cdot\{\partial^{\alpha-\beta}[(\partial^{\alpha-\beta}B)(\partial^\beta q)]\}\,dx \\
&= -\int_{\mathbf{R}^n} \langle D(\partial^\beta u), (\partial^{\alpha-\beta}B)\partial^\alpha q + (\partial^{2(\alpha-\beta)}B)\partial^\beta q \rangle\,dx \\
&= -\int_{\mathbf{R}^n} [\langle (\partial^{\alpha-\beta}B^T)D(\partial^\beta u), \partial^\alpha q \rangle + \langle D(\partial^\beta u), (\partial^{2(\alpha-\beta)}B)\partial^\beta q \rangle]\,dx.
\end{aligned}
$$

Thus, it follows that

$$
\begin{aligned}
\mathcal{I}_0^\alpha + \mathcal{I}_3^\alpha &= \int_{\mathbf{R}^n} \Big\{ \langle (A - BB^T)D(\partial^\alpha u), D(\partial^\alpha u) \rangle \\
&\quad + \Big| B^T D(\partial^\alpha u) + \partial^\alpha q - b(\partial^\alpha u) - \sum_{\substack{0 \leq \beta < \alpha \\ |\beta| = |\alpha|-1}} C_{\alpha\beta}(\partial^{\alpha-\beta}B^T)D(\partial^\beta u) \Big|^2 \\
&\quad - |b(\partial^\alpha u)|^2 - \Big| \sum_{\substack{0 \leq \beta < \alpha \\ |\beta| = |\alpha|-1}} C_{\alpha\beta}(\partial^{\alpha-\beta}B^T)D(\partial^\beta u) \Big|^2 \\
(5.9) \quad &\quad + 2 \sum_{\substack{0 \leq \beta < \alpha \\ |\beta| = |\alpha|-1}} C_{\alpha\beta}\, \langle (\partial^{\alpha-\beta}B^T)D(\partial^\beta u), B^T D(\partial^\alpha u) \rangle \\
&\quad - 2 \sum_{\substack{0 \leq \beta < \alpha \\ |\beta| = |\alpha|-1}} C_{\alpha\beta}\, \langle (\partial^{\alpha-\beta}B^T)D(\partial^\beta u), b(\partial^\alpha u) \rangle \\
&\quad + [\nabla\cdot(a - Bb) - 2c](\partial^\alpha u)^2 \\
&\quad - 2 \sum_{\substack{0 \leq \beta < \alpha \\ |\beta| = |\alpha|-1}} C_{\alpha\beta}\, \langle D(\partial^\beta u), (\partial^{2(\alpha-\beta)}B)(\partial^\beta q) \rangle \Big\} dx.
\end{aligned}
$$

Note that

$$\int_{\mathbf{R}^n} \left| B^T D(\partial^\alpha u) + \partial^\alpha q - b(\partial^\alpha u) \right.$$

$$\left. - \sum_{\substack{0 \le \beta < \alpha \\ |\beta| = |\alpha| - 1}} C_{\alpha\beta}(\partial^{\alpha-\beta} B^T) D(\partial^\beta u) \right|^2 dx$$

$$(5.10) \quad \ge \frac{1}{2} \int_{\mathbf{R}^n} \left| B^T D(\partial^\alpha u) + \partial^\alpha q \right|^2 dx$$

$$- \frac{1}{2} \int_{\mathbf{R}^n} \left| b(\partial^\alpha u) + \sum_{\substack{0 \le \beta < \alpha \\ |\beta| = |\alpha| - 1}} C_{\alpha\beta}(\partial^{\alpha-\beta} B^T) D(\partial^\beta u) \right|^2 dx$$

$$\ge \frac{1}{2} \int_{\mathbf{R}^n} \left| B^T D(\partial^\alpha u) + \partial^\alpha q \right|^2 dx - C|u|_\ell^2.$$

Next, by the symmetry condition (2.2), for $\beta < \alpha$, $|\beta| = |\alpha| - 1$, we have

$$\int_{\mathbf{R}^n} \langle (\partial^{\alpha-\beta} B^T) D(\partial^\beta u), B^T D(\partial^\alpha u) \rangle \, dx$$

$$= \int_{\mathbf{R}^n} \langle (B \partial^{\alpha-\beta} B^T) D(\partial^\beta u), D(\partial^\alpha u) \rangle \, dx$$

$$(5.11) \quad = \int_{\mathbf{R}^n} \frac{1}{2} \Big[\partial^{\alpha-\beta} \langle (B \partial^{\alpha-\beta} B^T) D(\partial^\beta u), D(\partial^\beta u) \rangle$$

$$- \langle \partial^{\alpha-\beta} (B \partial^{\alpha-\beta} B^T) D(\partial^\beta u), D(\partial^\beta u) \rangle \Big] dx$$

$$= -\frac{1}{2} \int_{\mathbf{R}^n} \langle \partial^{\alpha-\beta} (B \partial^{\alpha-\beta} B^T) D(\partial^\beta u), D(\partial^\beta u) \rangle \, dx \ge -C|u|_\ell^2.$$

Combining (5.6), (5.9)–(5.11) yields

$$\mathcal{I}^\alpha = \mathcal{I}_0^\alpha + \mathcal{I}_1^\alpha + \mathcal{I}_2^\alpha + \mathcal{I}_3^\alpha$$

$$(5.12) \quad \ge \int_{\mathbf{R}^n} \Big\{ \langle (A - BB^T) D(\partial^\alpha u), D(\partial^\alpha u) \rangle$$

$$+ \frac{1}{2} |B^T D(\partial^\alpha u) + \partial^\alpha q|^2 \Big\} dx - C(|u|_\ell^2 + |q|_{\ell-1}^2).$$

Now, we sum (5.12) up for all $|\alpha| \le \ell$ to get the following:

$$\Psi_\ell \overset{\Delta}{=} \sum_{|\alpha| \le \ell} \int_{\mathbf{R}^n} \Big\{ -2(\partial^\alpha u)\partial^\alpha(\mathcal{L}u + \mathcal{M}q) + |\partial^\alpha q|^2 \Big\} dx$$

$$(5.13) \quad \ge \frac{1}{2} \sum_{|\alpha| \le \ell} \int_{\mathbf{R}^n} \Big\{ \langle (A - BB^T) D(\partial^\alpha u), D(\partial^\alpha u) \rangle$$

$$+ |B^T D(\partial^\alpha u) + \partial^\alpha q|^2 \Big\} dx - C(|u|_\ell^2 + |q|_{\ell-1}^2).$$

Thus, it follows that

$$\Phi_\ell \overset{\Delta}{=} \sum_{|\alpha|\leq\ell} \int_{\mathbf{R}^n} \left\{ \langle (A - BB^T)D(\partial^\alpha u), D(\partial^\alpha u) \rangle \right.$$

(5.14)
$$\left. + |B^T D(\partial^\alpha u) + \partial^\alpha q|^2 \right\} dx$$

$$\leq C\left(\Psi_\ell + |u|_\ell^2 + |q|_{\ell-1}^2\right).$$

Note that

(5.15) $$|\partial^\alpha q|^2 \leq 2|B^T D(\partial^\alpha u) + \partial^\alpha q|^2 + 2|B^T D(\partial^\alpha u)|^2.$$

Using the parabolicity condition (1.6) and the definition of Φ_ℓ (see (5.14)), we have

(5.16) $$|q|_{\ell-1}^2 \leq C(\Phi_{\ell-1} + |u|_\ell^2).$$

Consequently, from (5.14), we obtain

(5.17) $$\Phi_\ell \leq C(\Psi_\ell + \Phi_{\ell-1} + |u|_\ell^2), \quad 1 \leq \ell \leq m.$$

On the other hand, for $\ell = 0$ (i.e., $\alpha = 0$), we have

$$\int_{\mathbf{R}^n} \left\{ -2u(\mathcal{L}u + \mathcal{M}q) + |q|^2 \right\} dx$$

$$= \int_{\mathbf{R}^n} \left\{ -2u\left[\frac{1}{2}\nabla\cdot[ADu] + \langle a, Du \rangle + cu \right. \right.$$

$$\left. \left. + \nabla\cdot[Bq] + \langle b, q \rangle \right] + |q|^2 \right\} dx$$

(5.18)
$$= \int_{\mathbf{R}^n} \left\{ \langle (A - BB^T)Du, Du \rangle + |B^T Du|^2 + |q|^2 \right.$$

$$+ 2\langle q, B^T Du \rangle - 2\langle bu, q \rangle$$

$$\left. - 2\langle B^T Du, bu \rangle + [\nabla\cdot(a - Bb) - 2c]u^2 \right\} dx$$

$$\geq \int_{\mathbf{R}^n} \left\{ \langle (A - BB^T)Du, Du \rangle + |B^T Du + q|^2 dx - C|u|_0^2. \right.$$

This implies

(5.19) $$\Phi_0 \leq \int_{\mathbf{R}^n} \left\{ -2u(\mathcal{L}u + \mathcal{M}q) + |q|^2 \right\} dx + C|u|_0^2.$$

Hence, it follows from (5.17) and (5.19) that

(5.20) $$\Phi_m \leq C(\Psi_m + |u|_m^2),$$

which is the same as (4.1).

In the case that (2.6) holds, we use the following estimate:

(5.21)
$$\int_{\mathbf{R}^n} \langle (\partial^{\alpha-\beta} B^T) D(\partial^\beta u), B^T D(\partial^\alpha u) \rangle \, dx$$
$$\geq -\varepsilon \int_{\mathbf{R}^n} |B^T D(\partial^\alpha u)|^2 dx - C \int_{\mathbf{R}^n} |D(\partial^\beta u)|^2 dx,$$

for small enough $\varepsilon > 0$ to get

(5.22)
$$\int_{\mathbf{R}^n} \Big\{ \langle (A - BB^T) D(\partial^\alpha u), D(\partial^\alpha u) \rangle$$
$$+ 2 \sum_{\substack{0 \leq \beta < \alpha \\ |\beta| = |\alpha| - 1}} C_{\alpha\beta} \langle (\partial^{\alpha-\beta} B^T) D(\partial^\beta u), B^T D(\partial^\alpha u) \rangle$$
$$\geq \frac{\varepsilon_0}{2} \int_{\mathbf{R}^n} \langle (A - BB^T) D(\partial^\alpha u), D(\partial^\alpha u) \rangle \, dx - C |u|_\ell^2.$$

Then, we still have (5.14) and finally have (5.20) which is the same as (4.1). In the case (2.7) holds, we use the following estimate:

(5.23)
$$\int_{\mathbf{R}^n} \langle (\partial^{\alpha-\beta} B^T) D(\partial^\beta u), B^T D(\partial^\alpha u) \rangle \, dx$$
$$= \int_{\mathbf{R}^n} \langle (\partial^{\alpha-\beta} B^T) D(\partial^\beta u), \partial^{\alpha-\beta} [B^T D(\partial^\beta u)]$$
$$- (\partial^{\alpha-\beta} B^T) D(\partial^\beta u) \rangle \, dx$$
$$= - \int_{\mathbf{R}^n} \Big\{ \langle (\partial^{2(\alpha-\beta)} B^T) D(\partial^\beta u), B^T D(\partial^\beta u) \rangle$$
$$+ \langle (\partial^{(\alpha-\beta)} B^T) D(\partial^\alpha u), B^T D(\partial^\beta u) \rangle$$
$$+ |(\partial^{\alpha-\beta} B^T) D(\partial^\beta u)|^2 \Big\} dx$$
$$\geq -\varepsilon \int_{\mathbf{R}^n} |(\partial^{\alpha-\beta} B^T) D(\partial^\alpha u)|^2 dx - C |u|_\ell^2,$$

for $\varepsilon > 0$ small enough to obtain (5.22) and finally to obtain (5.20).

Note that in the case (2.6) or (2.7) holds, we have (2.10). Then, (4.2) follows from (4.1) easily. Finally, if in addition, (1.9) also holds, then, (4.3) follows from (4.2). This completes the proof of Lemma 4.1.   □

§6. Comparison Theorems

In this section, we are going to present some comparison theorems on the solutions of different BSPDEs. For convenience, we consider BSPDEs of form (1.5). Let us denote (compare (3.1))

(6.1)
$$
\begin{cases}
\mathcal{L}u \overset{\Delta}{=} \dfrac{1}{2}\mathrm{tr}\,[AD^2u] - \langle a, Du \rangle - cu \\[2mm]
\mathcal{M}q \overset{\Delta}{=} \mathrm{tr}\,[B^T Dq] - \langle b, q \rangle, \\[2mm]
\overline{\mathcal{L}}u \overset{\Delta}{=} \dfrac{1}{2}\mathrm{tr}\,[\overline{A}D^2\overline{u}] - \langle \overline{a}, D\overline{u} \rangle - \overline{c}u, \\[2mm]
\overline{\mathcal{M}}q \overset{\Delta}{=} \mathrm{tr}\,[\overline{B}^T D\overline{q}] - \langle \overline{b}, \overline{q} \rangle.
\end{cases}
$$

We assume that $(H)_m$ holds for $\{A, B, a, b, c\}$ and $\{\overline{A}, \overline{B}, \overline{a}, \overline{b}, \overline{c}\}$. Consider the following BSPDEs:

(6.2)
$$
\begin{cases}
du = -\{\mathcal{L}u + \mathcal{M}q + f\}dt + \langle q, dW(t) \rangle, \quad (t, x) \in [0, T] \times \mathbb{R}^n, \\
u\big|_{t=T} = g.
\end{cases}
$$

(6.3)
$$
\begin{cases}
d\overline{u} = -\{\overline{\mathcal{L}}\overline{u} + \overline{\mathcal{M}}\overline{q} + \overline{f}\}dt + \langle \overline{q}, dW(t) \rangle, \quad (t, x) \in [0, T] \times \mathbb{R}^n, \\
\overline{u}\big|_{t=T} = \overline{g}.
\end{cases}
$$

Note that (6.2) and (3.2) are a little different since the operators $\mathcal{L}$ and $\mathcal{M}$ are defined a little differently. However, by the discussion at the end of §2, we know that Theorems 2.1, 2.2 and 2.3 hold for (6.1). Throughout this section, we assume that the parabolicity condition (1.6), the symmetry condition (2.2) and $(H)_m$ (for some $m \geq 1$) hold for (6.2) and (6.3). Then by Theorem 2.3, for any pairs (f, g) and $(\overline{f}, \overline{g})$ satisfying (2.14), there exist unique adapted weak solutions (u, q) and $(\overline{u}, \overline{q})$ to (6.2) and (6.3), respectively. We hope to establish some comparisons between u and $\overline{u}$ in various cases.

Our comparison results are all based on the following lemma.

Lemma 6.1. *Let (1.6), (2.2) and $(H)_m$ with $m \geq 1$ hold. Let (u, q) be the unique adapted weak solution of (6.2) corresponding to some (f, g) satisfying (2.14) for some $\lambda \geq 0$. Then there exists a constant $\mu \in \mathbb{R}$, such that*

(6.4)
$$
\begin{aligned}
E &\int_{\mathbb{R}^n} e^{-\lambda \langle x \rangle} \big|u(t, x)^-\big|^2 dx \\
&\leq e^{\mu(T-t)} E \int_{\mathbb{R}^n} e^{-\lambda \langle x \rangle} \big|g(x)^-\big|^2 dx \\
&\quad + E \int_t^T e^{\mu(s-t)} \int_{\mathbb{R}^n} e^{-\lambda \langle x \rangle} \big|f(s, x)^-\big|^2 dx ds, \qquad \forall t \in [0, T].
\end{aligned}
$$

Proof. We first assume that (f, g) satisfies (2.14) with $\lambda = 0$. Let $\varphi : \mathbb{R} \to [0, \infty)$ be defined as follows:

(6.5)
$$
\varphi(r) = \begin{cases}
r^2, & r \leq -1, \\
(6r^3 + 8r^4 + 3r^5)^2, & -1 \leq r \leq 0, \\
0, & r \geq 0.
\end{cases}
$$

We can directly check that φ is C^2 and

(6.6)
$$\begin{cases} \varphi(0) = \varphi'(0) = \varphi''(0) = 0, \\ \varphi(-1) = 1, \ \varphi'(-1) = -2, \ \varphi''(-1) = 2. \end{cases}$$

Next, for any $\varepsilon > 0$, we let $\varphi_\varepsilon(r) = \varepsilon^2 \varphi(\frac{r}{\varepsilon})$. Then, it holds

(6.7)
$$\begin{cases} \lim_{\varepsilon \to 0} \varphi_\varepsilon(r) = |r^-|^2, \quad \lim_{\varepsilon \to 0} \varphi'_\varepsilon(r) = -2r^-, \quad \text{uniformly}, \\ |\varphi''_\varepsilon(r)| \leq C, \quad \forall \varepsilon > 0, \ r \in \mathbb{R}; \\ \lim_{\varepsilon \to 0} \varphi''_\varepsilon(r) = \begin{cases} 2, & r < 0, \\ 0, & r > 0. \end{cases} \end{cases}$$

Denote

(6.8)
$$\widehat{a} = a - \frac{1}{2} \nabla \cdot A, \quad \widehat{b} = b - \nabla \cdot B.$$

Then by (1.3), we have

(6.9)
$$\begin{cases} \frac{1}{2} \operatorname{tr}[AD^2 u] + \langle a, Du \rangle = \frac{1}{2} \nabla \cdot [ADu] + \langle \widehat{a}, Du \rangle, \\ \operatorname{tr}[B^T Dq] + \langle b, q \rangle = \nabla \cdot (Bq) + \langle \widehat{b}, q \rangle. \end{cases}$$

Applying the Itô's formula to $\varphi_\varepsilon(u)$, we obtain (let $Q_t = [t, T] \times \mathbb{R}^n$)

$$
E \int_{\mathbb{R}^n} \varphi_\varepsilon(g(x)) dx - E \int_{\mathbb{R}^n} \varphi_\varepsilon(u(t,x)) dx
$$
$$
= E \int_{Q_t} \Big\{ \varphi'_\varepsilon(u) \big[-\frac{1}{2} \nabla \cdot (ADu) - \nabla \cdot (Bq) - \langle \widehat{a}, Du \rangle \\
- cu - \langle \widehat{b}, q \rangle - f \big] + \frac{1}{2} \varphi''_\varepsilon(u) |q|^2 \Big\} dx ds
$$
$$
= E \int_{Q_t} \Big\{ \frac{1}{2} \varphi''_\varepsilon(u) \big[\langle ADu, Du \rangle + 2 \langle B^T Du, q \rangle + |q|^2 \big] \\
- \varphi'_\varepsilon(u) \big[\langle \widehat{a}, Du \rangle + cu + \langle \widehat{b}, q \rangle + f \big] \Big\} dx ds
$$

(6.10)
$$
= E \int_{Q_t} \Big\{ \frac{1}{2} \varphi''_\varepsilon(u) \big[\langle (A - BB^T) Du, Du \rangle + |B^T Du + q - \widehat{b} u|^2 \big] \\
+ \frac{1}{2} \varphi''_\varepsilon(u) \big[-|\widehat{b}|^2 u^2 + 2 \langle B^T Du, \widehat{b} u \rangle + 2 \langle \widehat{b} u, q \rangle \big] \\
- \langle \widehat{a}, D\varphi_\varepsilon(u) \rangle - \varphi'_\varepsilon(u) [cu + \langle \widehat{b}, q \rangle + f] \Big\} dx ds
$$
$$
\geq E \int_{Q_t} \Big\{ -\frac{1}{2} \varphi''_\varepsilon(u) |\widehat{b}|^2 u^2 + \langle B\widehat{b}, D \int_0^u \varphi''_\varepsilon(r) r dr \rangle \\
+ [\varphi''_\varepsilon(u) u - \varphi'_\varepsilon(u)] \langle \widehat{b}, q \rangle + (\nabla \cdot \widehat{a}) \varphi_\varepsilon(u) \\
- \varphi'_\varepsilon(u) [cu + f] \Big\} dx ds.
$$

We note that

$$
(6.11) \qquad \int_0^u \varphi_\varepsilon''(r) r dr = \varphi_\varepsilon'(u) u - \varphi_\varepsilon(u),
$$

and

$$
(6.12) \qquad \lim_{\varepsilon \to 0} [\varphi_\varepsilon''(u) u - \varphi_\varepsilon'(u)] = 2u I_{(u \le 0)} + 2u^- = 0.
$$

Thus, let $\varepsilon \to 0$ in (6.10), we obtain

$$
\begin{aligned}
& E \int_{\mathbf{R}^n} |g(x)^-|^2 dx - E \int_{\mathbf{R}^n} |u(t,x)^-|^2 dx ds \\
& \ge E \int_{Q_t} \Big\{ - I_{(u \le 0)} |\widehat{b}|^2 u^2 - \nabla \cdot (B\widehat{b})[-2u^- u - |u^-|^2] \\
(6.13) \qquad & \qquad + (\nabla \cdot \widehat{a}) |u^-|^2 + 2u^- [cu + f] \Big\} dx ds \\
& \ge E \int_{Q_t} \Big\{ (-|\widehat{b}|^2 - \nabla \cdot (B\widehat{b}) + \nabla \cdot \widehat{a} - 2c) |u^-|^2 - 2u^- f^- \Big\} dx ds \\
& \ge -\mu E \int_{Q_t} |u^-|^2 dx ds - E \int_{Q_t} |f^-|^2 dx ds,
\end{aligned}
$$

where

$$
(6.14) \qquad \mu \overset{\Delta}{=} \sup_{t,x,\omega} \big[-\nabla \cdot \widehat{a} + \nabla \cdot (B\widehat{b}) + |\widehat{b}|^2 + 2c + 1 \big] < \infty.
$$

Then by Gronwall's inequality, we obtain (6.4) for the case $\lambda = 0$. The general case can be proved by using transformation (2.11) and working on (v, p) for the transformed equations. $\qquad \square$

Our main comparison result is the following.

Theorem 6.2. Let (1.6), (2.2) and $(H)_m$ hold for (6.2) and (6.3). Let (f, g) and $(\overline{f}, \overline{g})$ satisfy (2.14) with some $\lambda \ge 0$. Let (u, q) and $(\overline{u}, \overline{q})$ be adapted strong solutions of (6.2) and (6.3), respectively. Then for some $\mu > 0$,

$$
\begin{aligned}
& E \int_{\mathbf{R}^n} e^{-\lambda \langle x \rangle} |[u(t,x) - \overline{u}(t,x)]^-|^2 dx \\
& \le e^{\mu(T-t)} E \int_{\mathbf{R}^n} e^{-\lambda \langle x \rangle} |[g(x) - \overline{g}(x)]^-|^2 dx \\
(6.15) \qquad & \quad + E \int_t^T e^{\mu(s-t)} \int_{\mathbf{R}^n} e^{-\lambda \langle x \rangle} |[(\mathcal{L} - \overline{\mathcal{L}})\overline{u}(s,x) \\
& \qquad + (\mathcal{M} - \overline{\mathcal{M}})\overline{q}(s,x) + f(s,x) - \overline{f}(s,x)]^-|^2 dx ds, \\
& \qquad \qquad \forall t \in [0, T],
\end{aligned}
$$

In the case that

(6.16)
$$\begin{cases} g(x) - \overline{g}(x) \geq 0, & \forall x \in \mathbb{R}^n, \text{ a.s.} \\ (\mathcal{L} - \overline{\mathcal{L}})\overline{u}(t,x) + (\mathcal{M} - \overline{\mathcal{M}})\overline{q}(t,x) + f(t,x) - \overline{f}(t,x) \geq 0, \\ \qquad\qquad \forall (t,x) \in [0,T] \times \mathbb{R}^n, \text{ a.s.} \end{cases}$$

it holds

(6.17) $u(t,x) \geq \overline{u}(t,x), \qquad \forall (t,x) \in [0,T] \times \mathbb{R}^n, \text{ a.s.}$

This is the case, in particular, if $\mathcal{L} = \overline{\mathcal{L}}$, $\mathcal{M} = \overline{\mathcal{M}}$ *and*

(6.18)
$$\begin{cases} g(x) \geq \overline{g}(x), & \text{a.e. } x \in \mathbb{R}^n, \text{ a.s.} \\ f(t,x) \geq \overline{f}(t,x), & \text{a.e. } (t,x) \in [0,T] \times \mathbb{R}^n, \text{ a.s.} \end{cases}$$

Proof. It is clear that

(6.19)
$$\begin{cases} d(u - \overline{u}) = -\{\mathcal{L}(u - \overline{u}) + \mathcal{M}(q - \overline{q}) \\ \qquad\qquad + (\mathcal{L} - \overline{\mathcal{L}})\overline{u} + (\mathcal{M} - \overline{\mathcal{M}})\overline{q} + f - \overline{f}\}dt \\ \qquad\qquad + \langle q - \overline{q}, dW(t) \rangle, \\ (u - \overline{u})\big|_{t=T} = g - \overline{g}. \end{cases}$$

Then, (6.15) follows from (6.4). In the case (6.16) holds, (6.15) becomes

(6.20) $\displaystyle E \int_{\mathbb{R}^n} e^{-\lambda \langle x \rangle} \big| [u(t,x) - \overline{u}(t,x)]^- \big|^2 dx \leq 0, \quad \forall t \in [0,T].$

This yields (6.17). The last conclusion is clear.   □

Corollary 6.3. *Let the condition of Lemma 6.1 hold. Let*

(6.21)
$$\begin{cases} g(x) \geq 0, & \text{a.e. } x \in \mathbb{R}^n, \text{ a.s.} \\ f(t,x) \geq 0, & \text{a.e. } (t,x) \in [0,T] \times \mathbb{R}^n, \text{ a.s.} \end{cases}$$

and let (u,q) *be an adapted strong solution of (6.2). Then*

(6.22) $u(t,x) \geq 0, \qquad \text{a.e. } (t,x) \in [0,T] \times \mathbb{R}^n, \text{ a.s.}$

Proof. We take $\overline{\mathcal{L}} = \mathcal{L}$, $\overline{\mathcal{M}} = \mathcal{M}$, $\overline{f} \equiv 0$ and $\overline{g} \equiv 0$. Then $(\overline{u}, \overline{q}) = (0,0)$ is the unique adapted classical solution of (6.3) and (6.18) holds. Consequently, (6.22) follows from (6.17).   □

Let us make an observation on Theorem 6.2. Suppose $(\overline{u}, \overline{q})$ is an adapted strong solution of (6.3). Then (6.16) gives a condition on A, B, a, b, c, f and g, such that the solution (u,q) of the equation (6.2) satisfies (6.17). This has a very interesting interpretation (see Chapter 8). We now look at the cases that condition (6.16) holds.

Lemma 6.4. *Let* $\overline{A}$, $\overline{B}$, $\overline{a}$, $\overline{b}$ *and* $\overline{c}$ *be independent of* x. *Let* $\overline{f}$ *and* $\overline{g}$ *be convex in* x. *Let* $(\overline{u}, \overline{q})$ *be a strong solution of (3.1). Then,* $\overline{u}$ *is convex in* x *almost surely.*

Proof. First, we assume that f and g are smooth enough in x. Then, the corresponding solution $(\bar{u}, \bar{q})$ is smooth enough in x. Now, for any $\eta \in \mathbb{R}^n$, we define

$$\begin{cases} v(t,x) = \langle D^2 \bar{u}(t,x)\eta, \eta \rangle, \\ p(t,x) = (p^1(t,x), \cdots, p^d(t,x)), \qquad \forall (t,x) \in [0,T] \times \mathbb{R}^n, \text{ a.s.} \\ p^k(t,x) = \langle D^2 \bar{q}^k(t,x)\eta, \eta \rangle, \quad 1 \le k \le d, \end{cases}$$

Then, it holds

$$(6.23) \qquad \begin{cases} dv = [-\bar{\mathcal{L}}v - \overline{\mathcal{M}}p - \langle (D^2 \bar{f})\eta, \eta \rangle]dt + \langle p, dW(t) \rangle, \\ v|_{t=T} = \langle (D^2 \bar{g})\eta, \eta \rangle. \end{cases}$$

By Corollary 6.3 and the convexity of f and g (in x), we obtain

$$(6.24) \qquad \begin{aligned} \langle D^2 \bar{u}(t,x)\eta, \eta \rangle &= v(t,x) \ge 0, \\ &\forall (t,x) \in [0,T] \times \mathbb{R}^n, \ \eta \in \mathbb{R}^n, \text{ a.s.} \end{aligned}$$

This implies the convexity of $\bar{u}(t,x)$ in x almost surely. In the case that $\bar{f}$ and $\bar{g}$ are not necessarily smooth enough, we may make approximation. $\quad\square$

Proposition 6.5. *Let $\bar{A}$, $\bar{B}$, $\bar{a}$, $\bar{b}$ and $\bar{c}$ be independent of x. Let $\bar{f}$ and $\bar{g}$ be convex in x and nonnegative. Let $(\bar{u}, \bar{q})$ be a strong solution of (6.3). Let $\mathcal{M} = \overline{\mathcal{M}}$ and let*

$$(6.25) \qquad \begin{cases} A(t,x) = \bar{A}(t) + A_0(t,x), \\ c(t,x) = \bar{c}(t) + c_0(t,x), \\ f(t,x) = \bar{f}(t,x) + f_0(t,x), \qquad (t,x) \in [0,T] \times \mathbb{R}^n, \text{ a.s.} \\ g(x) = \bar{g}(x) + g_0(x), \end{cases}$$

with

$$(6.26) \qquad \begin{cases} A_0(t,x) \ge 0, \quad c_0(t,x) \ge 0, \\ f_0(t,x) \ge 0, \quad g_0(x) \ge 0, \end{cases} \qquad \forall (t,x) \in [0,T] \times \mathbb{R}^n, \text{ a.s.}$$

Then (6.16) is satisfied and thus (6.17) holds.

Proof. By Corollary 6.3 and Lemma 6.4, $\bar{u}$ is convex and nonnegative. Thus,

$$(\mathcal{L} - \bar{\mathcal{L}})\bar{u}(t,x) = \frac{1}{2}\mathrm{tr}\,[A_0 D^2 \bar{u}] + c_0 \bar{u} \ge 0.$$

Then (6.16) follows. $\quad\square$

Next, we have the following.

Proposition 6.6. *Let all the functions $\bar{A}$, $\bar{B}$, $\bar{a}$, $\bar{b}$, $\bar{c}$, $\bar{f}$ and $\bar{g}$ be deterministic. Let $\bar{u}$ be the solution of the following equation:*

$$(6.27) \qquad \begin{cases} \bar{u}_t = -\bar{\mathcal{L}}\bar{u} - \bar{f}, \qquad (t,x) \in [0,T] \times \mathbb{R}^n, \\ \bar{u}|_{t=T} = \bar{g}. \end{cases}$$

Further, we assume that $\bar{u}(t, x)$ is convex in x. Next, let (6.25) hold. Then (6.16) is satisfied and (6.17) holds.

Proof. In the present case, $(\bar{u}, 0)$ is an adapted strong solution of (6.3). Then similar to the proof of Proposition 6.5 and note $\bar{q} = 0$, we can obtain our assertion. $\square$

Note that in Proposition 6.6, B and b are arbitrary.

Chapter 6
Method of Continuation

In this chapter, we consider the solvability of the following FBSDE which is the same as (3.16) of Chapter 1 (We rewrite here for convenience):

$$(0.1) \quad \begin{cases} dX(t) = b(t, X(t), Y(t), Z(t))dt + \sigma(t, X(t), Y(t), Z(t))dW(t), \\ dY(t) = h(t, X(t), Y(t), Z(t))dt + Z(t)dW(t), \\ X(0) = x, \qquad Y(T) = g(X(T)). \end{cases}$$

Here, functions b, σ, h and g are allowed to be random, i.e., they can depend on $\omega \in \Omega$. For the notational simplicity, we have suppressed ω and we will do so below.

We have seen that for the case when all the coefficients are deterministic, one can use the Four Step Scheme to approach the problem (see Chapter 4), which involving the study of parabolic systems; in the case of random coefficients, in applying the Four Step Scheme, we need to study the solvability of BSPDEs (see Chapter 5). In this chapter, we are going to introduce a completely different method to approach the solvability of (0.1). Such a method is called the *method of continuation*.

§1. The Bridge

Recall that S^n is the set of all $(n \times n)$ symmetric matrices. In what follows, whenever A is a square matrix, (with λ being a scalar), by $A + \lambda$, we mean $A + \lambda I$. For any $A \in S^n$, by $A \geq \delta$, we mean that $A - \delta$ is positive semidefinite. The meaning of $A \leq -\delta$ is similar. For simplicity of notation, we will denote $M = \mathbb{R}^n \times \mathbb{R}^m \times \mathbb{R}^{m \times d}$; a generic point in M is denoted by $\theta = (x, y, z)$ with $x \in \mathbb{R}^n$, $y \in \mathbb{R}^m$ and $z \in \mathbb{R}^{m \times d}$. The norm in M is defined by

$$(1.1) \quad |\theta| \stackrel{\Delta}{=} \{|x|^2 + |y|^2 + |z|^2\}^{1/2}, \qquad \forall \theta \equiv (x, y, z) \in M,$$

where $|z|^2 \stackrel{\Delta}{=} \mathrm{tr}\,(zz^T)$. Similarly, we will use $\Theta = (X, Y, Z)$, and so on.

Now, let $T > 0$ be fixed and let

$$(1.2) \quad \begin{aligned} H[0, T] = &L^2_{\mathcal{F}}(0, T; W^{1,\infty}(M; \mathbb{R}^n \times \mathbb{R}^{n \times d} \times \mathbb{R}^m)) \\ &\times L^2_{\mathcal{F}_T}(\Omega; W^{1,\infty}(\mathbb{R}^n; \mathbb{R}^m)). \end{aligned}$$

Any generic element in $H[0, T]$ is denoted by $\Gamma \equiv (b, \sigma, h, g)$. Thus, $\Gamma \equiv (b, \sigma, h, g) \in H[0, T]$ if and only if

$$\begin{cases} b \in L^2_{\mathcal{F}}(0, T; W^{1,\infty}(M; \mathbb{R}^n)), \\ \sigma \in L^2_{\mathcal{F}}(0, T; W^{1,\infty}(M; \mathbb{R}^{n \times d})), \\ h \in L^2_{\mathcal{F}}(0, T; W^{1,\infty}(M; \mathbb{R}^m)), \\ g \in L^2_{\mathcal{F}_T}(\Omega; W^{1,\infty}(\mathbb{R}^n; \mathbb{R}^m)), \end{cases}$$

where the space $L_{\mathcal{F}}^2(0, T; W^{1,\infty}(M; \mathbb{R}^n))$, etc. are defined as in Chapter 1, §2. Further, we let

(1.3)
$$\begin{aligned}
\mathcal{H}[0, T] &= L_{\mathcal{F}}^2(0, T; \mathbb{R}^n) \times L_{\mathcal{F}}^2(0, T; \mathbb{R}^{n \times d}) \\
&\quad \times L_{\mathcal{F}}^2(0, T; \mathbb{R}^m) \times L_{\mathcal{F}_T}^2(\Omega; \mathbb{R}^m).
\end{aligned}$$

An element in $\mathcal{H}[0, T]$ is denoted by $\gamma \equiv (b_0, \sigma_0, h_0, g_0)$ with

$$\begin{cases}
b_0 \in L_{\mathcal{F}}^2(0, T; \mathbb{R}^n), \\
\sigma_0 \in L_{\mathcal{F}}^2(0, T; \mathbb{R}^{n \times d}), \\
h_0 \in L_{\mathcal{F}}^2(0, T; \mathbb{R}^m), \\
g_0 \in L_{\mathcal{F}_T}^2(\Omega; \mathbb{R}^m).
\end{cases}$$

We note that the range of the elements in $H[0, T]$ and $\mathcal{H}[0, T]$ are all in $\mathbb{R}^n \times \mathbb{R}^{n \times d} \times \mathbb{R}^m \times \mathbb{R}^m$. Hence, for any $\Gamma \equiv (b, \sigma, h, g) \in H[0, T]$ and $\gamma = (b_0, \sigma_0, h_0, g_0) \in \mathcal{H}[0, T]$, we can naturally define

(1.4) $\Gamma + \gamma = (b + b_0, \sigma + \sigma_0, h + h_0, g + g_0) \in H[0, T]$.

Now, for any $\Gamma \equiv (b, \sigma, h, g) \in H[0, T]$, $\gamma \equiv (b_0, \sigma_0, h_0, g_0) \in \mathcal{H}[0, T]$ and $x \in \mathbb{R}^n$, we associate them with the following FBSDE on $[0, T]$:

$(1.5)_{\Gamma,\gamma,x}$ $\begin{cases} dX(t) = \{b(t, \Theta(t)) + b_0(t)\}dt + \{\sigma(t, \Theta(t)) + \sigma_0(t)\}dW(t), \\ dY(t) = \{h(t, \Theta(t)) + h_0(t)\}dt + Z(t)dW(t), \\ X(0) = x, \qquad Y(T) = g(X(T)) + g_0, \end{cases}$

with $\Theta(t) \equiv (X(t), Y(t), Z(t))$. In what follows, sometimes, we will simply identify the FBSDEs $(1.5)_{\Gamma,\gamma,x}$ with (Γ, γ, x) or even with Γ (since γ and x are not essential in some sense). Let us recall the following definition.

Definition 1.1. A process $\Theta(\cdot) \equiv (X(\cdot), Y(\cdot), Z(\cdot)) \in \mathcal{M}[0, T]$ is called an *adapted solution* of $(1.5)_{\Gamma,\gamma,x}$, if the following holds for any $t \in [0, T]$, almost surely.

$(1.6)_{\Gamma,\gamma,x}$ $\begin{cases} X(t) = x + \displaystyle\int_0^t \{b(t, \Theta(s)) + b_0(s)\}ds \\ \qquad\quad + \displaystyle\int_0^t \{\sigma(t, \Theta(s)) + \sigma_0(s)\}dW(s), \\ Y(t) = g(X(T)) + g_0 - \displaystyle\int_t^T \{h(t, \Theta(s)) + h_0(s)\}ds \\ \qquad\quad - \displaystyle\int_t^T Z(s)dW(s). \end{cases}$

When $(1.5)_{\Gamma,\gamma,x}$ admits a unique adapted solution, we say that $(1.5)_{\Gamma,\gamma,x}$ is (*uniquely*) *solvable*.

We see that $(1.6)_{\Gamma,\gamma,x}$ is the integral form of $(1.5)_{\Gamma,\gamma,x}$. In what follows, we will not distinguish $(1.5)_{\Gamma,\gamma,x}$ and $(1.6)_{\Gamma,\gamma,x}$.

Definition 1.2. Let $T > 0$. A $\Gamma \in H[0,T]$ is said to be *solvable* if for any $x \in \mathbb{R}^n$ and $\gamma \in \mathcal{H}[0,T]$, equation $(1.5)_{\Gamma,\gamma,x}$ admits a unique adapted solution $\Theta(\cdot) \in \mathcal{M}[0,T]$. The set of all $\Gamma \in H[0,T]$ that is solvable is denoted by $\mathcal{S}[0,T]$. Any $\Gamma \in H[0,T] \setminus \mathcal{S}[0,T]$ is said to be *nonsolvable*.

Now, let us introduce the following notions, which will play the central role in this chapter.

Definition 1.3. Let $T > 0$ and $\Gamma \equiv (b, \sigma, h, g) \in H[0,T]$. A C^1 function
$$\Phi \equiv \begin{pmatrix} A & B^T \\ B & C \end{pmatrix} : [0,T] \to S^{n+m}, \text{ with } A : [0,T] \to S^n, B : [0,T] \to \mathbb{R}^{m \times n}$$
and $C : [0,T] \to S^m$, is called a *bridge* extending from Γ, (defined on $[0,T]$), if there exist some constants $K, \delta > 0$, such that

$$(1.7) \qquad \begin{cases} C(T) \leq 0, \qquad A(t) \geq 0, \quad \forall t \in [0,T], \\ \Phi(0) \leq K \begin{pmatrix} I & 0 \\ 0 & 0 \end{pmatrix}, \end{cases}$$

and either (1.8)–(1.9) or $(1.8)'$–$(1.9)'$ hold:

$$(1.8) \quad \left\langle \Phi(T) \begin{pmatrix} x - \overline{x} \\ g(x) - g(\overline{x}) \end{pmatrix}, \begin{pmatrix} x - \overline{x} \\ g(x) - g(\overline{x}) \end{pmatrix} \right\rangle \geq \delta |x - \overline{x}|^2, \quad \forall x, \overline{x} \in \mathbb{R}^n.$$

$$(1.9) \qquad \begin{aligned} &\left\langle \dot{\Phi}(t) \begin{pmatrix} x - \overline{x} \\ y - \overline{y} \end{pmatrix}, \begin{pmatrix} x - \overline{x} \\ y - \overline{y} \end{pmatrix} \right\rangle \\ &+ 2 \left\langle \Phi(t) \begin{pmatrix} x - \overline{x} \\ y - \overline{y} \end{pmatrix}, \begin{pmatrix} b(t,\theta) - b(t,\overline{\theta}) \\ h(t,\theta) - h(t,\overline{\theta}) \end{pmatrix} \right\rangle \\ &+ \left\langle \Phi(t) \begin{pmatrix} \sigma(t,\theta) - \sigma(t,\overline{\theta}) \\ z - \overline{z} \end{pmatrix}, \begin{pmatrix} \sigma(t,\theta) - \sigma(t,\overline{\theta}) \\ z - \overline{z} \end{pmatrix} \right\rangle \\ &\leq -\delta |x - \overline{x}|^2, \qquad \forall \theta, \overline{\theta} \in M, \text{ a.e. } t \in [0,T], \text{ a.s.} \end{aligned}$$

$$(1.8)' \qquad \left\langle \Phi(T) \begin{pmatrix} x - \overline{x} \\ g(x) - g(\overline{x}) \end{pmatrix}, \begin{pmatrix} x - \overline{x} \\ g(x) - g(\overline{x}) \end{pmatrix} \right\rangle \geq 0, \quad \forall x, \overline{x} \in \mathbb{R}^n.$$

$$(1.9)' \qquad \begin{aligned} &\left\langle \dot{\Phi}(t) \begin{pmatrix} x - \overline{x} \\ y - \overline{y} \end{pmatrix}, \begin{pmatrix} x - \overline{x} \\ y - \overline{y} \end{pmatrix} \right\rangle \\ &+ 2 \left\langle \Phi(t) \begin{pmatrix} x - \overline{x} \\ y - \overline{y} \end{pmatrix}, \begin{pmatrix} b(t,\theta) - b(t,\overline{\theta}) \\ h(t,\theta) - h(t,\overline{\theta}) \end{pmatrix} \right\rangle \\ &+ \left\langle \Phi(t) \begin{pmatrix} \sigma(t,\theta) - \sigma(t,\overline{\theta}) \\ z - \overline{z} \end{pmatrix}, \begin{pmatrix} \sigma(t,\theta) - \sigma(t,\overline{\theta}) \\ z - \overline{z} \end{pmatrix} \right\rangle \\ &\leq -\delta\{|y - \overline{y}|^2 + |z - \overline{z}|^2\}, \quad \forall \theta, \overline{\theta} \in M, \text{ a.e. } t \in [0,T], \text{ a.s.} \end{aligned}$$

If (1.7)–(1.9) (resp. (1.7) and $(1.8)'$–$(1.9)'$) hold, we call Φ a *type (I)* (resp. *type (II)*) *bridge* extending from Γ (defined on $[0,T]$). The set of all type

(I) and type (II) bridges extending from Γ (defined on $[0,T]$) are denoted by $\mathcal{B}_I(\Gamma;[0,T])$ and $\mathcal{B}_{II}(\Gamma;[0,T])$, respectively. Finally, we let

(1.10)
$$\begin{cases} \mathcal{B}(\Gamma;[0,T]) = \mathcal{B}_I(\Gamma;[0,T]) \bigcup \mathcal{B}_{II}(\Gamma;[0,T]), \\ \mathcal{B}^s(\Gamma;[0,T]) = \mathcal{B}_I(\Gamma;[0,T]) \bigcap \mathcal{B}_{II}(\Gamma;[0,T]). \end{cases}$$

Any element $\Phi \in \mathcal{B}^s(\Gamma;[0,T])$ is called a *strong bridge* extending from Γ (defined on $[0,T]$).

Definition 1.4. Let $T > 0$ and $\Gamma, \overline{\Gamma} \in H[0,T]$. We say that they are *linked by a direct bridge* if

(1.11)
$$\begin{aligned} &\{\mathcal{B}_I(\Gamma;[0,T]) \bigcap \mathcal{B}_I(\overline{\Gamma};[0,T])\} \\ &\bigcup \{\mathcal{B}_{II}(\Gamma;[0,T]) \bigcap \mathcal{B}_{II}(\overline{\Gamma};[0,T])\} \neq \phi; \end{aligned}$$

and we say that they are *linked by a bridge*, if there are $\Gamma_1, \cdots, \Gamma_k \in H[0,T]$, such that with $\Gamma_0 = \Gamma$ and $\Gamma_{k+1} = \overline{\Gamma}$, it holds

(1.12)
$$\begin{aligned} &\{\mathcal{B}_I(\Gamma_i;[0,T]) \bigcap \mathcal{B}_I(\Gamma_{i+1};[0,T])\} \\ &\bigcup \{\mathcal{B}_{II}(\Gamma_i;[0,T]) \bigcap \mathcal{B}_{II}(\Gamma_{i+1};[0,T])\} \neq \phi, \quad 0 \le i \le k. \end{aligned}$$

We may similarly define the notion that Γ and $\overline{\Gamma}$ are linked by a (direct) strong bridge.

§2. Method of Continuation

In this section, we are going to present the solvability of FBSDEs by the *method of continuation*. The notion of bridge plays an important role here.

§2.1. The solvability of FBSDEs linked by bridges

Let us state the following theorem.

Theorem 2.1. *Let $T > 0$ and $\Gamma_1, \Gamma_2 \in H[0,T]$ be linked by a bridge. Then, $\Gamma_1 \in \mathcal{S}[0,T]$ if and only if $\Gamma_2 \in \mathcal{S}[0,T]$.*

The above theorem tells us that if the FBSDE associated with Γ_1 is solvable, so is the one associated with Γ_2, provided Γ_1 and Γ_2 are linked by a bridge. In applications, if one wants to prove the solvability of the FBSDE associated with Γ_2, he/she can start with a known solvable FBSDE Γ_1, and try to construct a bridge linking Γ_1 and Γ_2. We will see a detailed construction of bridges in §5 for an interesting case.

Let us now explain the idea of proving Theorem 2.1. First of all, we make a simple reduction. By induction, to prove Theorem 2.1, it suffices to prove it for the case that $\Gamma_1 \equiv (b_1, \sigma_1, h_1, \gamma_1)$ and $\Gamma_2 \equiv (b_2, \sigma_2, h_2, g_2)$ are linked by a direct bridge. We now assume this. Next, for any $\gamma \equiv$

$(b_0(\cdot), \sigma_0(\cdot), h_0(\cdot), g_0) \in \mathcal{H}[0,T]$, $x \in \mathbb{R}^n$ and $\alpha \in [0,1]$, we consider the following FBSDE:

$$(2.1)^\alpha_{\gamma,x} \begin{cases} dX(t) = \big\{(1-\alpha)b_1(t, \Theta(t)) + \alpha b_2(t, \Theta(t)) + b_0(t)\big\}dt \\ \qquad\qquad + \big\{(1-\alpha)\sigma_1(t, \Theta(t)) + \alpha\sigma_2(t, \Theta(t)) + \sigma_0(t)\big\}dW(t), \\ dY(t) = \big\{(1-\alpha)h_1(t, \Theta(t)) + \alpha h_2(t, \Theta(t)) + h_0(t)\big\}dt \\ \qquad\qquad\qquad\qquad + Z(t)dW(t), \\ X(0) = x, \qquad Y(T) = (1-\alpha)g_1(X(T)) + \alpha g_2(X(T)) + g_0. \end{cases}$$

We may give the definition of the (adapted) solutions to above system $(2.1)^\alpha_{\gamma,x}$ similar to Definition 1.1. It is clear that $(2.1)^0_{\gamma,x}$ and $(2.1)^1_{\gamma,x}$ coincide with $(1.5)_{\Gamma_1,\gamma,x}$ and $(1.5)_{\Gamma_2,\gamma,x}$, respectively. Let us assume that $\Gamma_1 \in \mathcal{S}[0,T]$, i.e., $(2.1)^0_{\gamma,x}$ is uniquely solvable for any $\gamma \in \mathcal{H}[0,T]$ and $x \in \mathbb{R}^n$. We want to prove $\Gamma_2 \in \mathcal{S}[0,T]$, i.e., $(2.1)^1_{\gamma,x}$ is uniquely solvable for all $\gamma \in \mathcal{H}[0,T]$ and $x \in \mathbb{R}^n$. The essence of the *method of continuation* is contained in the following claim:

There exists a fixed step-length $\varepsilon_0 > 0$, such that if for some $\alpha \in [0,1)$, $(2.1)^\alpha_{\gamma,x}$ is uniquely solvable for any $\gamma \in \mathcal{H}[0,T]$ and $x \in \mathbb{R}^n$, then the same conclusion holds for α being replaced by $\alpha + \varepsilon \le 1$ with $\varepsilon \in [0, \varepsilon_0]$.

Once this has been proved, we can start with $(2.1)^\alpha_{\gamma,x}$ with $\alpha = 0$ which is solvable by our assumption, increase the parameter α step by step and finally reach $\alpha = 1$, which gives the unique solvability of $(2.1)^1_{\gamma,x}$.

In order to prove the above claim, the following a priori estimates for the adapted solutions of $(2.1)^\alpha_{\gamma,x}$ will be crucial.

Lemma 2.2. Let $\alpha \in [0,1]$. Let $\Theta(\cdot) \overset{\Delta}{=} (X(\cdot), Y(\cdot), Z(\cdot))$ and $\overline{\Theta}(\cdot) \overset{\Delta}{=} (\overline{X}(\cdot), \overline{Y}(\cdot), \overline{Z}(\cdot))$ be adapted solutions of $(2.1)^\alpha_{\gamma,x}$ and $(2.1)^\alpha_{\overline{\gamma},\overline{x}}$, respectively, with $\gamma = (b_0, \sigma_0, h_0, g_0)$, $\overline{\gamma} = (\overline{b}_0, \overline{\sigma}_0, \overline{h}_0, \overline{g}_0) \in \mathcal{H}[0,T]$ and $x, \overline{x} \in \mathbb{R}^n$. Then, the following estimate holds:

$$\begin{aligned} &\|\Theta(\cdot) - \overline{\Theta}(\cdot)\|^2_{\mathcal{M}[0,T]} \\ &\equiv E \sup_{t\in[0,T]} |X(t) - \overline{X}(t)|^2 + E \sup_{t\in[0,T]} |Y(t) - \overline{Y}(t)|^2 \end{aligned}$$

$$(2.2) \qquad\qquad\qquad\qquad + E \int_0^T |Z(t) - \overline{Z}(t)|^2 dt$$

$$\le C\Big\{|x - \overline{x}|^2 + E|g_0 - \overline{g}_0|^2 + E \int_0^T \big\{|b_0(t) - \overline{b}_0(t)|^2 \\ + |\sigma_0(t) - \overline{\sigma}_0(t)|^2 + |h_0(t) - \overline{h}_0(t)|^2\big\}dt\Big\}.$$

Since the proof of the above lemma is technical and lengthy, we would like to postpone it to the next subsection. Based on the above a priori estimate, we now prove the following result, which we call it the *continuation lemma*.

Lemma 2.3. *Let $\Gamma_1, \Gamma_2 \in H[0,T]$ be linked by a direct bridge. Then, there exists an absolute constant $\varepsilon_0 > 0$, such that if for some $\alpha \in [0,1]$, $(2.1)_{\gamma,x}^{\alpha}$ is uniquely solvable for any $\gamma \in \mathcal{H}[0,T]$ and $x \in \mathbb{R}^n$, then the same is true for $(2.1)_{\gamma,x}^{\alpha+\varepsilon}$ with $\varepsilon \in [0,\varepsilon_0]$, $\alpha + \varepsilon \le 1$.*

Proof. Let $\varepsilon_0 > 0$ be undetermined. Let $\varepsilon \in [0,\varepsilon_0]$. For $k \ge 0$, we successively solve the following systems for $\Theta^k(t) \stackrel{\Delta}{=} (X^k(t), Y^k(t), Z^k(t))$: (compare $(2.1)_{\gamma,x}^{\alpha+\varepsilon}$)

$$(2.3)_{\gamma,x}^{\alpha+\varepsilon} \begin{cases} \Theta^0(t) \stackrel{\Delta}{=} (X^0(t), Y^0(t), Z^0(t)) \equiv 0, \\ dX^{k+1}(t) = \big\{ (1-\alpha)b_1(t, \Theta^{k+1}(t)) + \alpha b_2(t, \Theta^{k+1}(t)) \\ \qquad - \varepsilon b_1(t, \Theta^k(t)) + \varepsilon b_2(t, \Theta^k(t)) + b_0(t) \big\} dt \\ \qquad + \big\{ (1-\alpha)\sigma_1(t, \Theta^{k+1}(t)) + \alpha \sigma_2(t, \Theta^{k+1}(t)) \\ \qquad - \varepsilon \sigma_1(t, \Theta^k(t)) + \varepsilon \sigma_2(t, \Theta^k(t)) + \sigma_0(t) \big\} dW(t), \\ dY^{k+1}(t) = \big\{ (1-\alpha)h_1(t, \Theta^{k+1}(t)) + \alpha h_2(t, \Theta^{k+1}(t)) \\ \qquad - \varepsilon h_1(t, \Theta^k(t)) + \varepsilon h_2(t, \Theta^k(t)) + h_0(t) \big\} dt \\ \qquad + Z^{k+1}(t) dW(t), \\ X^{k+1}(0) = x, \\ Y^{k+1}(T) = (1-\alpha)g_1(X^{k+1}(T)) + \alpha g_2(X^{k+1}(T)) \\ \qquad - \varepsilon g_1(X^k(T)) + \varepsilon g_2(X^k(T)) + g_0. \end{cases}$$

By our assumption, the above systems are uniquely solvable. We now apply Lemma 2.2 to $\Theta^{k+1}(\cdot)$ and $\Theta^k(\cdot)$. It follows that

$$(2.4) \quad \begin{aligned} \|\Theta^{k+1}&(\cdot) - \Theta^k(\cdot)\|_{\mathcal{M}[0,T]} \\ &\le C \Big\{ \varepsilon^2 E |X^k(T) - X^{k-1}(T)|^2 \\ &\qquad + \varepsilon^2 E \int_0^T |\Theta^k(t) - \Theta^{k-1}(t)|^2 dt \Big\} \\ &\le \varepsilon^2 C_0 \|\Theta^k(\cdot) - \Theta^{k-1}(\cdot)\|_{\mathcal{M}[0,T]}. \end{aligned}$$

We note that the constant $C_0 > 0$ appearing in (2.4) is independent of α and ε. Hence, if we choose $\varepsilon_0 > 0$ so that $\varepsilon_0^2 C_0 < 1/2$, then for any $\varepsilon \in [0,\varepsilon_0]$, we have the following estimate:

$$(2.5) \quad \|\Theta^{k+1}(\cdot) - \Theta^k(\cdot)\|_{\mathcal{M}[0,T]} \le \frac{1}{2} \|\Theta^k(\cdot) - \Theta^{k-1}(\cdot)\|_{\mathcal{M}[0,T]}, \quad \forall k \ge 1.$$

This implies that the sequence $\{\Theta^k(\cdot)\}$ is Cauchy in the Banach space $\mathcal{M}[0,T]$. Hence, it admits a limit. Clearly, this limit is an adapted solution to $(2.1)_{\gamma,x}^{\alpha+\varepsilon}$. Uniqueness follows from estimate (2.2) immediately. □

Now, we are ready to give a proof of our main result.

Proof of Theorems 2.1. We know that it suffices to consider the case that Γ_1 and Γ_2 are linked by a direct bridge. Let us assume that $(1.5)_{\Gamma_1,\gamma,x}$

is uniquely solvable for any $\gamma \in \mathcal{H}[0,T]$ and $x \in \mathbb{R}^n$. This means that $(2.1)^0_{\gamma,x}$ is uniquely solvable. By Lemma 2.3, we can then solve $(2.1)^\alpha_{\gamma,x}$ uniquely for any $\alpha \in [0,1]$. In particular, $(2.1)^1_{\gamma,x}$, which is $(1.5)_{\Gamma_2,\gamma,x}$, is uniquely solvable. This proves Theorem 2.1. □

Note that Lemma 2.2 has the following implication.

Corollary 2.4. *Let* $\Gamma \in H[0,T]$ *with* $\mathcal{B}(\Gamma;[0,T]) \neq \phi$. *Then, for any* $\gamma \in \mathcal{H}[0,T]$ *and* $x \in \mathbb{R}^n$, $(1.5)_{\Gamma,\gamma,x}$ *admits at most one adapted solution. Moreover, for any* $\gamma, \overline{\gamma} \in \mathcal{H}[0,T]$ *and* $x, \overline{x} \in \mathbb{R}^n$, *the stability estimate* (2.2) *holds for any adapted solutions* $\Theta(\cdot)$ *of* $(1.5)_{\Gamma,\gamma,x}$ *and* $\overline{\Theta}(\cdot)$ *of* $(1.5)_{\Gamma,\overline{\gamma},\overline{x}}$.

Proof. We take $\Gamma_1 = \Gamma_2 = \Gamma$ in Lemma 2.2. Then, (2.2) applies. □

From Corollary 2.4 we see that for the Γ associated with example (3.3) in Chapter 1, $\mathcal{B}(\Gamma;[0,T]) = \phi$ for $T = k\pi + \frac{3\pi}{4}$, $k \geq 0$.

§2.2. A priori estimate

In this subsection, we present a proof of the a priori estimate stated in Lemma 2.2.

Proof. Let Θ and $\overline{\Theta}$ be two adapted solutions of $(2.1)^\alpha_{\gamma,x}$ and $(2.1)^\alpha_{\overline{g},\overline{x}}$, respectively. Define $\widehat{\xi} = \xi - \overline{\xi}$ for $\xi = X, Y, Z, \Theta, b_0, \sigma_0, h_0, g_0$, $\widehat{x} = x - \overline{x}$, and

$$(2.6) \quad \begin{cases} \widehat{b}_i(t) = b_i(t, \Theta(t)) - b_i(t, \overline{\Theta}(t)), \\ \widehat{\sigma}_i(t) = \sigma_i(t, \Theta(t)) - \sigma_i(t, \overline{\Theta}(t)), \\ \widehat{h}_i(t) = h_i(t, \Theta(t)) - h_i(t, \overline{\Theta}(t)), \\ \widehat{g}_i(T) = g_i(X(T)) - g_i(\overline{X}(T)), \end{cases} \quad i = 1, 2,$$

Note that $\Gamma_i \in H[0,T]$ implies that all the functions b_i, σ_i, h_i, g_i are uniformly Lipschitz continuous. Suppose the common Lipschitz constant is $L > 0$. Applying Itô's formula to $|\widehat{X}(t)|^2$, we obtain that

$$(2.7) \quad \begin{aligned} |\widehat{X}(t)|^2 &= |\widehat{x}|^2 + 2 \int_0^t \langle \widehat{X}(s), (1-\alpha)\widehat{b}_1(s) + \alpha \widehat{b}_2(s) + \widehat{b}_0(s) \rangle \, ds \\ &\quad + \int_0^t |(1-\alpha)\widehat{\sigma}_1(s) + \alpha \widehat{\sigma}_2(s) + \widehat{\sigma}_0(s)|^2 ds \\ &\quad + 2 \int_0^t \langle \widehat{X}(s), [(1-\alpha)\widehat{\sigma}_1(s) + \alpha \widehat{\sigma}_2(s) + \widehat{\sigma}_0(s)] dW(s) \rangle \\ &\leq |\widehat{x}|^2 + C \int_0^t |\widehat{X}(s)|\{|\widehat{X}(s)| + |\widehat{Y}(s)| + |\widehat{Z}(s)| + |\widehat{b}_0(s)|\} ds \\ &\quad + C \int_0^t \{|\widehat{X}(s)| + |\widehat{Y}(s)| + |\widehat{Z}(s)| + |\widehat{\sigma}_0(s)|\}^2 ds \\ &\quad + 2 \int_0^t \langle \widehat{X}(s), [(1-\alpha)\widehat{\sigma}_1(s) + \alpha \widehat{\sigma}_2(s) + \widehat{\sigma}_0(s)] dW(s) \rangle, \end{aligned}$$

with some constant $C > 0$. As before, in what follows, C will be some generic constant, which can be different in different places. By taking the expectation and using Gronwall's inequality, we obtain

$$(2.8)\quad E|\widehat{X}(t)|^2 \le CE\Big\{|\widehat{x}|^2 + \int_0^T \{|\widehat{Y}(t)|^2 + |\widehat{Z}(t)|^2 + |\widehat{b}_0(t)|^2 + |\widehat{\sigma}_0(t)|^2\}dt\Big\},$$

with some constant $C = C(L,T)$. Next, applying Burkholder-Davis-Gundy's inequality to (2.7) (note (2.8)), one has that

$$(2.9)\quad \begin{aligned} E \sup_{t\in[0,T]} |\widehat{X}(t)|^2 \le C\Big\{&|\widehat{x}|^2 + \int_0^T \{|\widehat{Y}(t)|^2 + |\widehat{Z}(t)|^2 \\ &+ |\widehat{b}_0(t)|^2 + |\widehat{\sigma}_0(t)|^2\}dt\Big\}. \end{aligned}$$

On the other hand, by applying Itô's formula to $|\widehat{Y}(t)|^2$, we have

$$(2.10)\quad \begin{aligned} &|\widehat{Y}(t)|^2 + \int_t^T |\widehat{Z}(s)|^2 ds \\ &= |\widehat{Y}(T)|^2 - 2\int_t^T \langle \widehat{Y}(s), (1-\alpha)\widehat{h}_1(s) + \alpha\widehat{h}_2(s) + \widehat{h}_0(s)\rangle\, ds \\ &\quad - 2\int_t^T \langle \widehat{Y}(s), \widehat{Z}(s)dW(s)\rangle \\ &\le C\Big\{|\widehat{X}(T)|^2 + |\widehat{g}_0|^2 + \int_t^T \{|\widehat{X}(s)|^2 + |\widehat{Y}(s)|^2 + |\widehat{h}_0(s)|^2\}ds\Big\} \\ &\quad - \frac{1}{2}\int_t^T |\widehat{Z}(s)|^2 ds - 2\int_t^T \langle \widehat{Y}(s), \widehat{Z}(s)dW(s)\rangle. \end{aligned}$$

Similar to the procedure of getting (2.9), we obtain

$$(2.11)\quad \begin{aligned} E \sup_{t\in[0,T]} &|\widehat{Y}(t)|^2 + E\int_0^T |\widehat{Z}(t)|^2 dt \\ &\le CE\Big\{|\widehat{X}(T)|^2 + |\widehat{g}_0|^2 + \int_0^T \{|\widehat{X}(t)|^2 + |\widehat{h}_0(t)|^2\}dt\Big\}. \end{aligned}$$

We emphasize that the constants C appeared in (2.9) and (2.11) only depend on L and T. Also, in deriving these two estimates, only the condition $\Gamma_i \in H[0,T]$ has been used (and we have not used the bridge yet). Now, we apply Itô's formula to

$$\Big\langle \Phi(t)\begin{pmatrix}\widehat{X}(t)\\\widehat{Y}(t)\end{pmatrix}, \begin{pmatrix}\widehat{X}(t)\\\widehat{Y}(t)\end{pmatrix}\Big\rangle.$$

It follows that

$$
\begin{aligned}
E \left\langle \Phi(T) \begin{pmatrix} \widehat{X}(T) \\ \widehat{Y}(T) \end{pmatrix}, \begin{pmatrix} \widehat{X}(T) \\ \widehat{Y}(T) \end{pmatrix} \right\rangle - E \left\langle \Phi(0) \begin{pmatrix} x \\ \widehat{Y}(0) \end{pmatrix}, \begin{pmatrix} x \\ \widehat{Y}(0) \end{pmatrix} \right\rangle &
\end{aligned}
$$

$$
\begin{aligned}
= E \int_0^T \Big\{ & \left\langle \dot{\Phi}(t) \begin{pmatrix} \widehat{X}(t) \\ \widehat{Y}(t) \end{pmatrix}, \begin{pmatrix} \widehat{X}(t) \\ \widehat{Y}(t) \end{pmatrix} \right\rangle \\
& + 2 \left\langle \Phi(t) \begin{pmatrix} \widehat{X}(t) \\ \widehat{Y}(t) \end{pmatrix}, \begin{pmatrix} (1-\alpha)\widehat{b}_1(t) + \alpha \widehat{b}_2(t) + \widehat{b}_0(t) \\ (1-\alpha)\widehat{h}_1(t) + \alpha \widehat{h}_2(t) + \widehat{h}_0(t) \end{pmatrix} \right\rangle \\
& + \left\langle \Phi(t) \begin{pmatrix} (1-\alpha)\widehat{\sigma}_1(t) + \alpha \widehat{\sigma}_2(t) + \widehat{\sigma}_0(t) \\ \widehat{Z}(t) \end{pmatrix}, \right. \\
& \qquad \left. \begin{pmatrix} (1-\alpha)\widehat{\sigma}_1(t) + \alpha \widehat{\sigma}_2(t) + \widehat{\sigma}_0(t) \\ \widehat{Z}(t) \end{pmatrix} \right\rangle \Big\} dt.
\end{aligned}
$$

(2.12)

Let us separate two cases.

Case 1. Suppose $\Phi \in \mathcal{B}_I(\Gamma_i; [0,T])$ $(i = 1, 2)$. In this case, we have

$$
\begin{aligned}
F(\alpha) & \triangleq \left\langle \Phi(T) \begin{pmatrix} \widehat{X}(T) \\ (1-\alpha)\widehat{g}_1(T) + \alpha \widehat{g}_2(T) \end{pmatrix}, \right. \\
& \qquad \left. \begin{pmatrix} \widehat{X}(T) \\ (1-\alpha)\widehat{g}_1(T) + \alpha \widehat{g}_2(T) \end{pmatrix} \right\rangle \\
& = \langle A(T)\widehat{X}(T), \widehat{X}(T) \rangle + 2 \langle B(T)\widehat{X}(T), (1-\alpha)\widehat{g}_1(T) + \alpha \widehat{g}_2(T) \rangle \\
& \quad + \langle C(T)\{(1-\alpha)\widehat{g}_1(T) + \alpha \widehat{g}_2(T)\}, (1-\alpha)\widehat{g}_1(T) + \alpha \widehat{g}_2(T) \rangle \\
& = \alpha^2 \langle C(T)\{\widehat{g}_2(T) - \widehat{g}_1(T)\}, \{\widehat{g}_2(T) - \widehat{g}_1(T)\} \rangle \\
& \quad + \alpha\{\cdots\} + \{\cdots\} \geq \delta |\widehat{X}(T)|^2, \qquad \forall \alpha \in [0,1],
\end{aligned}
$$

(2.13)

where $\{\cdots\}$ are terms that do not depend on α. The above holds because $C(T) \leq 0$ implies that $F(\alpha)$ is concave in α, whereas (1.8) tells us that (recall $\Phi \in \mathcal{B}_I(\Gamma_i; [0,T])$, $i = 1, 2$)

$$
(2.14) \qquad\qquad F(0), \ F(1) \geq \delta |\widehat{X}(T)|^2.
$$

Then, (2.13) follows easily. Similarly, we have

$$
\begin{aligned}
f(\alpha) & \triangleq \left\langle \dot{\Phi}(t) \begin{pmatrix} \widehat{X}(t) \\ \widehat{Y}(t) \end{pmatrix}, \begin{pmatrix} \widehat{X}(t) \\ \widehat{Y}(t) \end{pmatrix} \right\rangle \\
& \quad + 2 \left\langle \Phi(t) \begin{pmatrix} \widehat{X}(t) \\ \widehat{Y}(t) \end{pmatrix}, \begin{pmatrix} (1-\alpha)\widehat{b}_1(t) + \alpha \widehat{b}_2(t) \\ (1-\alpha)\widehat{h}_1(t) + \alpha \widehat{h}_2(t) \end{pmatrix} \right\rangle \\
& \quad + \left\langle \Phi(t) \begin{pmatrix} (1-\alpha)\widehat{\sigma}_1(t) + \alpha \widehat{\sigma}_2(t) \\ \widehat{Z}(t) \end{pmatrix}, \right. \\
& \qquad \left. \begin{pmatrix} (1-\alpha)\widehat{\sigma}_1(t) + \alpha \widehat{\sigma}_2(t) \\ \widehat{Z}(t) \end{pmatrix} \right\rangle \\
& = \alpha^2 \langle A(t)\{\widehat{\sigma}_2(t) - \widehat{\sigma}_1(t)\}, \widehat{\sigma}_2(t) - \widehat{\sigma}_1(t) \rangle \\
& \quad + \alpha\{\cdots\} + \{\cdots\} \leq -\delta |\widehat{X}(t)|^2, \qquad \forall \alpha \in [0,1],
\end{aligned}
$$

(2.15)

since now $A(t) \geq 0$ which implies $f(\alpha)$ is convex in α. Then, we have

$$
\begin{aligned}
\text{Left side of (2.12)} = E\Big\{ &\langle A(T)\widehat{X}(T), \widehat{X}(T)\rangle \\
&+ 2\langle B(T)\widehat{X}(T), (1-\alpha)\widehat{g}_1(T) + \alpha\widehat{g}_2(T) + \widehat{g}_0\rangle \\
&+ \langle C(T)\{(1-\alpha)\widehat{g}_1(T) + \alpha\widehat{g}_2(T) + \widehat{g}_0\}, \\
&\qquad (1-\alpha)\widehat{g}_1(T) + \alpha\widehat{g}_2(T) + \widehat{g}_0\rangle \Big\}
\end{aligned}
$$

$$
\begin{aligned}
&- E\langle \Phi(0)\begin{pmatrix}\widehat{x}\\ \widehat{Y}(0)\end{pmatrix}, \begin{pmatrix}\widehat{x}\\ \widehat{Y}(0)\end{pmatrix}\rangle \\
&\geq \delta E|\widehat{X}(T)|^2 - 2|B(T)|E(|\widehat{X}(T)||\widehat{g}_0|) \\
&\quad - 2L|C(T)|E(|\widehat{X}(T)||\widehat{g}_0|) - |C(T)|E|\widehat{g}_0|^2 - K|\widehat{x}|^2 \\
&\geq \frac{\delta}{2}E|\widehat{X}(T)|^2 - C\{|\widehat{x}|^2 + E|\widehat{g}_0|^2\}.
\end{aligned}
$$

(2.16)

Here, the constant $C > 0$ only depends on K, L, δ, $|B(T)|$ and $|C(T)|$. Similarly, we have the following estimate for the right hand side of (2.13).

$$
\begin{aligned}
\text{Right side of (2.12)} \leq E\int_0^T \Big\{ &-\delta|\widehat{X}(t)|^2 dt \\
&+ 2\langle \Phi(t)\begin{pmatrix}\widehat{X}(t)\\ \widehat{Y}(t)\end{pmatrix}, \begin{pmatrix}\widehat{b}_0(t)\\ \widehat{h}_0(t)\end{pmatrix}\rangle \\
&+ 2\langle \Phi(t)\begin{pmatrix}(1-\alpha)\widehat{\sigma}_1(t) + \alpha\widehat{\sigma}_2(t)\\ \widehat{Z}(t)\end{pmatrix}, \begin{pmatrix}\widehat{\sigma}_0(t)\\ 0\end{pmatrix}\rangle \\
&+ \langle \Phi(t)\begin{pmatrix}\widehat{\sigma}_0(t)\\ 0\end{pmatrix}, \begin{pmatrix}\widehat{\sigma}_0(t)\\ 0\end{pmatrix}\rangle \Big\} dt
\end{aligned}
$$

(2.17)

$$
\begin{aligned}
&\leq -\frac{\delta}{2}E\int_0^T |\widehat{X}(t)|^2 dt + \varepsilon E\int_0^T \{|\widehat{Y}(t)|^2 + |\widehat{Z}(t)|^2\} dt i \\
&\quad + C_\varepsilon E\int_0^T \{|\widehat{b}_0(t)|^2 + |\widehat{\sigma}_0(t)|^2 + |\widehat{h}_0(t)|^2\} dt.
\end{aligned}
$$

with the constant $C_\varepsilon > 0$ only depending on the bounds of $|\Phi(t)|$, as well as δ, L and the undetermined small positive number $\varepsilon > 0$. Combining

(2.16)–(2.17) and note (2.11), we have

$$E|\widehat{X}(T)|^2 + E \int_0^T |\widehat{X}(t)|^2 dt$$

$$\leq C_\varepsilon \left\{ |\widehat{x}|^2 + E|\widehat{g}_0|^2 + E \int_0^T \{|\widehat{b}_0(t)|^2 + |\widehat{\sigma}_0(t)|^2 + |\widehat{h}_0(t)|^2\} dt \right\}$$

$$(2.18) \qquad + \frac{2\varepsilon}{\delta} E \int_0^T \{|\widehat{Y}(t)|^2 + |\widehat{Z}(t)|^2\} dt$$

$$\leq C_\varepsilon \left\{ |\widehat{x}|^2 + E|\widehat{g}_0|^2 + E \int_0^T \{|\widehat{b}_0(t)|^2 + |\widehat{\sigma}_0(t)|^2 + |\widehat{h}_0(t)|^2\} dt \right\}$$

$$+ \varepsilon \overline{C} E \left\{ |\widehat{X}(T)|^2 + |\widehat{g}_0|^2 + \int_0^T \{|\widehat{X}(t)|^2 + |\widehat{h}_0(t)|^2\} dt \right\},$$

with the constant $\overline{C}$ independent of $\varepsilon > 0$, and C_ε might be different from that appeared in (2.17). Thus, we may choose suitable $\varepsilon > 0$, such that

$$E|\widehat{X}(T)|^2 + E \int_0^T |\widehat{X}(t)|^2 dt$$

$$(2.19)$$

$$\leq CE \left\{ |\widehat{x}|^2 + |\widehat{g}_0|^2 + \int_0^T \{|\widehat{b}_0(t)|^2 + |\widehat{\sigma}_0(t)|^2 + |\widehat{h}_0(t)|^2\} dt \right\}.$$

Then, return to (2.11), we obtain

$$E \sup_{t \in [0,T]} |\widehat{Y}(t)|^2 + E \int_0^T |\widehat{Z}(t)|^2 dt$$

$$(2.20)$$

$$\leq CE \left\{ |\widehat{x}|^2 + |\widehat{g}_0|^2 + \int_0^T \{|\widehat{b}_0(t)|^2 + |\widehat{\sigma}_0(t)|^2 + |\widehat{h}_0(t)|^2\} dt \right\}.$$

Finally, by (2.9), we have

$$E \sup_{t \in [0,T]} |\widehat{X}(t)|^2 \leq CE \left\{ |\widehat{x}|^2 + |\widehat{g}_0|^2 \right.$$

$$(2.21)$$

$$\left. + \int_0^T \{|\widehat{b}_0(t)|^2 + |\widehat{\sigma}_0(t)|^2 + |\widehat{h}_0(t)|^2\} dt \right\}.$$

Hence, (2.2) follows from (2.20) and (2.21).

Case 2. Let $\Phi \in \mathcal{B}_{II}(\Gamma_i; [0,T])$ $(i = 1, 2)$ now. In this case, we still have (2.9), (2.11) and (2.12). Further, we have inequalities similar to (2.13) and (2.15) with $|\widehat{X}(T)|^2$ and $|\widehat{X}(t)|^2$ replaced by 0 and $|\widehat{Y}(t)|^2 + |\widehat{Z}(t)|^2$, respectively. Thus, it follows that

$$(2.22) \qquad \text{Left side of } (2.12) \geq -\varepsilon E|\widehat{X}(T)|^2 - C_\varepsilon \{|\widehat{x}|^2 + E|\widehat{g}_0|^2\},$$

with the constant $C_\varepsilon > 0$ depending on $K, L, \delta, |B(T)|, |C(T)|$, and the undetermined constant $\varepsilon > 0$. Whereas,

Right side of (2.12)

(2.23)
$$\leq -\frac{\delta}{2}E\int_0^T \{|\widehat{Y}(t)|^2 + |\widehat{Z}(t)|^2\}dt + \varepsilon E\int_0^T |\widehat{X}(t)|^2 dt$$
$$+ C_\varepsilon E\int_0^T \{|\widehat{b}_0(t)|^2 + |\widehat{\sigma}_0(t)|^2 + |\widehat{h}_0(t)|^2\}dt.$$

Now, combining (2.22)–(2.23) and using (2.9), we obtain (for suitable choice of $\varepsilon > 0$)

(2.24)
$$E\int_0^T \{|\widehat{Y}(t)|^2 + |\widehat{Z}(t)|^2\}dt$$
$$\leq CE\Big\{|\widehat{x}|^2 + |\widehat{g}_0|^2 + \int_0^T \{|\widehat{b}_0(t)|^2 + |\widehat{\sigma}_0(t)|^2 + |\widehat{h}_0(t)|^2\}dt\Big\}.$$

Finally, by (2.9) and (2.11) again, we obtain the estimate (2.2). □

§3. Some Solvable FBSDEs

In this section, we are going to prove the unique solvability of some FBSDEs by constructing appropriate bridges.

§3.1. A trivial FBSDE

We denote $\Gamma_0 = (0,0,0,0) \in H[0,T]$. The FBSDE associated with Γ_0 reads as (compare with $(1.5)_{\Gamma,\gamma,x}$)

(3.1)
$$\begin{cases} dX(t) = b_0(t)dt + \sigma_0(t)dW(t), \\ dY(t) = h_0(t)dt + Z(t)dW(t), \\ X(0) = x, \qquad Y(T) = g_0. \end{cases}$$

Clearly, (3.1) is trivially uniquely solvable for all $\gamma \equiv (b_0, \sigma_0, h_0, g_0) \in \mathcal{H}[0,T]$ and $x \in \mathbb{R}^n$. Thus, hereafter, we will refer to the FBSDE associated with Γ_0 as the *trivial* FBSDE. Now, let us present the following result.

Proposition 3.1. *Let $T > 0$ and $\Gamma_0 = \{0,0,0,0\} \in H[0,T]$. Then,*
$$\Phi \equiv \begin{pmatrix} A & B^T \\ B & C \end{pmatrix} \in \mathcal{B}^s(\Gamma_0; [0,T])$$ *if and only if*

(3.2)
$$\begin{cases} C(0) < 0, \qquad A(T) > 0, \\ \dot{\Phi}(t) < 0, \qquad \forall t \in [0,T]. \end{cases}$$

Proof. By Definition 1.3, we know that $\Phi \in \mathcal{B}^s(\Gamma_0; [0,T])$ if and only if (1.7)–(1.9) and (1.8)′–(1.9)′ hold. These are equivalent to the following:

(3.3)
$$\begin{cases} C(0) \leq -\delta, \ A(T) \geq \delta, \\ \dot{\Phi}(t) \leq -\delta, \qquad \forall t \in [0,T], \end{cases}$$

for some $\delta > 0$. We note that under condition $C(0) < 0$, the second inequality in (1.7) is always true for sufficiently large $K > 0$. Then, we see easily that $\Phi \in \mathcal{B}^s(\Gamma_0; [0, T])$ is characterized by (3.3) since $\delta > 0$ can be arbitrarily small. □

From the above, we also have the following characterization:

(3.4)
$$\mathcal{B}^s(\Gamma_0; [0, T]) = \Big\{ Q - \int_0^\cdot \Psi(s)ds \Big|$$
$$0 < \Psi(\cdot) = \begin{pmatrix} \Psi_1(\cdot) & \Psi_2(\cdot)^T \\ \Psi_2(\cdot) & \Psi_3(\cdot) \end{pmatrix} \in C([0, T]; S^{n+m}),$$
$$Q = \begin{pmatrix} Q_1 & Q_2^T \\ Q_2 & Q_3 \end{pmatrix} \in S^{n+m}, \ Q_3 < 0,$$
$$Q_1 - \int_0^T \Psi_1(s)ds > 0 \Big\}.$$

A useful consequence of Proposition 3.1 is the following.

Corollary 3.2. *Let $\Gamma \in H[0, T]$ admit a bridge $\Phi \in \mathcal{B}(\Gamma; [0, T])$ satisfying (3.2). Then, $\Gamma \in \mathcal{S}[0, T]$.*

Proof. Under our assumptions, it holds that

$$\Phi \in \mathcal{B}(\Gamma_0; [0, T]) \bigcap \mathcal{B}(\Gamma; [0, T]).$$

Since $\Gamma_0 \in \mathcal{S}[0, T]$, Theorem 2.1 applies. □

Next, we would like to discuss some concrete cases.

§3.2. Decoupled FBSDEs

Let $\Gamma \equiv (b, \sigma, h, g) \in H[0, T]$ such that

(3.5)
$$\begin{cases} b(t, x, y, z) \equiv b(t, x), \\ \sigma(t, x, y, z) \equiv \sigma(t, x), \end{cases} \quad \forall (t, x, y, z) \in [0, T] \times M.$$

We see that the associated FBSDE is decoupled, which is known to be solvable under usual Lipschitz conditions, by the result of Chapter 1, §4. The following result recovers this conclusion with some deeper insight.

Proposition 3.3. *Let $T > 0$, $\Gamma_0 \equiv (0, 0, 0, 0) \in H[0, T]$ and $\Gamma \equiv (b, \sigma, h, g) \in H[0, T]$ satisfying (3.5). Then,*

(3.6)
$$\mathcal{B}^s(\Gamma_0; [0, T]) \bigcap \mathcal{B}^s(\Gamma; [0, T]) \neq \phi.$$

Consequently, $\Gamma \in \mathcal{S}[0, T]$.

Proof. We take

(3.7)
$$\begin{cases} \Phi(t) = \begin{pmatrix} a(t)I & 0 \\ 0 & c(t)I \end{pmatrix}, \\ a(t) = A_0 e^{A_0(T-t)}, \quad c(t) = -C_0 e^{C_0 t}, \quad t \in [0, T], \end{cases}$$

where $A_0, C_0 > 0$ are undetermined constants. We first check that this $\Phi \in \mathcal{B}^s(\Gamma_0; [0, T])$. In fact,

(3.8)
$$\begin{cases} c(0) = -C_0 < 0, \qquad a(T) = A_0 > 0, \\ \dot{a}(t) = -A_0^2 e^{A_0(T-t)} < 0, \qquad t \in [0, T], \\ \dot{c}(t) = -C_0^2 e^{C_0 t} < 0, \qquad t \in [0, T]. \end{cases}$$

Thus, by Proposition 3.1, we see that $\Phi \in \mathcal{B}^s(\Gamma_0; [0, T])$. Next, we show that $\Phi \in \mathcal{B}^s(\Gamma; [0, T])$ for suitable choice of A_0 and C_0. To this end, we let L be the common Lipschitz constant for b, σ, h and g. We note that (3.8) implies (1.7). Thus, it is enough to further have

(3.9)
$$a(T) + L^2 c(T) \geq \delta,$$

and

(3.10)
$$\begin{aligned} &\dot{a}(t)|x - \overline{x}|^2 + \dot{c}(t)|y - \overline{y}|^2 + c(t)|z - \overline{z}|^2 \\ &\quad + 2a(t)\langle x - \overline{x}, b(t, x) - b(t, \overline{x}) \rangle + a(t)|\sigma(t, x) - \sigma(t, \overline{x})|^2 \\ &\quad + 2c(t)\langle y - \overline{y}, h(t, x, y, z) - h(t, \overline{x}, \overline{y}, \overline{z}) \rangle \\ &\leq -\delta\{|x - \overline{x}|^2 + |y - \overline{y}|^2 + |z - \overline{z}|^2\}, \\ &\qquad \forall t \in [0, T], \ x, \overline{x} \in \mathbb{R}^n, \ y, \overline{y} \in \mathbb{R}^m, \ z, \overline{z} \in \mathbb{R}^{m \times d}, \ \text{a.s.} \end{aligned}$$

Let us first look at (3.10). We note that

(3.11)
$$\begin{aligned} \text{Left side of (3.10)} &\leq \dot{a}(t)|x - \overline{x}|^2 + \dot{c}(t)|y - \overline{y}|^2 + c(t)|z - \overline{z}|^2 \\ &\quad + 2a(t)L|x - \overline{x}|^2 + a(t)L^2|x - \overline{x}|^2 \\ &\quad + 2|c(t)|L|y - \overline{y}|\{|x - \overline{x}| + |y - \overline{y}| + |z - \overline{z}|\} \\ &\leq \{\dot{a}(t) + 2a(t)L + a(t)L^2 + |c(t)|L\}|x - \overline{x}|^2 \\ &\quad + \{\dot{c}(t) + 3|c(t)|L + 2L^2|c(t)|\}|y - \overline{y}|^2 + \frac{c(t)}{2}|z - \overline{z}|^2. \end{aligned}$$

Hence, to have (3.10), it suffices to have the following:

(3.12)
$$\begin{cases} \dot{a}(t) + (2L + L^2)a(t) + L|c(t)| \leq -\delta, \\ \dot{c}(t) + (3L + 2L^2)|c(t)| \leq -\delta, \qquad \forall t \in [0, T]. \\ c(t) \leq -2\delta, \end{cases}$$

Now, we take $a(t)$ and $c(t)$ as in (3.7) and we require

(3.13)
$$\begin{aligned} \dot{c}(t) + (3L + 2L^2)|c(t)| &= -C_0(C_0 - 3L - 2L^2)e^{C_0 t}, \\ &\leq -C_0(C_0 - 3L - 2L^2) \leq -\delta, \qquad \forall t \in [0, T], \end{aligned}$$

and

(3.14)
$$c(t) = -C_0 e^{C_0 t} \leq -C_0 \leq -2\delta, \qquad \forall t \in [0, T].$$

These two are possible if $C_0 > 0$ is large enough. Next, for this fixed $C_0 > 0$, we choose $A_0 > 0$ as follows. We want

$$(3.15) \quad a(T) + c(T)L^2 = A_0 e^{A_0(T-t)} - C_0 L^2 e^{C_0 t} \geq A_0 - C_0 L^2 e^{C_0 T} \geq \delta,$$

and

$$(3.16) \quad \begin{aligned} \dot{a}(t) &+ (2L + L^2)a(t) + L|c(t)| \\ &= -A_0(A_0 - 2L - L^2)e^{A_0(T-t)} + LC_0 e^{C_0 t} \\ &\leq -A_0(A_0 - 2L - L^2) + LC_0 e^{C_0 T} \leq -\delta. \end{aligned}$$

These are also possible by choosing $A_0 > 0$ large enough. Hence, (3.9) and (3.12) hold and $\Phi \in \mathcal{B}^s(\Gamma; [0, T])$. □

From the above, we obtain that any decoupled FBSDE is solvable. In particular, any BSDE is solvable. Moreover, from Lemma 2.2, we see that the adapted solutions to such equations have the continuous dependence on the data.

The above proposition also tells us that decoupled FBSDEs are very "close" to the trivial FBSDE since they can be linked by some direct strong bridges of Γ_0.

§3.3. FBSDEs with monotonicity conditions

In this subsection, we are going to consider coupled FBSDEs which satisfy certain kind of monotonicity conditions. Let $\Gamma = (b, \sigma, h, g) \in H[0, T]$. We introduce the following conditions:

(M) Let $m \geq n$. There exists a matrix $B \in \mathbb{R}^{m \times n}$ such that for some $\beta > 0$, it holds that

$$(3.17) \quad \langle B(x - \overline{x}), g(x) - g(\overline{x}) \rangle \geq \beta |x - \overline{x}|^2, \qquad \forall x, \overline{x} \in \mathbb{R}^n, \text{ a.s.}$$

$$(3.18) \quad \begin{aligned} &\langle B^T[h(t, \theta) - h(t, \overline{\theta})], x - \overline{x} \rangle + \langle B[b(t, \theta) - b(t, \overline{\theta})], y - \overline{y} \rangle \\ &+ \langle B[\sigma(t, \theta) - \sigma(t, \overline{\theta})], z - \overline{z} \rangle \leq -\beta |x - \overline{x}|^2, \\ &\forall t \in [0, T], \ \theta, \overline{\theta} \in M, \text{ a.s.} \end{aligned}$$

(M)' Let $m \leq n$. There exists a matrix $B \in \mathbb{R}^{m \times n}$ such that for some $\beta > 0$, it holds that

$$(3.17)' \quad \langle B(x - \overline{x}), g(x) - g(\overline{x}) \rangle \geq 0, \qquad \forall x, \overline{x} \in \mathbb{R}^n, \text{ a.s.}$$

$$(3.18)' \quad \begin{aligned} &\langle B^T[h(t, \theta) - h(t, \overline{\theta})], x - \overline{x} \rangle + \langle B[b(t, \theta) - b(t, \overline{\theta})], y - \overline{y} \rangle \\ &+ \langle B[\sigma(t, \theta) - \sigma(t, \overline{\theta})], z - \overline{z} \rangle \leq -\beta(|y - \overline{y}|^2 + |z - \overline{z}|^2), \\ &\forall t \in [0, T], \ \theta, \overline{\theta} \in M, \text{ a.s.} \end{aligned}$$

Condition (3.17) means that the function $x \mapsto B^T g(x)$ is uniformly monotone on $\mathbb{R}^n$, and condition (3.18) implies that the function $\theta \mapsto$

$-(B^T h(t,\theta), Bb(t,\theta), B\sigma(t,\theta))$ is monotone on the space M. The meaning of $(3.17)'$ and $(3.18)'$ are similar. Here, we should point out that (3.17) implies $m \geq n$ and $(3.17)'$ implies $m \leq n$. Hence, (M) and (M)$'$ overlaps only for the case $m = n$.

We now prove the following.

Proposition 3.4. *Let $T > 0$ and $\Gamma \equiv (b, \sigma, h, g) \in H[0, T]$ satisfy (M) (resp. (M)$'$). Then, (3.6) holds. Consequently, $\Gamma \in \mathcal{S}[0, T]$.*

Proof. First, we assume (M) holds. Take

$$(3.19) \qquad \begin{cases} \Phi(t) = \begin{pmatrix} A(t) & B(t)^T \\ B(t) & C(t) \end{pmatrix} \\ A(t) = a(t)I \equiv \delta e^{T-t}I, \qquad t \in [0, T], \\ B(t) \equiv B, \\ C(t) = c(t)I \equiv -2\delta C_0 e^{C_0 t}I, \end{cases}$$

with $\delta, C_0 > 0$ being undetermined. Since

$$(3.20) \qquad \begin{cases} C(0) = -2\delta C_0 I < 0, \\ A(T) = \delta I > 0, \\ \dot{\Phi}(t) = \begin{pmatrix} -\delta e^{T-t}I & 0 \\ 0 & -2\delta C_0^2 e^{C_0 t} \end{pmatrix} < 0, \end{cases}$$

by Proposition 3.1, we see that $\Phi \in \mathcal{B}^s(\Gamma_0; [0, T])$. Next, we prove $\Phi \in \mathcal{B}^s(\Gamma; [0, T])$ for suitable choice of δ and C_0. Again, we let L be the common Lipschitz constant for b, σ, h and g. We will choose δ and C_0 so that

$$(3.21) \qquad a(T) + 2\beta + c(T)L^2 \geq \delta,$$

and

$$(3.22) \qquad \begin{aligned} &\dot{a}(t)|x|^2 + \dot{c}(t)|y|^2 + c(t)|z|^2 + 2La(t)|x|(|x| + |y| + |z|) \\ &+ 2L|c(t)|\,|y|(|x| + |y| + |z|) + L^2 a(t)(|x| + |y| + |z|)^2 \\ &\leq (2\beta - \delta)|x|^2 - \delta(|y|^2 + |z|^2), \qquad \forall (t, \theta) \in [0, T] \times M. \end{aligned}$$

It is not hard to see that under (3.17)–(3.18), (3.21) implies (1.8) and (3.22) implies (1.7) and $(1.9)'$ (Note (1.8) implies $(1.8)'$). We see that the left hand side of (3.22) can be controlled by the following:

$$(3.23) \qquad \begin{aligned} &\{\dot{a}(t) + Ka(t) + K|c(t)|\}|x|^2 + \{\dot{c}(t) + K|c(t)| + Ka(t)\}|y|^2 \\ &+ \{\frac{c(t)}{2} + Ka(t)\}|z|^2, \end{aligned}$$

for some constant $K > 0$. Then, for this fixed $K > 0$, we now choose δ and C_0. First of all, we require

$$(3.24) \qquad \frac{c(t)}{2} + Ka(t) = -\delta C_0 e^{C_0 t} + K\delta e^{T-t} \leq -\delta C_0 + K\delta e^T \leq -\delta,$$

and

$$(3.25) \quad \begin{aligned} \dot{c}(t) + K|c(t)| + Ka(t) &= -2\delta C_0^2 e^{C_0 t} + 2K C_0 \delta e^{C_0 t} + K\delta e^{T-t} \\ &\leq -2\delta C_0(C_0 - K) + K\delta e^T < -\delta. \end{aligned}$$

These two can be achieved by choosing $C_0 > 0$ large enough (independent of $\delta > 0$). Next, we require

$$(3.26) \quad \begin{aligned} \dot{a}(t) + Ka(t) + K|c(t)| &= -\delta e^{T-t} + K\delta e^{T-t} + 2\delta K C_0 e^{C_0 t} \\ &\leq -\delta + K\delta e^T + 2\delta K C_0 e^{C_0 T} \leq 2\beta - \delta, \end{aligned}$$

and

$$(3.27) \qquad a(T) + 2\beta + c(T)L^2 = \delta + 2\beta - 2\delta C_0 e^{C_0 T} L^2 \geq \delta.$$

Since $\beta > 0$, (3.26) and (3.27) can be achieved by letting $\delta > 0$ be small enough (note again that the choice of C_0 is independent of $\delta > 0$). Hence, we have (3.21) and (3.22), which proves $\Phi \in \mathcal{B}^s(\Gamma; [0, T])$.

Now, we assume (M)$'$ holds. Take (compare (3.19))

$$(3.28) \quad \begin{cases} \Phi(t) = \begin{pmatrix} A(t) & B(t)^T \\ B(t) & C(t) \end{pmatrix}, \\ A(t) = a(t)I \equiv \delta A_0 e^{A_0(T-t)} I, & \forall t \in [0, T], \\ B(t) \equiv B, \\ C(t) = c(t)I \equiv -\delta e^t I, \end{cases}$$

with $\delta, A_0 > 0$ being undetermined. Note that

$$(3.29) \quad \begin{cases} C(0) = -\delta I < 0, \\ A(T) = A_0 I > 0, \\ \dot{\Phi}(t) = \begin{pmatrix} -\delta A_0^2 e^{A_0(T-t)} I & 0 \\ 0 & -\delta e^t I \end{pmatrix} < 0. \end{cases}$$

Thus, by Proposition 3.1, we have $\Phi \in \mathcal{B}^s(\Gamma_0; [0, T])$. We now choose the constants δ and A_0. In the present case, we will still require (3.21) and the following instead of (3.22):

$$(3.30) \quad \begin{aligned} &\dot{a}(t)|x|^2 + \dot{c}(t)|y|^2 + c(t)|z|^2 + 2La(t)|x|(|x| + |y| + |z|) \\ &\quad + 2L|c(t)||y|(|x| + |y| + |z|) + L^2 a(t)(|x| + |y| + |z|)^2 \\ &\leq -\delta|x|^2 + (2\beta - \delta)\{|y|^2 + |z|^2\}, \quad \forall (t, \theta) \in [0, T] \times M. \end{aligned}$$

These two will imply the conclusion $\Phi \in \mathcal{B}^s(\Gamma; [0, T])$. Again the left hand side of (3.30) can be controlled by (3.23) for some constant $K > 0$. Now, we require

$$(3.31) \quad \begin{aligned} \dot{a}(t) + Ka(t) + K|c(t)| &= -\delta A_0^2 e^{A_0(T-t)} + \delta K A_0 e^{A_0(T-t)} + K\delta e^t \\ &\leq -\delta A_0(A_0 - K) + \delta K e^T \leq -\delta, \end{aligned}$$

and

(3.32)
$$a(T) + c(T)L^2 = \delta A_0 e^{A_0(T-t)} - \delta L^2 e^t$$
$$\geq \delta(A_0 - L^2 e^T) > \delta.$$

We can choose $A_0 > 0$ large enough (independent of $\delta > 0$) to achieve the above two. Next, we require

(3.33)
$$\frac{c(t)}{2} + Ka(t) \leq Ka(t) \leq \delta K A_0 e^{A_0 T} \leq 2\beta - \delta,$$

and

(3.34)
$$\dot{c}(t) + K|c(t)| + Ka(t) = -\delta e^t + K\delta e^t + K A_0 \delta e^{A_0(T-t)}$$
$$\leq \delta(K e^T + K A_0 e^{A_0 T}) \leq 2\beta - \delta.$$

These two can be achieved by choosing $\delta > 0$ small enough. Hence, we obtain (3.21) and (3.30), which gives $\Phi \in \mathcal{B}^s(\Gamma; [0, T])$.

It should be pointed out that the above FBSDEs with monotonicity conditions do not cover the decoupled case. Here is a simple example.

Let $n = m = 1$. Consider the following decoupled FBSDE:

(3.35)
$$\begin{cases} dX(t) = X(t)dt + dW(t), \\ dY(t) = X(t)dt + Z(t)dW(t), \\ X(0) = x, \quad Y(T) = X(T). \end{cases}$$

We can easily check that neither (M) nor (M)$'$ holds. But, (3.35) is uniquely solvable over any finite time duration $[0, T]$.

Remark 3.5. From the above, we see that decoupled FBSDEs and the FBSDEs with monotonicity conditions are two different classes of solvable FBSDEs. None of them includes the other. On the other hand, however, these two classes are proved to be linked by direct bridges to the trivial FBSDE (the one associated with $\Gamma_0 = (0, 0, 0, 0)$). Thus, in some sense, these classes of FBSDEs are very "closer" to the trivial FBSDE.

§4. Properties of the Bridges

In order to find some more solvable FBSDEs with the aid of bridges, we need to explore some useful properties that bridges enjoy.

Proposition 4.1. *Let* $T > 0$.

(i) *For any* $\Gamma \in H[0, T]$, *the set* $\mathcal{B}_I(\Gamma; [0, T])$ *is a convex cone whenever it is nonempty. Moreover,*

(4.1)
$$\mathcal{B}_I(\Gamma; [0, T]) = \mathcal{B}_I(\Gamma + \gamma; [0, T]), \qquad \forall \gamma \in \mathcal{H}[0, T].$$

(ii) *For any* $\Gamma_1, \Gamma_2 \in H[0, T]$, *it holds*

(4.2)
$$\mathcal{B}_I(\Gamma_1; [0, T]) \bigcap \mathcal{B}_I(\Gamma_2; [0, T]) \subseteq \bigcap_{\alpha, \beta > 0} \mathcal{B}_I(\alpha\Gamma_1 + \beta\Gamma_2; [0, T]).$$

Proof. (i) The convexity of $\mathcal{B}_I(\Gamma; [0, T])$ is clear since (1.7)–(1.9) are linear inequalities in Φ. Conclusion (4.1) also follows easily from the definition of the bridge.

(ii) The proof follows from (2.13), (2.15) and the fact that $\mathcal{B}_I(\Gamma; [0, T])$ is a convex cone. $\square$

It is clear that the same conclusions as Proposition 4.1 hold for $\mathcal{B}_{II}(\Gamma; [0, T])$ and $\mathcal{B}^s(\Gamma; [0, T])$.

As a consequence of (3.2), we see that if $\Gamma_1, \Gamma_2 \in H[0, T]$, then

(4.3)
$$\mathcal{B}_I(\alpha\Gamma_1 + \beta\Gamma_2; [0, T]) = \phi, \quad \text{for some } \alpha, \beta > 0,$$
$$\Rightarrow \quad \mathcal{B}_I(\Gamma_1; [0, T]) \bigcap \mathcal{B}_I(\Gamma_2; [0, T]) = \phi.$$

This means that for such a case, Γ_1 and Γ_2 are *not* linked by a direct bridge (of type (I)). Let us look at a concrete example. Let $\Gamma_i = (b_i, \sigma_i, h_i, g_i) \in H[0, T]$, $i = 1, 2, 3$, with

(4.4)
$$\begin{cases} \begin{pmatrix} b_1 \\ h_1 \end{pmatrix} = \begin{pmatrix} -\lambda & 0 \\ -1 & -\nu \end{pmatrix} \begin{pmatrix} x \\ y \end{pmatrix}, \qquad \begin{pmatrix} b_2 \\ h_2 \end{pmatrix} = \begin{pmatrix} \lambda & 1 \\ 0 & \nu \end{pmatrix} \begin{pmatrix} x \\ y \end{pmatrix}, \\ \begin{pmatrix} b_3 \\ h_3 \end{pmatrix} = \begin{pmatrix} 0 & 1 \\ -1 & 0 \end{pmatrix} \begin{pmatrix} x \\ y \end{pmatrix}, \qquad \begin{matrix} \sigma_1 = \sigma_2 = \sigma_3 = 0, \\ g_1 = g_2 = g_3 = -x, \end{matrix} \end{cases}$$

with $\lambda, \nu \in \mathbb{R}$. Clearly, it holds

(4.5)
$$\Gamma_3 = \Gamma_1 + \Gamma_2.$$

By the remark right after Corollary 2.4, we know that $\mathcal{B}(\Gamma_3; [0, T]) = \phi$. Thus, it follows from (3.5) and (4.3) that Γ_1 and Γ_2 are not linked by a direct bridge. However, we see that the FBSDE associated with Γ_1 is decoupled and thus it is uniquely solvable (see Chapter 1). In §5, we will show that for suitable choice of λ and ν, $\Gamma_2 \in \mathcal{S}[0, T]$. Hence, we find two elements in $\mathcal{S}[0, T]$ that are *not* linked by a direct bridge. This means Γ_1 and Γ_2 are *not* very "close".

Next, for any $b_1, b_2 \in L^2_{\mathcal{F}}(0, T; W^{1,\infty}(M; \mathbb{R}^n))$, we define

(4.6)
$$\|b_1 - b_2\|_0(t)$$
$$= \operatorname*{esssup}_{\omega \in \Omega} \sup_{\theta, \bar{\theta} \in M} \frac{\left| b_1(t, \theta; \omega) - b_1(t, \bar{\theta}; \omega) - b_2(t, \theta; \omega) + b_2(t, \bar{\theta}; \omega) \right|}{|\theta - \bar{\theta}|}.$$

We define $\|h_1 - h_2\|_0(t)$ and $\|\sigma_1 - \sigma_2\|_0(t)$ similarly. For $g_1, g_2 \in L^2_{\mathcal{F}_T}(\Omega; W^{1,\infty}(\mathbb{R}^n; \mathbb{R}^m))$, we define

(4.7)
$$\|g_1 - g_2\|_0$$
$$= \operatorname*{esssup}_{\omega \in \Omega} \sup_{x, \bar{x} \in \mathbb{R}^n} \frac{\left| g_1(x; \omega) - g_1(\bar{x}; \omega) - g_2(x; \omega) + g_2(\bar{x}; \omega) \right|}{|x - \bar{x}|}.$$

Then, for any $\Gamma_i = (b_i, \sigma_i, h_i, g_i) \in H[0, T]$ $(i = 1, 2)$, set

$$
(4.8) \qquad
\begin{aligned}
\|\Gamma_1 - \Gamma_2\|_0(t) &= \|b_1 - b_2\|_0(t) + \|\sigma_1 - \sigma_2\|_0(t) \\
&\quad + \|h_1 - h_2\|_0(t) + \|g_1 - g_2\|_0.
\end{aligned}
$$

Note that $\|\cdot\|_0(t)$ is just a family of semi-norms (parameterized by $t \in [0, T]$). As a matter of fact, $\|\Gamma_1 - \Gamma_2\|_0(t) = 0$ for all $t \in [0, T]$ if and only if

$$
(4.9) \qquad\qquad\qquad \Gamma_2 = \Gamma_1 + \gamma,
$$

for some $\gamma \in \mathcal{H}[0, T]$.

Theorem 4.2. *Let $T > 0$ and $\Gamma \in H[0, T]$. Let $\Phi \in \mathcal{B}^s(\Gamma; [0, T])$. Then, there exists an $\varepsilon > 0$, such that for any $\Gamma' \in H[0, T]$ with*

$$
(4.10) \qquad\qquad \|\Gamma - \Gamma'\|_0(t) < \varepsilon, \qquad \forall t \in [0, T],
$$

we have $\Phi \in \mathcal{B}^s(\Gamma'; [0, T])$.

Proof. Let $\Gamma = (b, \sigma, h, g)$ and $\Gamma' = (b', \sigma', h', g')$. Suppose $\Phi \in \mathcal{B}^s(\Gamma; [0, T])$. Then, for some $K, \delta > 0$, (1.7)–(1.9) and (1.8)′–(1.9)′ hold. Now, we denote (for any $\theta, \overline{\theta} \in M$)

$$
(4.11) \qquad
\begin{cases}
\widehat{x} = x - \overline{x}, \ \widehat{\theta} = \theta - \overline{\theta}, \\
\widehat{b} = b(t, \theta) - b(t, \overline{\theta}), \ \widehat{\sigma} = \sigma(t, \theta) - \sigma(t, \overline{\theta}), \\
\widehat{h} = h(t, \theta) - h(t, \overline{\theta}), \ \widehat{g} = g(x) - g(\overline{x}), \\
\widehat{b}' = b'(t, \theta) - b'(t, \overline{\theta}), \ \widehat{\sigma}' = \sigma'(t, \theta) - \sigma'(t, \overline{\theta}), \\
\widehat{h}' = h'(t, \theta) - h'(t, \overline{\theta}), \ \widehat{g}' = g'(x) - g'(\overline{x}).
\end{cases}
$$

Then one has

$$
(4.12) \qquad |\widehat{g}' - \widehat{g}| = |g'(x) - g'(\overline{x}) - g(x) + g(\overline{x})| \le \|g' - g\|_0 |\widehat{x}|.
$$

Similarly, we have

$$
(4.13) \qquad
\begin{cases}
|\widehat{b}' - \widehat{b}| \le \|b' - b\|_0(t)|\widehat{\theta}|, \\
|\widehat{\sigma}' - \widehat{\sigma}| \le \|\sigma' - \sigma\|_0(t)|\widehat{\theta}|, \\
|\widehat{h}' - \widehat{h}| \le \|h' - h\|_0(t)|\widehat{\theta}|.
\end{cases}
$$

Hence, it follows that

$$
\begin{aligned}
\langle \Phi(T) \begin{pmatrix} \widehat{x} \\ \widehat{g}' \end{pmatrix}, \begin{pmatrix} \widehat{x} \\ \widehat{g}' \end{pmatrix} \rangle &= \langle \Phi(T) \begin{pmatrix} \widehat{x} \\ \widehat{g} \end{pmatrix}, \begin{pmatrix} \widehat{x} \\ \widehat{g} \end{pmatrix} \rangle \\
&+ 2 \langle \Phi(T) \begin{pmatrix} \widehat{x} \\ \widehat{g} \end{pmatrix}, \begin{pmatrix} 0 \\ \widehat{g}' - \widehat{g} \end{pmatrix} \rangle \\
&+ \langle \Phi(T) \begin{pmatrix} 0 \\ \widehat{g}' - \widehat{g} \end{pmatrix}, \begin{pmatrix} 0 \\ \widehat{g}' - \widehat{g} \end{pmatrix} \rangle \\
&\geq \delta |\widehat{x}|^2 + 2 \langle B(T)\widehat{x}, \widehat{g}' - \widehat{g} \rangle + \langle C(T)(\widehat{g}' + \widehat{g}), \widehat{g}' - \widehat{g} \rangle \\
&\geq \left\{ \delta - 2|B(T)|\|g' - g\|_0 - |C(T)|\|g' + g\|_0 \|g' - g\|_0 \right\} |\widehat{x}|^2 \\
&\geq \frac{\delta}{2} |\widehat{x}|^2,
\end{aligned}
$$
(4.14)

provided $\|g' - g\|_0$ is small enough. Similarly, we have the following:

$$
\begin{aligned}
\langle \dot{\Phi}(t) \begin{pmatrix} \widehat{x} \\ \widehat{y} \end{pmatrix}, \begin{pmatrix} \widehat{x} \\ \widehat{y} \end{pmatrix} \rangle &+ 2 \langle \Phi(t) \begin{pmatrix} \widehat{x} \\ \widehat{y} \end{pmatrix}, \begin{pmatrix} \widehat{b}' \\ \widehat{h}' \end{pmatrix} \rangle \\
&+ \langle \Phi(t) \begin{pmatrix} \widehat{\sigma}' \\ \widehat{z} \end{pmatrix}, \begin{pmatrix} \widehat{\sigma}' \\ \widehat{z} \end{pmatrix} \rangle \\
\leq -\delta |\widehat{\theta}|^2 &+ 2 \langle \Phi(t) \begin{pmatrix} \widehat{x} \\ \widehat{y} \end{pmatrix}, \begin{pmatrix} \widehat{b}' - \widehat{b} \\ \widehat{h}' - \widehat{h} \end{pmatrix} \rangle \\
&+ 2 \langle \Phi(t) \begin{pmatrix} \widehat{\sigma} \\ \widehat{z} \end{pmatrix}, \begin{pmatrix} \widehat{\sigma}' - \widehat{\sigma} \\ 0 \end{pmatrix} \rangle \\
&+ \langle \Phi(t) \begin{pmatrix} \widehat{\sigma}' - \widehat{\sigma} \\ 0 \end{pmatrix}, \begin{pmatrix} \widehat{\sigma}' - \widehat{\sigma} \\ 0 \end{pmatrix} \rangle \\
\leq -\delta |\widehat{\theta}|^2 &+ 2 \langle A(t)\widehat{x} + B(t)^T \widehat{y}, \widehat{b}' - \widehat{b} \rangle \\
&+ 2 \langle B(t)\widehat{x} + C(t)\widehat{y}, \widehat{h}' - \widehat{h} \rangle \\
&+ 2 \langle B(t)^T \widehat{z}, \widehat{\sigma}' - \widehat{\sigma} \rangle + \langle A(t)(\widehat{\sigma}' + \widehat{\sigma}), \widehat{\sigma}' - \widehat{\sigma} \rangle \\
\leq \Big\{ -\delta &+ 2\big(|A(t)| + |B(t)|\big)\|b' - b\|_0(t) \\
&+ 2\big(|B(t)| + |C(t)|\big)\|h' - h\|_0(t) \\
&+ 2|B(t)|\|\sigma' - \sigma\|_0(t) + |A(t)|\|\sigma' + \sigma\|_0(t)\|\sigma' - \sigma\|_0(t) \Big\} |\widehat{\theta}|^2.
\end{aligned}
$$
(4.15)

Then, our assertion follows.   □

The above result tells us that if the equation associated with Γ is solvable and Γ admits a strong bridge, then all the equations "nearby" are solvable. This is a kind of stability result.

Remark 4.3. We see from (4.14) and (4.15) that the condition (4.10) can be replaced by

$$(4.16) \quad \begin{cases} 2\big(|B(T)| + |C(T)|\big)\|g' + g\|_0\big)\|g' - g\|_0 < \delta, \\[2mm] \sup_{t \in [0,T]} \Big\{ 2\big(|A(t)| + |B(t)|\big)\|b' - b\|_0(t) \\[2mm] \qquad\qquad + 2\big(|B(t)| + |C(t)|\big)\|h' - h\|_0(t) \\[2mm] \qquad\qquad + \big[2|B(t)| + |A(t)|\big]\|\sigma' + \sigma\|_0(t)\big]\|\sigma' - \sigma\|_0(t) \Big\} < \delta, \end{cases}$$

where $\delta > 0$ is the one appeared in the definition of the bridge (see Definition 1.3). Actually, (4.16) can further be replaced by the following even weaker conditions:

$$(4.17) \quad \begin{cases} 2\,\langle\, B(T)\widehat{x}, \widehat{g}' - \widehat{g}\,\rangle + \langle\, C(T)(\widehat{g}' + \widehat{g}), \widehat{g}' - \widehat{g}\,\rangle > -\delta|\widehat{x}|^2, \\[2mm] \qquad\qquad\qquad \forall x, \overline{x} \in \mathbb{R}^n, \\[2mm] \sup_{t \in [0,T]} \Big\{ 2\,\langle\, A(t)\widehat{x} + B(t)^T\widehat{y}, \widehat{b}' - \widehat{b}\,\rangle \\[2mm] \quad + 2\,\langle\, B(t)\widehat{x} + C(t)\widehat{y}, \widehat{h}' - \widehat{h}\,\rangle + 2\,\langle\, B(t)^T\widehat{z}, \widehat{\sigma}' - \widehat{\sigma}\,\rangle \\[2mm] \quad + \langle\, A(t)(\widehat{\sigma}' + \widehat{\sigma}), \widehat{\sigma}' - \widehat{\sigma}\,\rangle \Big\} < \delta|\widehat{\theta}|^2, \quad \forall \theta, \overline{\theta} \in M. \end{cases}$$

The above means that if the perturbation is made not necessarily small but in the right direction, the solvability will be kept. This observation will be useful later.

To conclude this section, we present the following simple proposition.

Proposition 4.4.   Let $T > 0$, $\Gamma \equiv (b, \sigma, h, g) \in H[0,T]$ and $\Phi \in \mathcal{B}_I(\Gamma; [0,T])$. Let $\beta \in \mathbb{R}$ and

$$(4.18) \quad \begin{cases} \widetilde{\Phi}(t) = e^{2\beta t}\Phi(t), \qquad t \in [0,T], \\[2mm] \widetilde{\Gamma} = (b - \beta x, \sigma, h - \beta y, g) \in H[0,T]. \end{cases}$$

Then, $\widetilde{\Phi} \in \mathcal{B}_I(\widetilde{\Gamma}; [0,T])$.

The proof is immediate. Clearly, the similar conclusion holds if we replace $\mathcal{B}_I(\Gamma; [0,T])$ by $\mathcal{B}_{II}(\Gamma; [0,T])$, $\mathcal{B}(\Gamma; [0,T])$ or $\mathcal{B}^s(\Gamma; [0,T])$.

§5. Construction of Bridges

In this section, we are going to present some more results on the solvability of FBSDEs by constructing certain bridges.

§5.1. A general consideration

Let us start with the following linear FBSDE:

$$(5.1) \quad \begin{cases} d\begin{pmatrix} X(t) \\ Y(t) \end{pmatrix} = \Big\{ A\begin{pmatrix} X(t) \\ Y(t) \end{pmatrix} + \begin{pmatrix} b_0(t) \\ h_0(t) \end{pmatrix} \Big\}dt + \begin{pmatrix} \sigma_0(t) \\ Z(t) \end{pmatrix}dW(t), \\[3mm] \qquad\qquad\qquad\qquad\qquad t \in [0,T], \\[3mm] X(0) = x, \qquad Y(T) = GX(T) + g_0, \end{cases}$$

where $\mathcal{A} \in \mathbb{R}^{(n+m)\times(n+m)}$, $G \in \mathbb{R}^{m\times n}$, $\gamma \equiv (b_0, \sigma_0, h_0, g_0) \in \mathcal{H}[0, T]$ (see (1.3)) and $x \in \mathbb{R}^n$. We have the following result.

Lemma 5.1. *Let $T > 0$. Then, the two-point boundary value problem (5.1) is uniquely solvable for all $\gamma \in \mathcal{H}[0, T]$ if and only if*

$$(5.2) \qquad \det\left\{ (-G, I)e^{\mathcal{A}t} \begin{pmatrix} 0 \\ I \end{pmatrix} \right\} > 0, \quad t \in [0, T].$$

Proof. Let

$$(5.3) \qquad \begin{pmatrix} \xi(t) \\ \eta(t) \end{pmatrix} = \begin{pmatrix} I & 0 \\ -G & I \end{pmatrix} \begin{pmatrix} X(t) \\ Y(t) \end{pmatrix}.$$

Then we have the linear FBSDE for (ξ, η) as follows:

$$(5.3) \qquad \begin{cases} d\begin{pmatrix} \xi(t) \\ \eta(t) \end{pmatrix} = \left\{ \begin{pmatrix} I & 0 \\ -G & I \end{pmatrix} \mathcal{A} \begin{pmatrix} I & 0 \\ G & I \end{pmatrix} \begin{pmatrix} \xi(t) \\ \eta(t) \end{pmatrix} \right. \\ \qquad\qquad + \begin{pmatrix} b_0(t) \\ h_0(t)Gb_0(t) \end{pmatrix} \Big\} dt \\ \qquad\qquad + \begin{pmatrix} \sigma_0(t) \\ Z(t) - G\sigma_0(t) \end{pmatrix} dW(t), \quad t \in [0, T], \\ \xi(0) = x, \qquad \eta(T) = g_0, \end{cases}$$

Clearly, the solvability of (5.3) is equivalent to that of (5.1). By Theorem 3.7 of Chapter 2, we obtain that (5.3) is solvable for all $\gamma \in \mathcal{H}[0, T]$ if and only if (3.16) and (3.19) of Chapter 2 hold. In the present case, these two conditions are the same as (5.2). This proves the result. □

Now, let us relate the above result to the notion of bridge. From Theorem 2.1, we know that if Γ_1 and Γ_2 are linked by a bridge, then Γ_1 and Γ_2 have the same solvability. On the other hand, for any given Γ, Corollary 2.4 tells us that if Γ admits a bridge, then, the FBSDE associated with Γ admits at most one adapted solution. The existence, however, is not claimed. The following result tells us something concerning the existence. This result will be useful below.

Proposition 5.2. *Let $T_0 > 0$ and $\Gamma = (b, 0, h, g)$ with*

$$(5.4) \qquad \begin{pmatrix} b(t, \theta) \\ h(t, \theta) \end{pmatrix} = \mathcal{A} \begin{pmatrix} x \\ y \end{pmatrix}, \qquad g(x) = Gx, \qquad \forall(t, \theta) \in [0, T_0] \times M.$$

Then $\Gamma \in \mathcal{S}[0, T]$ for all $T \in (0, T_0]$ if $\mathcal{B}(\Gamma; [0, T]) \neq \phi$ for all $T \in (0, T_0]$.

Proof. Since $\mathcal{B}(\Gamma; [0, T]) \neq \phi$, by Corollary 2.4, (5.1) admits at most one solution. By taking $\gamma \equiv (b_0, \sigma_0, h_0, g_0) = 0$ and $x = 0$, we see that the resulting homogeneous equation only admits the zero solution. This is equivalent to that (5.1) with the nonhomogeneous terms being zero only admits the zero solution. On the other hand, in this case, the solution of (5.1) is given by

$$(5.5) \qquad \begin{pmatrix} X(t) \\ Y(t) \end{pmatrix} = e^{\mathcal{A}t} \begin{pmatrix} x \\ Y(0) \end{pmatrix}, \qquad t \in [0, T],$$

with the condition

$$(5.6) \qquad 0 = (-G, I) \begin{pmatrix} X(T) \\ Y(T) \end{pmatrix} = (-G, I) e^{\mathcal{A}T} \begin{pmatrix} 0 \\ I \end{pmatrix} Y(0).$$

We require that (5.6) leads to $Y(0) = 0$. Thus, it is necessary that the left hand side of (5.2) is non-zero for $t = T$. Since $T \in (0, T_0]$ is arbitrary, we must have (5.2). Then, by Lemma 5.1, we have $\Gamma \in \mathcal{S}[0, T]$. □

Let us now look at some class of nonlinear FBSDEs. Recall the semi-norms $\|\cdot\|_0(t)$ defined by (4.8).

Theorem 5.3. *Let $T_0 > 0$, $\mathcal{A} \in \mathbb{R}^{(n+m) \times (n+m)}$ and $\Gamma \equiv (b, 0, h, g)$ be defined by (5.4). Suppose (5.2) holds for $T = T_0$ and that $\mathcal{B}^s(\Gamma; [0, T]) \neq \phi$ for all $T \in (0, T_0]$. Then for any $T \in (0, T]$, there exists an $\varepsilon > 0$, such that for all $\beta \in \mathbb{R}$ and $\overline{\Gamma} \equiv (\overline{b}, \overline{\sigma}, \overline{h}, \overline{g}) \in H[0, T]$ with*

$$(5.7) \qquad \|\overline{\Gamma}\|_0(t) < \varepsilon, \qquad t \in [0, T],$$

the following FBSDE:

$$(5.8) \qquad \begin{cases} d \begin{pmatrix} X(t) \\ Y(t) \end{pmatrix} = \left\{ (\mathcal{A} + \beta I) \begin{pmatrix} X(t) \\ Y(t) \end{pmatrix} + \begin{pmatrix} \overline{b}(t, \Theta(t)) \\ \overline{h}(t, \Theta(t)) \end{pmatrix} \right\} dt \\ \qquad\qquad + \begin{pmatrix} \overline{\sigma}(t, \Theta(t)) \\ Z(t) \end{pmatrix} dW(t), \qquad t \in [0, T], \\ X(0) = x, \qquad Y(T) = GX(T) + \overline{g}(X(T)), \end{cases}$$

admits a unique adapted solution $\Theta \equiv (X, Y, Z) \in \mathcal{M}[0, T]$.

Proof. We note that if

$$(5.9) \qquad \widetilde{b}(t, \theta) = e^{\beta t} \overline{b}(t, e^{-\beta t} \theta), \qquad \forall (t, \theta) \in [0, T] \times M,$$

then,

$$(5.10) \qquad \|\widetilde{b}\|_0(t) = \|\overline{b}\|_0(t), \qquad \forall t \in [0, T].$$

Similar conclusion holds for $\overline{\sigma}$, $\overline{h}$ and $\overline{g}$ if we define $\widetilde{\sigma}$, $\widetilde{h}$ and $\widetilde{g}$ similar to (5.9). On the other hand, if $\Theta(t) \equiv (X(t), Y(t), Z(t))$ is an adapted solution of (5.8) with $\beta = 0$, then $\widetilde{\Theta}(t) \overset{\Delta}{=} e^{\beta t} \Theta(t)$ is an adapted solution of (5.8). Thus, we need only consider the case $\beta = 0$ in (5.8). Then, by Theorems 2.1, 4.2 and Proposition 5.2, we obtain our conclusion immediately. □

We note that FBSDEs (5.8) is nonlinear and the Lipschitz constants of the coefficients could be large. Also, (5.8) is not necessarily decoupled nor with monotonicity conditions. Thus, Theorem 5.3 gives the unique solvability of a (new) class of nonlinear FBSDEs, which is not covered by the classes discussed before. On the other hand, by Remark 4.3, we see that condition (5.7) can be replaced by something like (4.16), or even (4.17). This further enlarges the class of FBSDEs covered by (5.8).

We note that the key assumption of Theorem 5.3 is that $\Gamma \equiv (b, 0, h, g)$ given by (5.4) admits a strong bridge. Thus, the major problem left is whether we can construct a (strong) bridge for Γ. In the rest of this section, we will concentrate on this issue.

We now consider the construction of the strong bridges for $\Gamma = (b, 0, h, g)$ given by (5.4). From the definition of strong bridge, we can check that $\Phi \in \mathcal{B}^s(\Gamma; [0, T])$ if it is the solution to the following differential equation for some constants $K, \overline{K}, \delta, \varepsilon > 0$,

(5.11)
$$
\begin{cases}
\dot{\Phi}(t) + \mathcal{A}^T \Phi(t) + \Phi(t)\mathcal{A} = -\delta I, & t \in [0, T], \\
\Phi(0) = \begin{pmatrix} K & 0 \\ 0 & -\overline{K} \end{pmatrix},
\end{cases}
$$

satisfying the following additional conditions:

(5.12)
$$
\begin{cases}
(I, 0)\Phi(t)\begin{pmatrix} I \\ 0 \end{pmatrix} \geq 0, \quad (0, I)\Phi(t)\begin{pmatrix} 0 \\ I \end{pmatrix} \leq -\varepsilon I, \quad \forall t \in [0, T], \\
(I, G^T)\Phi(T)\begin{pmatrix} I \\ G \end{pmatrix} \geq \varepsilon I.
\end{cases}
$$

On the other hand, we find that the solution to (5.11) is given by

(5.13)
$$
\Phi(t) = e^{-\mathcal{A}^T t}\begin{pmatrix} K & 0 \\ 0 & -\overline{K} \end{pmatrix} e^{-\mathcal{A}t} - \delta \int_0^t e^{-\mathcal{A}^T s} e^{-\mathcal{A}s} ds,
$$
$$
t \in [0, T].
$$

Thus, in principle, if we can find constants $K, \overline{K}, \delta, \varepsilon > 0$, such that (5.12) holds with $\Phi(t)$ given by (5.13), then we obtain a strong bridge $\Phi(\cdot)$ for Γ and Theorem 5.3 applies.

§5.2. A one dimensional case

In this subsection, we are going to carry out a detailed construction of strong bridges for a case of $n = m = d = 1$ based on the general consideration of the previous subsection. The corresponding class of solvable FBSDEs will also be determined.

Let $\Gamma = (b, 0, h, g)$ be given by

(5.14)
$$
\begin{cases}
\begin{pmatrix} b(t, x, y, z) \\ h(t, x, y, z) \end{pmatrix} = \mathcal{A}\begin{pmatrix} x \\ y \end{pmatrix} \equiv \begin{pmatrix} -\lambda & \mu \\ 0 & 0 \end{pmatrix}\begin{pmatrix} x \\ y \end{pmatrix}, \\
g(x) = -gx,
\end{cases}
$$

for all $(t, x, y, z) \in [0, \infty) \times \mathbb{R}^3$, with $\lambda, \mu, g \in \mathbb{R}$ being constants satisfying the following:

$$(5.15) \qquad \lambda, \mu, g > 0, \qquad \frac{1}{2} + \frac{3g\mu}{2\lambda} - g^2 \geq 0.$$

We point out that conditions (5.15) for the constants λ, μ, g are not necessarily the best. We prefer not to get into the most generality to avoid some complicated computation. Let us now carry out some calculations. First of all

$$(5.16) \qquad e^{At} = \begin{pmatrix} e^{-\lambda t} & \frac{\mu}{\lambda}(1 - e^{-\lambda t}) \\ 0 & 1 \end{pmatrix}, \qquad \forall t \geq 0.$$

Thus, for all $t \geq 0$,

$$(5.17) \qquad \begin{aligned} & e^{-A^T t} \begin{pmatrix} K & 0 \\ 0 & -\overline{K} \end{pmatrix} e^{-At} \\ &= \begin{pmatrix} e^{\lambda t} & 0 \\ \frac{\mu}{\lambda}(1 - e^{\lambda t}) & 1 \end{pmatrix} \begin{pmatrix} K & 0 \\ 0 & -\overline{K} \end{pmatrix} \begin{pmatrix} e^{\lambda t} & \frac{\mu}{\lambda}(1 - e^{\lambda t}) \\ 0 & 1 \end{pmatrix} \\ &= \begin{pmatrix} Ke^{2\lambda t} & \frac{K\mu}{\lambda}(e^{\lambda t} - e^{2\lambda t}) \\ \frac{K\mu}{\lambda}(e^{\lambda t} - e^{2\lambda t}) & -\overline{K} + \frac{K\mu^2}{\lambda^2}(1 - e^{\lambda t})^2 \end{pmatrix}, \end{aligned}$$

and

$$(5.18) \qquad \begin{aligned} \int_0^t e^{-A^T s} e^{-As} ds &= \int_0^t \begin{pmatrix} e^{2\lambda s} & \frac{\mu}{\lambda}(e^{\lambda s} - e^{2\lambda s}) \\ \frac{\mu}{\lambda}(e^{\lambda s} - e^{2\lambda s}) & 1 + \frac{\mu^2}{\lambda^2}(1 - e^{\lambda s})^2 \end{pmatrix} ds \\ &= \begin{pmatrix} \frac{1}{2\lambda}(e^{2\lambda t} - 1) & -\frac{\mu}{2\lambda^2}(e^{\lambda t} - 1)^2 \\ -\frac{\mu}{2\lambda^2}(e^{\lambda t} - 1)^2 & \frac{\mu^2}{2\lambda^3}(e^{\lambda t} - 2)^2 + \frac{\lambda^2 + \mu^2}{\lambda^2}t - \frac{\mu^2}{2\lambda^3} \end{pmatrix}. \end{aligned}$$

We let $\overline{K} > 0$ be undetermined and choose

$$(5.19) \qquad K = \frac{3}{4\lambda}, \qquad \delta = 1.$$

Then, according to (5.13), we define

$$\Phi(t) = \begin{pmatrix} A(t) & B(t) \\ B(t) & C(t) \end{pmatrix}$$

with

$$
\begin{cases}
\begin{aligned}
A(t) &= Ke^{2\lambda t} - \frac{\delta}{2\lambda}(e^{2\lambda t} - 1) = \left(K - \frac{\delta}{2\lambda}\right)e^{2\lambda t} + \frac{\delta}{2\lambda} \\
&= \frac{1}{4\lambda}(e^{2\lambda t} + 2), \\
B(t) &= \frac{K\mu}{\lambda}(e^{\lambda t} - e^{2\lambda t}) + \frac{\delta\mu}{2\lambda^2}(e^{\lambda t} - 1)^2 \\
&= -\frac{\mu}{\lambda}\left(K - \frac{\delta}{2\lambda}\right)e^{2\lambda t} + \frac{\mu}{\lambda}\left(K - \frac{\delta}{\lambda}\right)e^{\lambda t} + \frac{\delta\mu}{2\lambda^2} \\
&= -\frac{\mu}{4\lambda^2}(e^{2\lambda t} + e^{\lambda t} - 2), \\
C(t) &= -\overline{K} + \frac{K\mu^2}{\lambda^2}(e^{\lambda t} - 1)^2 - \frac{\delta\mu^2}{2\lambda^3}(e^{\lambda t} - 2)^2 \\
&\quad - \frac{\delta(\lambda^2 + \mu^2)}{\lambda^2}t + \frac{\delta\mu^2}{2\lambda^3} \\
&= \frac{\mu^2}{\lambda^2}\left(K - \frac{\delta}{2\lambda}\right)e^{2\lambda t} - \frac{2\mu^2}{\lambda^2}\left(K - \frac{\delta}{\lambda}\right)e^{\lambda t} - \frac{\mu^2}{\lambda^2}\left(K - \frac{3\delta}{2\lambda}\right) \\
&\quad - \overline{K} - \frac{\delta(\lambda^2 + \mu^2)}{\lambda^2}t \\
&= \frac{\mu^2}{4\lambda^3}(e^{2\lambda t} + 2e^{\lambda t} + 3) - \overline{K} - \frac{\lambda^2 + \mu^2}{\lambda^2}t.
\end{aligned}
\end{cases}
\tag{5.20}
$$

From (5.12), we need the following: ($\varepsilon > 0$ is undetermined)

$$
\begin{cases}
A(t) \geq 0, \quad C(t) \leq -\varepsilon, \qquad \forall t \in [0, T], \\
A(T) - 2gB(T) + g^2 C(T) \geq \varepsilon.
\end{cases}
\tag{5.21}
$$

Let us now look at these requirements separately.

First of all, it is clear true that $A(t) \geq 0$ for all $t \in [0, T]$. Next, $C(t) \leq -\varepsilon$ for all $t \in [0, T]$, if and only if

$$
\overline{K} \geq \varepsilon + \frac{\mu^2}{4\lambda^3}(e^{2\lambda t} + 2e^{\lambda t} + 3) - \frac{\lambda^2 + \mu^2}{\lambda^2}t \stackrel{\Delta}{=} f(t), \qquad t \in [0, T].
\tag{5.22}
$$

Since $f''(t) \geq 0$ for all $t \in [0, \infty)$, the function $f(t)$ is convex. Thus, (5.22) holds if and only if

$$
\overline{K} \geq f(0) \vee f(T).
\tag{5.23}
$$

Finally, we need

$$\varepsilon \le A(T) - 2gB(T) + g^2 C(T)$$

$$= \frac{1}{4\lambda}(e^{2\lambda T} + 2) + \frac{g\mu}{2\lambda^2}(e^{2\lambda T} + e^{\lambda T} - 2)$$

(5.24)

$$+ \frac{g^2\mu^2}{4\lambda^3}(e^{2\lambda T} + 2e^{\lambda T} + 3) - g^2\overline{K} - \frac{g^2(\lambda^2 + \mu^2)}{\lambda^2}T$$

$$= \frac{1}{4\lambda}\left(1 + \frac{g\mu}{\lambda}\right)^2 e^{2\lambda T} + \frac{g\mu}{2\lambda^2}\left(1 + \frac{g\mu}{\lambda}\right)e^{\lambda T} - \frac{g^2(\lambda^2 + \mu^2)}{\lambda^2}T$$

$$+ \frac{1}{2\lambda} - \frac{g\mu}{\lambda^2} + \frac{3g^2\mu^2}{4\lambda^3} - g^2\overline{K}.$$

Thus, we need (note (5.23))

$$F(T) \overset{\Delta}{=} \frac{1}{4\lambda}\left(1 + \frac{g\mu}{\lambda}\right)^2 e^{2\lambda T} + \frac{g\mu}{2\lambda^2}\left(1 + \frac{g\mu}{\lambda}\right)e^{\lambda T} - \frac{g^2(\lambda^2 + \mu^2)}{\lambda^2}T$$

(5.25)

$$+ \frac{1}{2\lambda} - \frac{g\mu}{\lambda^2} + \frac{3g^2\mu^2}{4\lambda^3} - \varepsilon$$

$$\ge g^2\overline{K} \ge g^2(f(0) \vee f(T)).$$

We now separate two cases (with $f(T)$ and $f(0)$, respectively). First of all, for $f(T)$, we want

(5.26)

$$0 \le F(T) - g^2 f(T)$$

$$= \frac{1}{4\lambda}\left(1 + \frac{2g\mu}{\lambda}\right)e^{2\lambda T} + \frac{g\mu}{2\lambda^2}e^{\lambda T} + \frac{1}{2\lambda} - \frac{g\mu}{\lambda^2} - \varepsilon(1 + g^2) \overset{\Delta}{=} \widehat{F}(T).$$

We see that $T \mapsto \widehat{F}(T)$ is monotone increasing. Thus, to have the above, it suffices to have

(5.27)

$$0 \le \widehat{F}(0) = \frac{3}{4\lambda} - \varepsilon(1 + g^2).$$

Hence, in what follows, we take

(5.28)

$$\varepsilon = \frac{3}{4\lambda(1 + g^2)}.$$

Then, (5.26) holds. Next, we claim that under (5.15) and (5.28), the following holds.

(5.29)

$$F(T) - g^2 f(0) \ge 0.$$

In fact, by the choice of ε and by (5.27),

(5.30)

$$F(0) - g^2 f(0) = \widehat{F}(0) = 0.$$

On the other hand,

(5.31) $$F'(T) = \frac{1}{2}\left(1 + \frac{g\mu}{\lambda}\right)^2 e^{2\lambda T} + \frac{g\mu}{2\lambda}\left(1 + \frac{g\mu}{\lambda}\right)e^{\lambda T} - \frac{g^2(\lambda^2 + \mu^2)}{\lambda^2}.$$

Thus, by (5.15), it follows that

$$(5.32) \qquad F'(0) = \left(\frac{1}{2} + \frac{3g\mu}{2\lambda} - g^2\right) \geq 0.$$

Then, by $F''(T) \geq 0$, together with (5.30) and (5.32), we must have (5.29). Hence, we obtain (5.25). This shows that a strong bridge $\Phi(t)$ has been constructed with K, δ and ε being given by (5.19) and (5.28), respectively, and we may take

$$(5.33) \qquad \overline{K} = f(0) \vee f(T).$$

It is interesting that the $\Phi(\cdot)$ constructed in the above is not in $\mathcal{B}(\Gamma_0; [0, T])$ for any $T > 0$ since $\dot{A}(t) > 0$. On the other hand, we note that both $A(t)$ and $B(t)$ are independent of T. However, due to the fact that $\overline{K}$ depending on T, $C(t)$ depends on T. But, we claim that there exists a constant $c_0 > 0$, only depending on λ, μ, g (independent of T), such that

$$(5.34) \qquad \begin{cases} -c_0 - f(T) \leq C(t) \leq -\dfrac{3}{4\lambda(1 + g^2)}, & t \in [0, T], \\ -c_0 \leq C(T) \leq -\dfrac{3}{4\lambda(1 + g^2)}, \end{cases}$$

where $f(t)$ is defined by (5.22). In fact, by (5.20), (5.22), (5.28) and (5.33), we have

$$(5.35) \qquad C(t) = f(t) - f(0) \vee f(T) - \frac{3}{4\lambda(1 + g^2)}.$$

Clearly, $C(t)$ is convex. Thus,

$$(5.36) \qquad C(t) \leq C(0) \vee C(T) = -\frac{3}{4\lambda(1 + g^2)}, \qquad \forall t \in [0, T].$$

On the other hand, by the fact that $f(t)$ is strictly convex and $\lim_{t \to \infty} f(t) = \infty$, we see that there exists a unique $T_0 > 0$, only depending on λ and μ, such that

$$(5.37) \qquad C(t) \geq f(T_0) - f(0) \vee f(T) - \frac{3}{4\lambda(1 + g^2)}, \qquad t \in [0, T].$$

This proves the first relation in (5.34). Next, we see easily that there exists a unique $T_1 > T_0$, such that $f(T_1) = f(0)$, and

$$(5.38) \qquad \begin{cases} f(t) \leq f(0), & \forall t \in [0, T_1], \\ f(t) > f(0), & \forall t \in (T_1, \infty). \end{cases}$$

Hence, we obtain

$$(5.39) \qquad C(T) \geq f(T_0) - f(0) - \frac{3}{4\lambda(1 + g^2)}.$$

This proves the second relation in (5.34).

Now, from Remark 4.3 and Theorem 5.3, we know that the following FBSDEs is solvable on $[0, T]$.

$$
(5.40)\quad
\begin{cases}
dX(t) = \{(\beta - \lambda)X(t) + \mu Y(t) + \bar{b}(t, X(t), Y(t), Z(t))\}dt \\
\qquad\qquad + \bar{\sigma}(t, X(t), Y(t), Z(t))dW(t), \\
dY(t) = \{\beta Y(t) + \bar{h}(t, X(t), Y(t), Z(t))\}dt + Z(t)dW(t), \\
X(0) = x, \qquad Y(T) = -gX(T) + \bar{g}(X(T)),
\end{cases}
$$

where $\lambda, \mu, g > 0$ satisfying (5.15), $\beta \in \mathbb{R}$, and $\bar{\Gamma} \equiv (\bar{b}, \bar{\sigma}, \bar{h}, \bar{g}) \in H[0, T]$ satisfying

$$
(5.41)\quad
\begin{cases}
2|B(T)|\|\bar{g}\|_0 + |C(T)|\|\bar{g}\|_0^2 < \varepsilon \wedge 1, \\
\sup_{t\in[0,T]} \Big\{2(|A(t)| + |B(t)|)\|\bar{b}\|_0(t) + 2(|B(t)| + |C(t)|)\|\bar{h}\|_0(t) \\
\qquad\qquad + 2|B(t)|\|\bar{\sigma}\|_0(t) + |A(t)|\|\bar{\sigma}\|_0(t)^2\Big\} < \varepsilon \wedge 1,
\end{cases}
$$

with $A(\cdot)$, $B(\cdot)$ and $C(\cdot)$ given by (5.20) and $\varepsilon > 0$ given by (5.28). If we use (4.17), then, (5.41) can be relaxed to the following:

$$
(5.42)\quad
\begin{cases}
2B(T)\widehat{x\bar{g}} + C(T)(\widehat{\bar{g}} - 2g\widehat{x})\widehat{\bar{g}} > -(\varepsilon \wedge 1)|\widehat{x}|^2, \qquad \forall x, \bar{x} \in \mathbb{R}, \\
\sup_{t\in[0,T]} \Big\{2(A(t)\widehat{x} + B(t)^T\widehat{y})\widehat{\bar{b}} + 2(B(t)\widehat{x} + C(t)\widehat{y})\widehat{\bar{h}} \\
\qquad\qquad + 2B(t)\widehat{z\bar{\sigma}} + A(t)\widehat{\bar{\sigma}}^2\Big\} < (\varepsilon \wedge 1)|\widehat{\theta}|^2, \quad \forall\theta, \bar{\theta} \in M.
\end{cases}
$$

If $\bar{b}$, $\bar{\sigma}$, $\bar{h}$ and $\bar{g}$ are differentiable, then, we see that (5.42) is equivalent to the following:

$$
(5.43)\quad
\begin{cases}
2B(T)\bar{g}_x(x) + C(T)(\bar{g}_x(x) - 2g)\bar{g}_x(x) > -(\varepsilon \wedge 1), \qquad \forall x \in \mathbb{R}, \\
\begin{pmatrix} A(t) & B(t) & 0 \\ B(t) & C(t) & 0 \\ 0 & 0 & B(t) \end{pmatrix} (\nabla\bar{b}(t, \theta), \nabla\bar{h}(t, \theta), \nabla\bar{\sigma}(t, \theta)) \\
\quad + \Big\{\begin{pmatrix} A(t) & B(t) & 0 \\ B(t) & C(t) & 0 \\ 0 & 0 & B(t) \end{pmatrix} (\nabla\bar{b}(t, \theta), \nabla\bar{h}(t, \theta), \nabla\bar{\sigma}(t, \theta))\Big\}^T \\
\quad + A(t)\nabla\bar{\sigma}(t, \theta)\{\nabla\bar{\sigma}(t, \theta)\}^T < \varepsilon \wedge 1, \quad \forall(t, \theta) \in [0, T] \times M,
\end{cases}
$$

where $\nabla\bar{b}(t, \theta) = (\bar{b}_x(t, \theta), \bar{b}_y(t, \theta), \bar{b}_z(t, \theta))^T$, and so on. Some direct computation shows that the first relation in (5.43) is equivalent to the following:

$$
(5.44)\quad
\begin{aligned}
-r(T) &\stackrel{\Delta}{=} -\sqrt{\frac{\varepsilon \wedge 1}{|C(T)|} + \Big(\frac{B(T)}{C(T)} - g\Big)^2} - \frac{B(T)}{C(T)} + g \leq \bar{g}_x(x) \\
&< \sqrt{\frac{\varepsilon \wedge 1}{|C(T)|} + \Big(\frac{B(T)}{C(T)} - g\Big)^2} - \frac{B(T)}{C(T)} + g, \qquad \forall x \in \mathbb{R}.
\end{aligned}
$$

By (5.34), we know that $C(T)$ is bounded uniformly in T, while, $B(T) \to -\infty$ as $T \to \infty$ (see (5.20)). Thus, by some calculation, we see that

$$(5.45) \qquad -\sqrt{\frac{\varepsilon \wedge 1}{|C(T)|}} \geq -r(T) \downarrow -\infty, \qquad \text{as } T \to \infty,$$

and $\bar{g}$ need only to satisfy the following:

$$(5.46) \qquad -r(T) \leq \bar{g}_x(x) \leq 0, \qquad \forall t \in \mathbb{R}.$$

Clearly, the larger the T, the weaker the restriction of (5.46). The second condition in (5.43) is also checkable (although it is a little more complicated than the first one). It is not hard to see that the choice of functions $\bar{b}$ and $\bar{\sigma}$ are independent of T as $A(t)$ and $B(t)$ do not depend on T. However, since $C(t)$ depends on T, by some direct calculation, we see that in order FBSDE (5.40) is solvable for all $T > 0$, we have to restrict ourselves to the case that $\bar{h}(t, \theta) = \bar{h}(t, y)$. Clearly, even with such a restriction, (5.40) is still a very big class of FBSDEs, which are not necessarily decoupled, nor monotone. Also, $\bar{\sigma}$ is allowed to be degenerate. We omit the exact statement of the explicit conditions on $\bar{b}, \bar{\sigma}$ and $\bar{h}$ under which (5.40) is solvable to avoid some lengthy computation. Instead, to conclude our discussion, let us finally look at the following FBSDE:

$$(5.47) \qquad \begin{cases} dX(t) = \{(\beta - \lambda)X(t) + \mu Y(t) + \bar{b}(t, X(t), Y(t), Z(t))\}dt \\ \qquad\qquad + \bar{\sigma}(t, X(t), Y(t), Z(t))dW(t), \\ dY(t) = \{\beta Y(t) + h_0(t)\}dt + Z(t)dW(t), \\ X(0) = x, \qquad Y(T) = -gX(T) + g_0, \end{cases}$$

with $\lambda, \mu, g > 0$ satisfying (5.15) and

$$(5.48) \qquad \sup_{t \in [0,\infty)} \Big\{ 2\big(|A(t)| + |B(t)|\big)\|\bar{b}\|_0(t) + 2|B(t)| \,\|\bar{\sigma}\|_0(t)$$
$$+ |A(t)| \,\|\bar{\sigma}\|_0(t)^2 \Big\} < \varepsilon \wedge 1.$$

This is a special case of (5.40) in which $\bar{h} \equiv h_0$ and $\bar{g} \equiv g_0$. Then, by the above analysis, we know that (5.47) is uniquely solvable over any finite time duration $[0, T]$. Condition (5.48) can be carried out explicitly as follows:

$$\{2(e^{2\lambda t} + 2) + \frac{2\mu^2}{\lambda}(e^{2\lambda t} + e^{\lambda t} - 2)\}\|\bar{b}\|_0(t)$$

$$(5.49) \qquad + \frac{2\mu^2}{\lambda}(2^{2\lambda t} + e^{\lambda t} - 2)\|\bar{\sigma}\|_0(t)$$

$$+ (e^{2\lambda t} + 2)\|\bar{\sigma}\|_0(t)^2 < \min\{4\lambda, \frac{3}{1+g^2}\}, \qquad t \in [0, \infty).$$

It is clear that although (5.47) is a special case of (5.40), it is still very general and in particular, it is not necessarily decoupled nor monotone. Also, if we regard (5.47) as a nonlinear perturbation of (5.1) (with $m =$

$n = d = 1$ and (5.14) holds), then the perturbation is not necessarily small (for t not large).

Chapter 7

Forward-Backward SDEs with Reflections

In this chapter we study FBSDEs with boundary conditions. In the simplest case when the FBSDE is decoupled, it is reduced to a combination of a well-understood (forward) reflected diffusion and a newly developed reflected backward SDE. However, the extension of such FBSDEs to the general coupled case is quite delicate. In fact, none of the methods that we have seen in the previous chapters seems to be applicable, due to the presence of the reflecting process. Therefore, the route we take in this chapter to reach the existence and uniqueness of the adapted solution is slightly different from those we have seen before.

§1. Forward SDEs with Reflections

Let $\mathcal{O}$ be a closed convex domain in $\mathbb{R}^n$. Define for any $x \in \partial\mathcal{O}$ the set of inward normals to $\mathcal{O}$ at x by

(1.1) $$\mathcal{N}_x = \{\gamma : |\gamma| = 1, \text{ and } \langle \gamma, x - y \rangle \leq 0, \ \forall y \in \mathcal{O}\}.$$

It is clear that if the boundary $\partial\mathcal{O}$ is smooth (say, C^1), then for any $x \in \partial\mathcal{O}$, the set $\mathcal{N}_x$ contains only one vector, that is, the unit inner normal vector at x. We denote $BV([0,T]; \mathbb{R}^n)$ to be the set of all $\mathbb{R}^n$-valued functions of bounded variation; and for $\eta \in BV([0,T]; \mathbb{R}^n)$, we denote $|\eta|(T)$ to be the total variation of η on $[0,T]$.

A general form of (forward) SDEs with reflection (FSDER, for short) is the following:

(1.2) $$X(t) = x + \int_0^t b(s, X(s))ds + \int_0^t \sigma(s, X(s))dW(s) + \eta(t).$$

Here the b and σ are functions of $(t, x, \omega) \in [0,T] \times \mathbb{R}^n \times \Omega$ (with ω being suppressed, as usual); and $\eta \in BV_{\mathcal{F}}([0,T]; \mathbb{R}^m)$, the set of all $\{\mathcal{F}_t\}_{t\geq0}$-adapted processes η with paths in $BV([0,T]; \mathbb{R}^m)$.

Definition 1.1. A pair of continuous, $\{\mathcal{F}_t\}_{t\geq0}$-adapted processes $(X, \eta) \in L^2_{\mathcal{F}}([0,T]; \mathbb{R}^n) \times BV_{\mathcal{F}}([0,T]; \mathbb{R}^n)$ is called a solution to the FSDER (1.2) if

1) $X(t) \in \mathcal{O}$, $\forall t \in [0,T]$, a.s.;

2) $\eta(t) = \int_0^t 1_{\{X(s)\in\partial\mathcal{O}\}}\gamma(s)d|\eta|(s)$, where $\gamma(s) \in \mathcal{N}_{X(s)}$, $0 \leq s \leq t \leq T$, $d|\eta|$-a.e.;

3) equation (1.2) is satisfied almost surely.

A widely used tool for solving an FSDER is the following (deterministic) function-theoretic technique known as the *Skorohod Problem*: Let the domain $\mathcal{O}$ be given,

Problem $SP(\cdot\,;\mathcal{O})$: Let $\psi \in C([0,T];\mathbb{R}^n)$ with $\psi(0) \in \mathcal{O}$ be given. Find a pair $(\varphi,\eta) \in C([0,T];\mathbb{R}^n) \times BV([0,T];\mathbb{R}^n)$ such that

1) $\varphi(t) = \psi(t) + \eta(t)$, $\forall t \in [0,T]$, and $\varphi(0) = \psi(0)$;

2) $\varphi(t) \in \mathcal{O}$, for $t \in [0,T]$;

3) $|\eta|(t) = \int_0^t 1_{\{\varphi(s) \in \partial\mathcal{O}\}} d|\eta|(s)$;

4) there exists a measurable function $\gamma : [0,T] \mapsto \mathbb{R}^n$, such that $\gamma(t) \in \mathcal{N}_{\varphi(t)}$ $(d|\eta|$ a.s.$)$ and $\eta(t) = \int_0^t \gamma(s) d|\eta|(s)$.

A pair (φ,η) satisfying the above 1)–4) is called a *solution* of the $SP(\psi;\mathcal{O})$.

It is known that under various technical conditions on the domain $\mathcal{O}$ and its boundary, for any $\psi \in C([0,T];\mathbb{R}^n)$ there exists a unique solution to $SP(\psi;\mathcal{O})$. In particular, these conditions are satisfied when $\mathcal{O}$ is convex and with smooth boundary, which will be the case considered throughout this chapter. Therefore we can consider a well-defined mapping $\Gamma : C([0,T];\mathbb{R}^n) \mapsto C([0,T];\mathbb{R}^n)$ such that $\Gamma(\psi)(t) = \varphi(t)$, $t \in [0,T]$, where (φ,η) is the (unique) solution to $SP(\psi;\mathcal{O})$. We will call Γ the *solution mapping* of the $SP(\cdot\,;\mathcal{O})$.

An elegant feature of the solution mapping Γ is that it may have a *Lipschitz property*: for some constant $K > 0$ that is independent of T, such that for $\psi_i \in C([0,T],\mathbb{R}^n)$, $i = 1,2$, it holds that

(1.3) $$|\Gamma(\psi_1)(\cdot) - \Gamma(\psi_2)(\cdot)|_T^* \le K|\psi_1(\cdot) - \psi_2(\cdot)|_T^*,$$

where $|\xi|_t^*$ denotes the sup-norm on $[0,t]$ for $\xi \in C([0,T];\mathbb{R}^n)$. Consequently, if (φ_i,η_i), $i = 1,2$ are solutions to $SP(\psi_i;\mathcal{O})$, $i = 1,2$, respectively, then for some constant K independent of T,

(1.4) $$|\varphi_1(\cdot) + \varphi_2(\cdot)|_T^* + |\eta_1(\cdot) - \eta_2(\cdot)|_T^* \le K|\psi_1(\cdot) - \psi_2(\cdot)|_T^*.$$

In what follows we call a (convex) domain $\mathcal{O} \subseteq \mathbb{R}^n$ *regular* if the solution mapping of the corresponding $SP(\cdot\,;\mathcal{O})$ satisfies (1.3). The simplest but typical example of a regular domain is the "half space" $\mathcal{O} = \mathbb{R}_+^n \stackrel{\Delta}{=} \{(x_1, \cdots, x_n) \in \mathbb{R}^n : x_n \ge 0\}$. With a standard localization technique, one can show that a convex domain with smooth boundary is also regular. A much deeper result of Dupuis and Ishii [1] shows that a convex polyhedron is regular, which can be extended to a class of convex domains with piecewise smooth boundaries. We should note that proving the regularity of a given domain is in general a formidable problem with independent interest of its own. To simplify presentation, however, in this chapter we consider only the case when the domains are regular, although the result we state below should hold true for a much larger class of (convex) domains, with proofs more complicated than what we present here.

We shall make use of the following assumptions.

(A1) (i) for fixed $x \in \mathbb{R}^n$, $b(\cdot,x,\cdot)$ and $\sigma(\cdot,x,\cdot)$ are $\{\mathcal{F}_t\}_{t \ge 0}$-progressively measurable;

(ii) there exists constant $K > 0$, such that for all $(t, \omega) \in [0, T] \times \Omega$ and $x, x' \in \mathbb{R}^n$, it holds that

(1.5)
$$|b(t, x, \omega) - b(t, x', \omega)| \leq K|x - x'|;$$
$$|\sigma(t, x, \omega) - \sigma(t, x', \omega)| \leq K|x - x'|.$$

Theorem 1.2. *Suppose that $\mathcal{O} \subseteq \mathbb{R}^n$ is a regular, convex domain; and that (A1) holds. Then the SDER (1.2) has a unique strong solution.*

Proof. Let Γ be the solution mapping to $SP(\cdot; \mathcal{O})$. Consider the following SDE (without reflection):

(1.6)
$$\widetilde{X}(t) = x + \int_0^t \widetilde{b}(s, \widetilde{X}(\cdot))ds + \int_0^t \widetilde{\sigma}(s, \widetilde{X}(\cdot))dW(s),$$

where for $y(\cdot) \in C([0, T]; \mathbb{R}^n)$,

$$\widetilde{b}(t, y(\cdot), \omega) = b(t, \Gamma(y)(t), \omega); \quad \widetilde{\sigma}(t, y(\cdot), \omega) = \sigma(t, \Gamma(y)(t), \omega).$$

Note that for any $\{\mathcal{F}_t\}_{t \geq 0}$-adapted, continuous process Y, the processes $\widetilde{b}(\cdot, Y(\cdot), \cdot)$ and $\widetilde{\sigma}(\cdot, Y(\cdot), \cdot)$, are all $\{\mathcal{F}_t\}_{t \geq 0}$-progressively measurable. Further, the regularity of the domain $\mathcal{O}$ implies that there exists a constant $K_0 > 0$ depending only on the Lipschitz constant of Γ and K in (A1), such that for any $\{\mathcal{F}_t\}_{t \geq 0}$-adapted, continuous processes Y and Y', it holds that

$$|\widetilde{b}(s, Y(\cdot, \omega), \omega) - \widetilde{b}(s, Y'(\cdot, \omega), \omega)|_t^* \leq K_0 |Y(s, \omega) - Y'(s, \omega))|_t^*;$$
$$|\widetilde{\sigma}(s, Y(\cdot, \omega), \omega) - \widetilde{\sigma}(s, Y'(\cdot, \omega), \omega)|_t^* \leq K_0 |Y(s, \omega) - Y'(s, \omega))|_t^*,$$

for all $(t, \omega) \in [0, T] \times \mathcal{O}$. Therefore, by the standard theory of SDEs (cf. e.g., Protter [1]), we know that the SDE (1.6) has a unique strong solution $\widetilde{X}$.

Next, we define a process $X(t) = \Gamma(\widetilde{X})(t)$, $t \in [0, T]$. Then by definition of the Skorohod problem, we see that there exists a process η such that (X, η) satisfies the conditions 1)–3) of Definition 1.1. Consequently, for all $t \in [0, T]$, we have

$$X(t) = \widetilde{X}(t) + \eta(t)$$
$$= x + \int_0^t \widetilde{b}(s, \widetilde{X}(\cdot))ds + \int_0^t \widetilde{\sigma}(s, \widetilde{X}(\cdot))dW(s) + \eta(t)$$
$$= x + \int_0^t b(s, X(s))ds + \int_0^t \sigma(s, X(s))dW(s) + \eta(t).$$

In other words, (X, η) is a solution to the SDER (1.5). The uniqueness follows easily from the construction of the solution and the Lipschitz property (1.3) and (1.4). The proof is complete. $\qquad \square$

§2. Backward SDEs with Reflections

In this section we study the reflected BSDEs (BSDERs, for short). For clearer notation we will call the domain in which a BSDE lives by $\mathcal{O}_2$, to distinguish it from those in the previous section. A slight difference is that we shall allow $\mathcal{O}_2$ to "move" when time varies, and even randomly. Namely, we shall consider a family of closed, convex domains $\{\mathcal{O}_2(t, \omega) : (t, \omega) \in [0, T] \times \Omega\}$ in $\mathbb{R}^m$ satisfying certain conditions. Let $\xi \in \mathcal{O}_2(T, \omega)$ be given, we consider the following SDE:

$$(2.1) \quad Y(t) = \xi + \int_t^T h(s, Y(s), Z(s))ds - \int_t^T Z(s)dW(s) + \zeta(T) - \zeta(t).$$

Analogous to the FSDER, we define the adapted solution to a BSDER as follows:

Definition 2.1. A triplet of processes $(Y, Z, \zeta) \in L_{\mathcal{F}}^2(\Omega; C([0, T]; \mathbb{R}^m)) \times L_{\mathcal{F}}^2(0, T; \mathbb{R}^{m \times d}) \times BV_{\mathcal{F}}([0, T]; \mathbb{R}^m)$ is called a solution to (2.1) if
   (1) $Y(t, \omega) \in \mathcal{O}_2(t, \omega)$, for all $t \in [0, T]$, **P**-a.e. ω;
   (2) for any $\{\mathcal{F}_t\}_{t \geq 0}$-adapted, RCLL process $V(t)$ such that $V(t) \in \mathcal{O}_2(t, \cdot)$, $\forall t \in [0, T]$, a.s., it holds that $\langle Y(t) - V(t), d\zeta(t) \rangle \leq 0$, as a signed measure.

We note that Definition 2.1 more or less requires that the domains $\{\mathcal{O}_2(\cdot, \cdot)\}$ be "measurable" (or even "progressively measurable") in (t, ω) in a certain sense, which we now describe. Let $y \in \mathbb{R}^m$ and $A \subseteq \mathbb{R}^m$ be any closed set, we define the *projection operator* Pr with respect to A, denoted $Pr(\cdot; A)$, by

$$(2.2) \qquad\qquad Pr(y; A) = y - \frac{1}{2}\nabla_y d^2(y, A), \qquad y \in \mathbb{R}^m;$$

where $d(\cdot, \cdot)$ is the usual distance function:

$$(2.3) \qquad\qquad d(y, A) \overset{\Delta}{=} \inf\{|y - x| : x \in A\}.$$

For each $y \in \mathbb{R}^m$, we define $\beta(t, y, \omega) = Pr(y; \mathcal{O}_2(t, \omega))$. Throughout this chapter we shall assume the following technical condition.

(A2) (i) For every fixed $y \in \mathbb{R}^m$, the process $(t, \omega) \mapsto \beta(t, y, \omega)$ is $\{\mathcal{F}_t\}_{t \geq 0}$-progressively measurable;
   (ii) for fixed $y \in \mathbb{R}^m$, it holds that

$$(2.4) \qquad\qquad E\int_0^T |\beta(t, y, \cdot)|^2 dt < \infty.$$

Before we go any further, let us look at some examples.

Example 2.2. Let $\mathcal{H}_m$ be the collection of all compact subsets of $\mathbb{R}^m$, endowed with the *Hausdorff* metric d^*, that is,

$$(2.5) \quad d^*(A, B) = \max\{\sup_{x \in A} d(x, B), \ \sup_{y \in B} d(y, A)\}, \qquad \forall A, B \in \mathcal{H}_m.$$

It is well-known that $(\mathcal{H}_m, d^*)$ is a complete metric space. Now suppose that $\mathcal{O}_2 \overset{\Delta}{=} \{\mathcal{O}_2(t, \omega) : (t, \omega) \in [0, T] \times \Omega\} \subseteq (\mathcal{H}_m, d^*)$, then we can view $\mathcal{O}_2$ as an

$(\mathcal{H}_m, d^*)$-valued process, and thus assume that it is $\{\mathcal{F}_t\}_{t \geq 0}$-progressively measurable. Noting that for fixed $y \in \mathbb{R}^m$, the mapping $A \mapsto d(y, A)$ is a continuous mapping from $(\mathcal{H}_m, d^*)$ to $\mathbb{R}$, as

$$|d(y, A) - d(y, B)| \leq d^*(A, B), \qquad \forall y \in \mathbb{R}^m, \ \forall A, B \in \mathcal{H}_m,$$

the composition function $(t, \omega) \mapsto d^2(y, \mathcal{O}(t, \omega))$ is $\{\mathcal{F}_t\}_{t \geq 0}$-progressively measurable as well, which then renders $\nabla_y d^2(y, \mathcal{O}_2(\cdot, \cdot))$ an $\{\mathcal{F}_t\}_{t \geq 0}$-progressively measurable process, for any fixed $y \in \mathbb{R}^m$. Consequently, $\mathcal{O}_2$ satisfies (A2)-(i).

Next, using elementary inequality $|d(z_1, A) - d(z_2, A)| \leq |z_1 - z_2|$, $\forall z_1, z_2 \in \mathbb{R}^m$, $\forall A \subseteq \mathbb{R}^m$ one shows that

$$|\nabla_y d^2(y, \mathcal{O}_2(t, \omega))| \leq 2d(y, \mathcal{O}_2(t, \omega)).$$

Assumption (A2)-(ii) is easily satisfied provided $d(y, \mathcal{O}_2(\cdot, \cdot)) \in L^2([0, T] \times \Omega)$, which is always the case if, for example, $0 \in \mathcal{O}_2(t, \omega)$ for all (t, ω), or, more generally, $\mathcal{O}_2(t, \omega)$ has a selection in $L^2_{\mathcal{F}}(0, T; \Omega)$.   □

Example 2.3. As a special case of Example 2.2, the following moving domains are often seen in applications. Let $\{\mathcal{O}(t, x) : (t, x) \in [0, T] \times \mathbb{R}^n\}$ be a family of convex, compact domains in $\mathbb{R}^m$ such that

   (i) the mapping $(t, x) \mapsto \mathcal{O}(t, x)$ is continuous as a function from $[0, T] \times \mathbb{R}^n$ to $(\mathcal{H}_m, d^*)$.

   (ii) for each (t, x), $0 \in \mathcal{O}(t, x)$; and there exists a constant $C > 0$ such that

$$\sup_{t \in [0, T]} \ d^*(\mathcal{O}(t, x), \mathcal{O}(t, 0)) \leq C|x|.$$

Let $X \in L^2_{\mathcal{F}}(\Omega; C([0, T]; \mathbb{R}^n))$, and define $\mathcal{O}_2(t, \omega) \overset{\Delta}{=} \mathcal{O}(t, X(t, \omega))$, $(t, \omega) \in [0, T] \times \Omega$. We leave it to the readers to check that $\mathcal{O}_2$ satisfies (A2).   □

Example 2.4. Continuing from the previous examples, let us assume that $m = 1$ and $\mathcal{O}(t, x) = [L(t, x), U(t, x)]$, where $-\infty < L(t, x) < 0 < U(t, x) < \infty$ for all $(t, x) \in [0, T] \times \mathbb{R}^n$. Suppose that the functions L and U are both uniformly Lipschitz in x, uniformly in $t \in [0, T]$. Then a simple calculation using the definition of the Hausdorff metric shows that

$$d^*(\mathcal{O}(t, 0), \mathcal{O}(t, x)) = \max\{|L(t, x) - L(t, 0)|, \ |U(t, x) - U(t, 0)|\} \leq C|x|.$$

Thus $\mathcal{O}_2$ satisfies (A2), thanks to the previous example.   □

Let us now turn our attention to the well-posedness of the BSDER (2.1). We shall make use of the following standing assumptions on coefficient $h : [0, T] \times \mathbb{R}^m \times \mathbb{R}^{m \times d} \times \Omega \mapsto \mathbb{R}^m$ and the domain $\{\mathcal{O}_2(t, \omega)\}$.

(**A3**) (i) for each $(y, z) \in \mathbb{R}^m \times \mathbb{R}^{m \times d}$, $h(\cdot, y, z, \cdot)$ is an $\{\mathcal{F}_t\}_{t \geq 0}$-progressively measurable process; and for fixed $(t, z) \in [0, T] \times \mathbb{R}^{m \times d}$ and a.e. $\omega \in \Omega$, $h(t, \cdot, z, \omega)$ is continuous;

(ii) $E \int_0^T |h(t,0,0)|^2 dt < \infty$;

(iii) there exist $\alpha \in \mathbb{R}$ and $k_2 > 0$, such that for all $t \in [0,T]$, $y, y' \in \mathbb{R}^m$, and $z, z' \in \mathbb{R}^{m \times d}$, it holds $\mathbf{P}$-a.s. that

$$\begin{cases} \langle y - y', h(t,y,z) - h(t,y',z) \rangle \leq \alpha |y - y'|^2; \\ |h(t,y,z) - h(t,y,z')| \leq k_2 |z - z'|; \\ |h(t,y,z) - h(t,0,z)| \leq k_2(1 + |y|). \end{cases}$$

(iv) The domains $\{\mathcal{O}_2(t, \cdot)\}$ is "non-increasing". In other words, it holds that

$$\mathcal{O}(t, \omega) \subseteq \mathcal{O}(s, \omega), \qquad \forall t \geq s, \text{ a.s.}$$

Our main result of this section is the following theorem.

Theorem 2.5. *Suppose that (A2) and (A3) are in force. Then the BSDER (2.1) has a unique (strong) solution. Furthermore, the process ζ_t is absolutely continuous with respect to Lebesgue measure, and for any process V_t such that $V_t(\omega) \in \mathcal{O}_2(t, \omega)$, $\forall t \in [0,T]$, a.s. , it holds that*

$$(2.6) \qquad \langle \frac{d\zeta_t}{dt}, Y_t - V_t \rangle \leq 0, \qquad \forall t \in [0,T], \text{ a.s.}$$

Remark 2.6. Suppose $m = 1$ and $\mathcal{O}_2 = [L, U]$, for appropriate processes L and U. Denote by $\zeta = \zeta^+ - \zeta^-$, $\zeta_0^+ = \zeta_0^- = 0$, the minimal decomposition of ζ as a difference of two non-decreasing processes. By replacing V in (2.6) by

$$V_t^L = L_t \mathbf{1}_{\{\frac{d\zeta_t}{dt} \geq 0\}} + Y_t \mathbf{1}_{\{\frac{d\zeta_t}{dt} < 0\}},$$
$$V_t^U = U_t \mathbf{1}_{\{\frac{d\zeta_t}{dt} \leq 0\}} + Y_t \mathbf{1}_{\{\frac{d\zeta_t}{dt} > 0\}}, \qquad t \in [0,T],$$

respectively, we obtain

$$(2.7) \qquad \langle Y_t - L_t, d\zeta_t^+ \rangle = 0, \quad \langle Y_t - U_t, d\zeta_t^- \rangle = 0, \quad \forall t \in [0,T], \text{ a.s.}$$

Proof of Theorem 2.5. Since the proof is quite lengthy, we shall split it into several lemmas. To begin with, let us first recall the notion of *Yosida approximation*, which is another typical route of attacking the existence and uniqueness of an SDE with reflection other than using Skorohod problem.

Let φ be any *proper*, lower semicontinuous (l.s.c., for short), convex function (by *proper* we mean that φ is not identically equal to $+\infty$). Let $\mathcal{D}(\varphi) = \{x : \varphi(x) < \infty\}$. We define the *subdifferential* of φ, denoted by $\partial\varphi$, as

$$\partial\varphi(y) \overset{\Delta}{=} \{x^* \in \mathbb{R}^m : \langle x^*, y - x \rangle \geq 0, \ \forall x \in \overline{\mathcal{D}(\varphi)}\}.$$

In what follows we denote $A \overset{\Delta}{=} \partial\varphi$. Define, for each $\varepsilon > 0$, a function

$$(2.8) \qquad \varphi_\varepsilon(y) \overset{\Delta}{=} \inf_{x \in \mathbb{R}^m} \left\{ \frac{1}{2\varepsilon} |y - x|^2 + \varphi(x) \right\}.$$

Since $\mathbb{R}^m$ is a Hilbert space, and φ is a l.s.c. proper convex mapping, the the following result can be found in standard text (cf. Barbu [1, Chapter II]):

Lemma 2.7. (i) The function φ_ε is (Fréchet) differentiable.

(ii) The Fréchet differential of φ_ε, denoted by $D\varphi_\varepsilon$, satisfies $D\varphi_\varepsilon = A_\varepsilon$, where A_ε is the Yosida approximation of A, define by

(2.9) $A_\varepsilon(y) = \dfrac{1}{\varepsilon}(y - J_\varepsilon(y))$,   where $J_\varepsilon(y) = (I + \varepsilon A)^{-1}(y)$.

(iii) $|J_\varepsilon(x) - J_\varepsilon(y)| \le |x - y|$; $|A_\varepsilon(x) - A_\varepsilon(y)| \le \frac{1}{\varepsilon}|x - y|$,

(iv) $A_\varepsilon(y) \in \partial\varphi(J_\varepsilon(y))$.

(v) $|A_\varepsilon(y)| \nearrow_{\varepsilon \to 0} \begin{cases} |A^0(x)|, & \text{if } x \in \mathcal{O}; \\ +\infty, & \text{otherwise,} \end{cases}$ where $A^0(y) \overset{\Delta}{=} Pr_{\partial\varphi(y)}(0)$,

$y \in \mathbb{R}^m$.   □

Let us now specify a l.s.c. proper convex function to fit our discussion. For any convex, closed subset $\mathcal{O} \subseteq \mathbb{R}^m$, we define its *indicator* function, denoted by $\varphi := I_{\mathcal{O}}$ to be

$$\varphi(y) \overset{\Delta}{=} \begin{cases} 0 & y \in \mathcal{O}; \\ +\infty & y \notin \mathcal{O}, \end{cases}$$

In this case, $\mathcal{D}(\varphi) = \mathcal{O}$. Now by definitions (2.8) and (2.9), we have

$$\varphi_\varepsilon(y) = \inf_{x \in \mathcal{O}} \frac{1}{2\varepsilon}|y - x|^2 = \frac{1}{2\varepsilon}d^2(y, \mathcal{O}),$$

$$A_\varepsilon(y) = D\varphi_\varepsilon(y) = \frac{1}{2\varepsilon}\nabla d^2(y, \mathcal{O}) = \frac{1}{\varepsilon}(y - Pr(y, \mathcal{O})),$$

Consequently, we have

(2.10) $\begin{cases} J_\varepsilon(y) = Pr(y; \mathcal{O}), & \forall \varepsilon > 0; \\ A_\varepsilon(y) = 0, & \forall y \in \mathcal{O}, \ \forall \varepsilon > 0; \\ A^0(y) = 0, & \forall y \in \mathcal{O}. \end{cases}$

Further, we replace $\mathcal{O}$ by the $(\mathcal{H}_m, d^*)$-valued process $\{\mathcal{O}_2\}$, then

(2.11) $\begin{aligned} \varphi_\varepsilon(t, y, \omega) &= \frac{1}{2\varepsilon}I_{\mathcal{O}_2(t,\omega)}(y), \quad \forall \varepsilon > 0; \\ J_\varepsilon(t, y, \omega) &= (I + \varepsilon A(t, \cdot, \omega))^{-1}(y); \\ A_\varepsilon(t, y, \omega) &= \frac{1}{\varepsilon}(y - J_\varepsilon(t, y, \omega)). \end{aligned}$

By (2.10) we know that $J_\varepsilon(t, y, \omega) = Pr(y, \mathcal{O}_2(t, \omega))$, and by assumption (A2) we have that for every $\varepsilon > 0$, $J_\varepsilon(\cdot, y, \cdot) \in L^2_{\mathcal{F}}(0, T; \mathbb{R}^m)$ for all $y \in \mathbb{R}^m$.

Let us now consider the following approximation of (2.1):

$$Y^\varepsilon(t) = \xi + \int_t^T h(s, Y^\varepsilon(s), Z^\varepsilon(s))ds - \int_t^T Z^\varepsilon(s)dW(s)$$

(2.12)
$$- \int_t^T A_\varepsilon(Y^\varepsilon(s))ds,$$

where A_ε is the Yosida approximation of $A(t,\omega) = \partial I_{\mathcal{O}_2(t,\omega)}$ defined by (2.11). Since A_ε is uniform Lipschitz for each fixed ε, by Lemma 2.7-(iii) and by slightly modifying the arguments in Chapter 1, §4 to cope with the current situation where α in (A3) is allowed to be negative, one shows that (2.12) has a unique strong solution $(Y^\varepsilon, Z^\varepsilon)$ satisfying

(2.13) $$E\left\{ \sup_{0 \le t \le T} |Y^\varepsilon(t)|^2 + \int_0^T \|Z^\varepsilon(t)\|^2 dt \right\} < \infty.$$

We will first show that as $\varepsilon \to 0$, $(Y^\varepsilon, Z^\varepsilon)$ converges in a certain sense, then show that the limit will give the solution of (2.12). To begin with, we need some elementary estimates.

Lemma 2.8. *Suppose that condition (A3) holds, and that $\xi \in L^2_{\mathcal{F}_T}(\Omega)$. Then there exists a constant $C > 0$, independent of ε, such that the following estimates hold*

(2.14)
$$\begin{cases} E\left\{ \sup_{t \in [0,T]} |Y^\varepsilon(t)|^2 + \int_0^T |Z^\varepsilon(t)|^2 dt \right\} \le C; \\ E\left\{ \int_0^T |A_\varepsilon(t, Y^\varepsilon(t))|^2 dt \right\} \le C. \end{cases}$$

Proof. The proof of the first inequality is quite similar to those we have seen many times before, with the help of the properties of Yosida approximations listed in §2.2, we only prove the second one. First note that since $\mathcal{O}_2$ is convex, so is $\varphi_\varepsilon(t, \cdot, \omega)$ (recall (2.11)). We have the following inequality (suppressing ω):

(2.15) $$\varphi_\varepsilon(t, y) + \langle D\varphi_\varepsilon(t, y), \tilde{y} - y \rangle \le \varphi_\varepsilon(t, \tilde{y}), \qquad \forall(t, y), \text{ a.s.}$$

Now let $t = t_0 < t_1 < \cdots < t_n = T$ be any partition of $[t, T]$. Then (2.15) leads to that

(2.16)
$$\varphi_\varepsilon(t_i, Y^\varepsilon(t_i)) + \langle D\varphi_\varepsilon(t_i, Y^\varepsilon(t_i)), Y^\varepsilon(t_{i+1}) - Y^\varepsilon(t_i) \rangle$$
$$\le \varphi_\varepsilon(t_i, Y^\varepsilon(t_{i+1})) \le \varphi_\varepsilon(t_{i+1}, Y^\varepsilon(t_{i+1})), \qquad \text{a.s.},$$

where the last inequality is due to Assumption (A3)-iv). Summing both sides of (2.16) up and letting the mesh size of the partition $\max_i |t_{i+1} - t_i| \to 0$ we obtain that

(2.17) $$\varphi_\varepsilon(t, Y^\varepsilon(t)) + \int_t^T \langle D\varphi_\varepsilon(s, Y^\varepsilon(s)), dY^\varepsilon(s) \rangle \le \varphi_\varepsilon(T, \xi) = 0.$$

Thus, recall the equation for Y^ε we have

$$\varphi_\varepsilon(t, Y^\varepsilon(t)) + \frac{1}{\varepsilon} \int_t^T |D\varphi_\varepsilon(Y^\varepsilon(s))|^2 ds$$

(2.18)
$$\leq \varphi_\varepsilon(T, \xi) + \int_t^T \langle D\varphi_\varepsilon(s, Y^\varepsilon(s)), h(s, Y^\varepsilon(s), Z^\varepsilon(s)) \rangle ds$$

$$- \int_t^T \langle D\varphi_\varepsilon(Y^\varepsilon(s)), Z^\varepsilon dW_s \rangle.$$

By Cauchy-Schwartz inequality and (A3)-(iii),

$$\langle D\varphi_\varepsilon(t, y), h(t, y, z) \rangle \leq \frac{1}{2\varepsilon}|D\varphi_\varepsilon(t, y)|^2 + \varepsilon C(1 + \|z\|^2 + |y|^2), \quad \forall(t, y, z).$$

We now recall that $\varphi_\varepsilon \geq 0$; $\xi \in \mathcal{O}_2(T, \cdot)$ (i.e., $\varphi_\varepsilon(T, \xi) = 0$); and $A_\varepsilon(t, y, \omega) = D\varphi_\varepsilon(t, y, \omega)$. Using the first inequality of this lemma we obtain that

$$E \int_t^T |A_\varepsilon(t, Y^\varepsilon(s))|^2 ds = E \int_t^T |D\varphi_\varepsilon(Y^\varepsilon(s))|^2 ds$$

$$\leq C\left(1 + E \sup_{t \in [0,T]} |Y^\varepsilon(t)|^2 + E \int_0^T \|Z^\varepsilon(t)\|^2 dt\right) \leq \widetilde{C},$$

where $\widetilde{C} > 0$ is some constant independent of ε. Thus, by a slightly abuse of notations on the constant C, we obtain the desired estimate. □

Lemma 2.9. *Suppose that the assumptions of Lemma 2.8 hold. Then there exists a constant $C > 0$, such that for any $\varepsilon, \delta > 0$, it holds that*

(2.19) $E\left\{ \sup_{t \in [0,T]} |Y^\varepsilon(t) - Y^\delta(t)|^2 + \int_0^T |Z^\varepsilon(t) - Z^\delta(t)|^2 dt\right\} \leq (\varepsilon + \delta)C.$

Proof. Applying Itô's formula we get

$$|Y^\varepsilon(t) - Y^\delta(t)|^2 + \int_t^T \|Z^\varepsilon(s) - Z^\delta(s)\|^2 ds$$

$$+ 2\int_t^T \langle A_\varepsilon(s, Y^\varepsilon(s)) - A_\delta(s, Y^\delta(s)), Y^\varepsilon(s) - Y^\delta(s) \rangle ds$$

(2.20)
$$= 2\int_t^T \langle h(s, Y^\varepsilon(s), Z^\varepsilon(s)) - h(s, Y^\delta(s), Z^\delta(s)), Y^\varepsilon(s) - Y^\delta(s) \rangle ds$$

$$- 2\int_t^T \langle Y^\varepsilon(s) - Y^\delta(s), [Z^\varepsilon(s) - Z^\delta(s)]dW(s) \rangle.$$

Since $A_\varepsilon(t, y, \omega) \in \partial\varphi(J_\varepsilon(y))$, we have by definition that

$$\langle A_\varepsilon(t, y, \omega), J_\varepsilon(t, y, \omega) - x \rangle \geq 0, \qquad \forall x \in \mathcal{O}_2(t, \omega).$$

In particular for any $\widetilde{y} \in \mathbb{R}^m$, and any $\delta > 0$, $J_\delta(t, \widetilde{y}, \omega) \in \mathcal{O}_2(t, \omega)$ and therefore

$$\langle A_\varepsilon(t, y, \omega), J_\varepsilon(t, y, \omega) - J_\delta(t, \widetilde{y}, \omega) \rangle \geq 0, \qquad \forall\, \widetilde{y} \in \mathbb{R}^m, \text{ a.e. } \omega \in \Omega.$$

Similarly,

$$\langle A_\delta(t, \widetilde{y}, \omega), J_\delta(t, \widetilde{y}, \omega) - J_\varepsilon(t, y, \omega) \rangle \geq 0, \qquad \forall\, y \in \mathbb{R}^m, \text{ a.e. } \omega \in \Omega.$$

Consequently, we have (suppressing ω)

$$
\begin{aligned}
& \langle A_\varepsilon(t, y) - A_\delta(t, \widetilde{y}), y - \widetilde{y} \rangle \\
&= \langle A_\varepsilon(t, y), [y - J_\varepsilon(t, y)] + [J_\varepsilon(t, y) - J_\delta(t, \widetilde{y})] + J_\delta(t, \widetilde{y}) - \widetilde{y} \rangle \\
&\quad + \langle A_\delta(t, \widetilde{y}), [\widetilde{y} - J_\delta(t, \widetilde{y})] + [J_\delta(t, \widetilde{y}) - J_\varepsilon(t, y)] + J_\varepsilon(t, y) - y \rangle \\
&\geq - \langle A_\varepsilon(t, y), \delta A_\delta(t, \widetilde{y}) \rangle - \langle A_\delta(t, \widetilde{y}), \varepsilon A_\varepsilon(t, y) \rangle \\
&= -(\varepsilon + \delta) \langle A_\varepsilon(t, y), A_\delta(t, \widetilde{y}) \rangle.
\end{aligned}
$$
(2.21)

Also, some standard arguments using Schwartz inequality lead to that

$$2 \langle h(t, y, z) - h(t, \widetilde{y}, \widetilde{z}), y - \widetilde{y} \rangle \leq \frac{1}{2} \|z - \widetilde{z}\|^2 + C|y - \widetilde{y}|^2.$$
(2.22)

Combining (2.20)—(2.22) and using the Burkholder and Gronwall inequalities we obtain, for some constant $C > 0$,

$$
E\Big\{ \sup_{t \in [0,T]} |Y^\varepsilon(t) - Y^\delta(t)|^2 + \int_0^T \|Z^\varepsilon(t) - Z^\delta(t)\|^2 dt \Big\}
$$

$$
\leq (\varepsilon + \delta) E \int_0^T \Big| \langle A_\varepsilon(t, Y^\varepsilon(t)), A_\delta(t, Y^\delta(t)) \rangle \Big| dt
$$

$$
\leq (\varepsilon + \delta) \Big\{ E \int_0^T |A_\varepsilon(Y^\varepsilon(t))|^2 dt \cdot E \int_0^T |A_\delta(Y^\delta(t))|^2 dt \Big\}^{\frac{1}{2}} \leq (\varepsilon + \delta)C,
$$

thanks to (2.14). This proves the Lemma. $\square$

As a direct consequence of Lemma 2.8, we see that if we send ε to zero along an arbitrary sequence $\{\varepsilon_n\}$, then there exist processes $Y \in L^2_{\mathcal{F}}(\Omega; C([0,T]; \mathbb{R}^m))$, $Z \in L^2_{\mathcal{F}}(\Omega \times [0,T]; \mathbb{R}^m))$, independent of the choice of the sequence $\{e_n\}$ chosen, such that

$$(Y^n, Z^n) \triangleq (Y^{\varepsilon_n}, Z^{\varepsilon_n}) \to (Y, Z), \qquad \text{as } n \to \infty,$$

strongly in $L^2_{\mathcal{F}}(\Omega; C([0,T]; \mathbb{R}^m)) \times L^2_{\mathcal{F}}(\Omega \times [0,T]; \mathbb{R}^m)$.

Furthermore, by Lemma 2.8 and the equation (2.12), it follows that for some $\eta \in L^2_{\mathcal{F}}(0,T; \mathbb{R}^m)$, $\zeta \in L^2_{\mathcal{F}}(\Omega; C([0,T]; \mathbb{R}^m))$, and possibly along a subsequence which we still denote by $\{\varepsilon_n\}$, it holds that

$$
\begin{cases}
A_{\varepsilon_n}(Y^{\varepsilon_n}(\cdot)) \to -\eta(\cdot), & \text{weakly in } L^2_{\mathcal{F}}(0,T; \mathbb{R}^m); \\
E\Big\{ \sup_{0 \leq t \leq T} \Big| \int_0^t A_{\varepsilon_n}(Y^{\varepsilon_n}(s)) ds + \zeta(t) \Big|^2 \Big\} \to 0, & \text{as } n \to \infty.
\end{cases}
$$

Here, we use $-\eta$ and $-\zeta$ to match the signs in (2.1) and (2.12). Obviously, we see that the limiting processes Y, Z, and ζ will satisfy the SDE (2.1), and the proof of Theorem 2.6 will be complete after we prove the following lemma.

Lemma 2.10. *Suppose that the process* (Y, Z), η, *and* ζ *are defined as before. Then* (Y, Z, ζ) *satisfies (2.11), such that*

(i) $E|\zeta|(T) = E \int_0^T |\eta(t)| dt < \infty;$

(ii) $Y(t) \in \mathcal{O}_2(t, \cdot)$, $\forall t \in [0, T]$, *a.s.;*

(iii) for any RCLL, $\{\mathcal{F}_t\}_{t \geq 0}$*-adapted process* V, $\langle Y(t) - V(t), \eta(t) \rangle \leq 0$, *a.s., as a signed measure.*

Proof. (i) We first show that ζ has absolutely continuous paths almost surely and that $\dot{\zeta} = \eta$. To see this, note that η is the weak limit of $A_{\varepsilon_n}(Y^{\varepsilon_n})$'s. By Mazur's theorem, there exists an convex combination of $A_{\varepsilon_n}(Y^{\varepsilon_n})$'s, denoted by $\tilde{A}_{\varepsilon_n}(Y^{\varepsilon_n})$, such that $\tilde{A}_{\varepsilon_n}(Y^{\varepsilon_n}) \to \eta$, strongly in $L^2_{\mathcal{F}}(\Omega \times [0, T]; \mathbb{R}^m))$. Note that for this sequence of convex combinations of the sequence $A_{\varepsilon_n}(Y^{\varepsilon_n})$, we also have

$$E\left\{ \sup_{0 \leq t \leq T} \left| \int_0^t \tilde{A}_{\varepsilon_n}(Y^{\varepsilon_n}(s)) ds + \zeta(t) \right|^2 \right\} \to 0, \qquad \text{as } n \to \infty.$$

Thus the uniqueness of the limit implies that $\zeta(t) = \int_0^t \eta(s) ds$, $\forall t \in [0, T]$. Furthermore, since $L^2_{\mathcal{F}}(\Omega) \subseteq L^1(\Omega)$, we derive (i) immediately.

(ii) In what follows we denote $d(y, t, \omega) = d(y, \mathcal{O}_2(t, \omega))$. Since $\mathcal{O}_2(t, \omega)$ is convex for fixed (t, ω), $d(\cdot, \mathcal{O}_2(t, \omega))$ is a convex function. Further, since $\mathcal{O}_2$ has smooth boundary, one derives from (2.9) that

$$d(y, t, \omega) = |y - Pr(y, \mathcal{O}_2(t, \omega))| = |y - J_\varepsilon(y)| = \varepsilon |A_\varepsilon(y, t, \omega)|.$$

for all $y \in \mathbb{R}^m$, and $t \in [0, T]$, P-a.s.. Hence by part (i), we see that

$$
\begin{aligned}
E \int_0^T d(Y^\varepsilon(t), t, \omega) dt &\leq \varepsilon E \int_0^T |A_\varepsilon(Y^\varepsilon(t))| dt \\
&\leq \varepsilon \sqrt{T} E \left\{ \int_0^T |A_\varepsilon(Y^\varepsilon(t))|^2 dt \right\}^{\frac{1}{2}} \to 0.
\end{aligned}
$$
(2.23)

Next, define for each $(t, \omega) \in [0, T] \times \Omega$ the *conjugate* function of $d(\cdot, t, \omega)$ by

$$(2.24) \qquad G(z, t, \omega) \overset{\triangle}{=} \inf_y \{d(y, t, \omega) - \langle z, y \rangle\},$$

and define the *effective domain* of G by

$$(2.25) \qquad \mathcal{D}^G(t, \omega) = \{z \in \mathbb{R} : G(z, t, \omega) > -\infty\}.$$

Since $d(\cdot, t, \omega)$ is convex and continuous everywhere, it must be identical to its biconjugate function, or equivalently, its closed convex hull (see

Hiriart-Urruty-Lemaréchal [1]). Consequently, the following conjugate relation holds:

$$(2.26) \qquad d(y,t,\omega) = \sup_{z \in \mathcal{D}^G(t,\omega)} \{G(z,t,\omega) + \langle z,y \rangle\};$$

and both the infimum of (2.24) and the supremum of (2.26) are achieved for every fixed (t,ω). Now for fixed (t,ω), and any $z_0 \in \mathcal{D}^G(t,\omega)$, we let $y_0 = y_0(t,\omega)$ be the minimizer in (2.24). Then

$$d(y_0,t,\omega) - \langle y_0, z_0 \rangle = G(z_0,t,\omega) \le d(y,t,\omega) - \langle y, z_0 \rangle, \qquad \forall y \in \mathbb{R}^n,$$

and hence

$$\langle y - y_0, z_0 \rangle \le d(y,t,\omega) - d(y_0,t,\omega), \qquad \forall y \in \mathbb{R}^n.$$

Since it is easily checked that $d(\cdot,t,\omega)$ is uniformly Lipschitz with Lipschitz constant 1, we deduce from above that $|z_0| \le 1$. Namely $\mathcal{D}^G(t,\omega) \subseteq [-1,1]$.

Now let Y be the limit process of Y^{ε_n}, we apply a measurable selection theorem to obtain a (bounded) $\{\mathcal{F}_t\}_{t \ge 0}$-adapted process R, such that $R(t,\omega) \in \mathcal{D}^G(t,\omega) \subseteq [-1,1]$, $\forall t$, a.s.; and

$$(2.27) \qquad \begin{cases} d(Y(t,\omega),t,\omega) = G(R(t,\omega),t,\omega) + \langle R(t,\omega), Y(t,\omega) \rangle, \\ d(Y^{\varepsilon_n}(t,\omega),t,\omega) \ge G(R(t,\omega),t,\omega) + \langle R(t,\omega), Y^{\varepsilon_n}(t,\omega) \rangle, \end{cases}$$

Therefore, recall that $Y^{\varepsilon_n} \to Y$, we have

$$\begin{aligned} E \int_0^T d(Y(t),t,\cdot)dt &= E \int_0^T \left\{ G(Y(t),t,\cdot) + \langle R(t), Y(t) \rangle \right\} dt \\ &= \lim_{n \to \infty} E \int_0^T \left\{ G(Y(t),t,\cdot) + \langle R(t), Y^n(t) \rangle \right\} dt \\ &\le \lim_{n \to \infty} E \int_0^T d(Y^{\varepsilon_n}(t),t,\cdot)dt = 0, \end{aligned}$$

thanks to (2.23). That is, $E \int_0^T d(Y(t),t,\cdot)dt = 0$, which implies that $Y(t,\omega) \in \mathcal{O}_2(t,\omega)$, $dt \times d\mathbf{P}$-a.e. Thus the conclusion follows from the continuity of the paths of Y.

(iii) Let $V(t)$ be any $\{\mathcal{F}_t\}_{t \ge 0}$-adapted process such that $V(t,\omega) \in \mathcal{O}_2(t,\omega)$, $\forall t \in [0,T]$, P-a.s. For every $\varepsilon > 0$, and $t \in [0,T]$, consider

$$(2.28) \qquad \Lambda_\varepsilon(t) = E \int_0^t \langle J_\varepsilon(Y^\varepsilon(s)) - V(s), A_\varepsilon(Y^\varepsilon(s)) \rangle ds.$$

Since $V(t) \in \mathcal{O}_2(t,\cdot)$, for all t, and $A_\varepsilon(Y^\varepsilon(t)) \in \partial I_{\mathcal{O}_2(t,\cdot)}(J_\varepsilon(Y^\varepsilon(t)))$ (see Lemma 2.7-(iv)), we have

$$\langle J_\varepsilon(Y^\varepsilon(t)) - V(t), A_\varepsilon(t,Y^\varepsilon(t)) \rangle \ge 0.$$

Namely, $\Lambda_\varepsilon(t) \geq 0$, $\forall \varepsilon > 0$ and $t \in [0, T]$. On the other hand, since

$$
\begin{aligned}
\Lambda_\varepsilon(t) = E \int_0^t & \Big\{ \langle J_\varepsilon(Y^\varepsilon(s)) - Y^\varepsilon(s), A_\varepsilon(s, Y^\varepsilon(s)) \rangle \\
& + \langle Y^\varepsilon(s) - V(s), A_\varepsilon(s, Y^\varepsilon(s)) \rangle \Big\} ds
\end{aligned}
$$

(2.29)

$$
= E \int_0^t \Big\{ - \varepsilon |A_\varepsilon(s, Y^\varepsilon(s))|^2 + \langle Y^\varepsilon(s) - V(s), A_\varepsilon(s, Y^\varepsilon(s)) \rangle \Big\} ds
$$

Now using the uniform boundedness (2.14) and the weak convergence of $\{A_{\varepsilon_n}(\cdot, Y^{\varepsilon_n})(\cdot))\}$, and the fact that Y^{ε_n} converges to Y strongly in $L^2_{\mathcal{F}}(\Omega; C([0, T]; \mathbb{R}^m))$, one derives easily by sending $n \to \infty$ in (2.29) that

$$
0 \leq E \int_0^t \langle Y(s) - V(s), -\eta(s) \rangle \, ds, \quad \forall t \in [0, T].
$$

Or equivalently,

$$
\langle Y(t) - V(t), \eta(t) \rangle = \langle Y(t) - V(t), \frac{d\zeta}{dt}(t) \rangle \leq 0, \quad \forall t \in [0, T], \quad \text{a.s.}
$$

as a (random) signed measure. Thus completes the proof of Lemma 2.10.

$\square$

§3. Reflected Forward-Backward SDEs

We are now ready to formulate forward-backward SDEs with reflection (FBSDER, for short). Let $\mathcal{O}_1$ be a closed, convex domain in $\mathbb{R}^n$, and $\mathcal{O}_2 = \{\mathcal{O}_2(t, \omega) : (t, \omega) \in [0, T] \times \mathbb{R}^n \times \Omega\}$ be a family of closed, convex domains in $\mathbb{R}^m$. Let $x \in \mathcal{O}_1$, and $g : \mathbb{R}^n \times \Omega \mapsto \mathbb{R}^m$ be a given $\mathcal{F}_T$-measurable random field satisfying

(3.1) $g(x, \omega) \in \mathcal{O}_2(T, \omega), \quad \forall (x, \omega).$

Consider the following FBSDER:

(3.2)
$$
\begin{cases}
X_t = x + \displaystyle\int_0^t b(s, X_s, Y_s, Z_s) ds + \int_0^t \sigma(s, X_s, Y_s, Z_s) dW_s + \eta_t; \\[2mm]
Y_t = g(X_T) + \displaystyle\int_t^T h(s, X_s, Y_s, Z_s) ds - \int_t^T Z_s dW_s + \zeta_T - \zeta_t;
\end{cases}
$$

Definition 3.1. A quintuple of processes (X, Y, Z, η, ζ) is called an adapted solution of the FBSDER (3.2) if

1) $(X, Y) \in L^2_{\mathcal{F}}(\Omega, C(0, T; \mathbb{R}^n \times \mathbb{R}^m))$, $Z \in L^2_{\mathcal{F}}(0, T; \mathbb{R}^{m \times d})$, $(\eta, \zeta) \in BV_{\mathcal{F}}(0, T; \mathbb{R}^n \times \mathbb{R}^m)$;
2) $X_t \in \mathcal{O}_1$, $Y_t \in \mathcal{O}_2(t, \cdot)$, $\forall t \in [0, T]$, a.s.;
3) $|\eta|_t = \int_0^t \mathbf{1}_{\{X_s \in \partial \mathcal{O}_1\}} d|\eta|_s$; $\eta_t = \int_0^t \gamma_s d|\eta|_s$, $\forall t \in [0, T]$, a.s., for some progressively measurable process γ such that $\gamma_s \in \mathcal{N}_{X_s}(\mathcal{O}_1)$, $d|\eta|$-a.e.;
4) for all RCLL and progressively measurable processes U such that $U_t \in \mathcal{O}_2(t, \cdot)$, $\forall t \in [0, T]$, a.s., one has $\langle Y_t - U_t, d\zeta_t \rangle \leq 0$, $\forall t \in [0, T]$, a.s.;

5) (X, Y, Z, η, ζ) satisfies the SDE (3.2) almost surely.

In light of assumptions (A1)-(A3), we will assume the following

(A4) (i) $\mathcal{O}_1$ has smooth boundary;

(ii) $\mathcal{O}_2(t, \omega) \subseteq \mathcal{O}_2(s, \omega), \forall t \geq s$, a.s.; and for fixed $y \in \mathbb{R}^m$, the mapping $(t, \omega) \mapsto \beta(t, y, \omega) \stackrel{\Delta}{=} Pr(y; \mathcal{O}_2(t, \omega))$ belongs to $L_{\mathcal{F}}^2([0, T]; \mathbb{R}^m)$.

(iii) The coefficients b, h, σ, and g are random fields defined on $[0, T] \times \mathbb{R}^n \times \mathbb{R}^m \times \mathbb{R}^{m \times d}$ such that for fixed (x, y, z), the processes $b(\cdot, x, y, z, \cdot)$, $h(\cdot, x, y, z, \cdot)$, and $\sigma(\cdot, x, y, z, \cdot)$ are $\{\mathcal{F}_t\}_{t \geq 0}$-progressively measurable, and $g(x, \cdot)$ is $\mathcal{F}_T$-measurable.

(iv) For fixed (t, x, z) and a.e. ω, $h(t, x, \cdot, z, \omega)$ is continuous, and there exists a constant $K > 0$ such that $|h(t, x, y, z, \omega)| \leq K(1 + |x| + |y|)$, for all (t, x, y, z, ω). Moreover,

$$E \int_0^T |b(t, 0, 0, 0)|^2 dt + E \int_0^T |\sigma(t, 0, 0, 0)|^2 dt + E|g(0)|^2 < \infty.$$

(v) There exist constants $k_i \geq 0, i = 1, 2$ and $\gamma \in \mathbb{R}$ such that for all $(t, \omega) \in [0, T] \times \Omega$ and $\mathbf{x} \stackrel{\Delta}{=} (x, y, z), \mathbf{x}_i \stackrel{\Delta}{=} (x_i, y_i, z_i) \in \mathbb{R}^n \times \mathbb{R}^m \times \mathbb{R}^{m \times d}$, $i = 1, 2$, and $\mathbf{x}^0 \stackrel{\Delta}{=} (x, y)$ for $\mathbf{x} = (x, y, z)$.

- $|b(t, \mathbf{x}_1, \omega) - b(t, \mathbf{x}_2, \omega)| \leq K|\mathbf{x}_1 - \mathbf{x}_2|$;
- $\langle h(t, x, y_1, z, \omega) - h(t, x, y_2, z, \omega), y_1 - y_2 \rangle \leq \gamma|y_1 - y_2|^2$;
- $|h(t, x_1, y, z_1, \omega) - h(t, x_2, y, z_2, \omega)| \leq K(|x_1 - x_2| + \|z_1 - z_2\|)$;
- $\|\sigma(t, \mathbf{x}_1, \omega) - \sigma(t, \mathbf{x}_2, \omega)\|^2 \leq K^2|\mathbf{x}_1^0 - \mathbf{x}_2^0|^2 + k_1^2\|z_1 - z_2\|^2$;
- $|g(x_1, \omega) - g(x_2, \omega)| \leq k_2|x_1 - x_2|$.

We should note that if $k_1 = k_2 = 0$, then σ and g are independent of z, just as the many cases we considered before. Therefore, the FBSDE considered in this chapter is more general. We note also that the method presented here should also work when there is no reflection involved (e.g., $\mathcal{O}_1 = \mathbb{R}^n$, $\mathcal{O}_2 \equiv \mathbb{R}^m$).

§3.1. A priori estimates

We first establish a new type of a priori estimates that is different from what we have seen in the previous chapters. To simplify notations we shall denote, for $t \in [0, T)$, $\mathbf{H}(t, T) = L_{\mathcal{F}}^2(t, T; \mathbb{R})$, and let $\mathbf{H}^{\mathbf{c}}(t, T)$ be the subset of $\mathbf{H}(t, T)$ consisting of all continuous processes. For any $\lambda \in \mathbb{R}$, define an equivalent norm on $\mathbf{H}(t, T)$ by:

$$\|\xi\|_{t, \lambda} \stackrel{\Delta}{=} \left\{ E \int_t^T e^{-\lambda s} |\xi(s)|^2 ds \right\}^{\frac{1}{2}}.$$

Then $\mathbf{H}_\lambda(t, T) \stackrel{\Delta}{=} \{\xi \in \mathbf{H}(t, T) : \|\xi\|_{t, \lambda} < \infty\} = \mathbf{H}(t, T)$. We shall also use the following norm on $\mathbf{H}^{\mathbf{c}}(t, T)$:

$$|\xi|_{t, \lambda, \beta} \stackrel{\Delta}{=} e^{-\lambda T} E|\xi_T|^2 + \beta\|\xi\|_{t, \lambda}^2, \qquad \xi \in \mathbf{H}^{\mathbf{c}}(t, T), \lambda \in \mathbb{R}, \beta > 0,$$

and denote $\mathbf{H}_{\lambda,\beta}(t,T)$ to be the completion of $\mathbf{H}^c(t,T)$ under norm $|\cdot|_{t,\lambda,\beta}$. Then for any λ and β, $\mathbf{H}_{\lambda,\beta}(t,T)$ is a Banach space. Further, if $t = 0$, we simply denote $\|\cdot\|_\lambda \stackrel{\Delta}{=} \|\cdot\|_{0,\lambda}$; $|\cdot|^2_{\lambda,\beta} \stackrel{\Delta}{=} |\cdot|^2_{0,\lambda,\beta}$; $\mathbf{H} = \mathbf{H}(0,T)$; $\mathbf{H}^c = \mathbf{H}^c(0,T)$; $\mathbf{H}_\lambda = \mathbf{H}_\lambda(t,T)$, and $\mathbf{H}_{\lambda,\beta} = \mathbf{H}_{\lambda,\beta}(t,T)$.

Moreover, the following functions will be frequently used in this section: for $\lambda \in \mathbb{R}$ and $t \in [0,T]$,

$$(3.3) \qquad A(\lambda,t) = e^{-(\lambda \wedge 0)t}; \quad B(\lambda,t) = \frac{1 - e^{-\lambda t}}{\lambda} = t \int_0^1 e^{-\lambda t \theta} d\theta.$$

It is easy to see that, for all $\lambda \in \mathbb{R}^n$, $B(\lambda,\cdot)$ is a nonnegative, increasing function, $A(\lambda,t) \geq 1$; and $B(\lambda,0) = 0$, $A(\lambda,0) = 1$.

Lemma 3.2. *Let (A4) hold. Let (X,Y,Z,η,ζ) and (X',Y',Z',η',ζ') be two solutions to the FBSDER (3.2), and let $\widehat{\xi} \stackrel{\Delta}{=} \xi - \xi'$, where $\xi = X,Y,Z,\eta,\zeta$, respectively.*

(i) Let $\lambda \in \mathbb{R}, C_1, C_2 > 0$, and let $\bar{\lambda}_1 = \lambda - K(2 + C_1^{-1} + C_2^{-1}) - K^2$. Then, for all $\lambda' \in \mathbb{R}$,

$$(3.4) \quad \begin{aligned} &e^{-\lambda t} E|\widehat{X}_t|^2 + (\bar{\lambda}_1 - \lambda') \int_0^t e^{-\lambda \tau} e^{-\lambda'(t-\tau)} E|\widehat{X}_\tau|^2 d\tau \\ &\leq \int_0^t e^{-\lambda \tau} e^{-\lambda'(t-\tau)} \{K(C_1 + K)E|\widehat{Y}_\tau|^2 + (KC_2 + k_1^2)E|\widehat{Z}_\tau|^2\} d\tau. \end{aligned}$$

(ii) Let $\lambda \in \mathbb{R}$ and $C_3, C_4 > 0$, and let $\bar{\lambda}_2 = -\lambda - 2\gamma - K(C_3^{-1} + C_4^{-1})$. Then, for all $\lambda' \in \mathbb{R}$,

$$(3.5) \quad \begin{aligned} &e^{-\lambda t} E|\widehat{Y}_t|^2 + (\bar{\lambda}_2 - \lambda') \int_t^T e^{-\lambda \tau} e^{-\lambda'(\tau-t)} E|\widehat{Y}_\tau|^2 d\tau \\ &+ (1 - KC_4) \int_t^T e^{-\lambda \tau} e^{-\lambda'(\tau-t)} E|\widehat{Z}_\tau|^2 d\tau \\ &\leq k_2^2 e^{-\lambda T} e^{-\lambda'(T-t)} E|\widehat{X}_T|^2 + KC_3 \int_t^T e^{-\lambda \tau} e^{-\lambda'(\tau-t)} |\widehat{X}_\tau|^2 d\tau \end{aligned}$$

Consequently, if $KC_4 = 1 - \alpha$ for some $\alpha \in (0,1)$, then

$$(3.6) \quad e^{-\lambda T} E|\widehat{X}_T|^2 + \bar{\lambda}_1 \|\widehat{X}\|^2_\lambda \leq K(C_1 + K)\|\widehat{Y}\|^2_\lambda + (KC_2 + k_1^2)\|\widehat{Z}\|^2_\lambda.$$

$$(3.7) \quad \|\widehat{X}\|^2_\lambda \leq B(\bar{\lambda}_1, T)[K(C_1 + K)\|\widehat{Y}\|^2_\lambda + (KC_2 + k_1^2)\|\widehat{Z}\|^2_\lambda].$$

$$(3.8) \quad \|\widehat{Y}\|^2_\lambda \leq B(\bar{\lambda}_2, T)[k_2^2 e^{-\lambda T} E|\widehat{X}_T|^2 + KC_3 \|\widehat{X}\|^2_\lambda],$$

$$(3.9) \quad \|\widehat{Z}\|^2_\lambda \leq \frac{A(\bar{\lambda}_2, T)}{\alpha}[k_2^2 e^{-\lambda T} E|\widehat{X}_T|^2 + KC_3 \|\widehat{X}\|^2_\lambda].$$

Proof. We first show (3.4). Let $t \in (0,T]$, λ, λ' be arbitrarily given, and consider the function $F_t(s,x) \stackrel{\Delta}{=} e^{-\lambda s} e^{-\lambda'(t-s)} |x|^2$, for $(s,x) \in [0,t] \times \mathbb{R}^n$.

Applying Itô's formula to $F_t(s, \widehat{X}_s)$ from 0 to t, and then taking expectation we have

$$
e^{-\lambda t} E|\widehat{X}_t|^2 + (\lambda - \lambda') E \int_0^t e^{-\lambda \tau} e^{-\lambda'(t-\tau)} |\widehat{X}_\tau|^2 d\tau
$$

$$
= \int_0^t e^{-\lambda \tau} e^{-\lambda'(t-\tau)} \Big\{ 2 \langle \widehat{X}_\tau, b(\tau, X_\tau, Y_\tau, Z_\tau) - b(\tau, X'_\tau, Y'_\tau, Z'_\tau) \rangle
$$

$$
+ \| \sigma(\tau, X_\tau, Y_\tau, Z_\tau) - \sigma(\tau, X'_\tau, Y'_\tau, Z'_\tau) \|^2 \Big\} d\tau
$$

$$
+ 2E \int_0^t e^{-\lambda \tau} e^{-\lambda'(t-\tau)} \langle \widehat{X}_\tau, d\widehat{\eta}_\tau \rangle.
$$

Since $X_t, X'_t \in \overline{\mathcal{O}}_1$, $\forall t \in [0,T]$, a.s., we derive from Definition 3.1-(3) that $e^{-\lambda t} e^{-\lambda'(t-\tau)} \langle \widehat{X}_t, d\widehat{\eta}_\tau \rangle \le 0$ (as a signed measure), $\forall s \in [0,T]$, a.s.. Therefore, repeatedly applying the Schwartz inequality and the inequality $2ab \le ca^2 + c^{-1}b^2$, $\forall c > 0$, using the definition of $\bar{\lambda}_1$, together with some elementary computation with the help of (A4), we derive (3.4).

To prove (3.5), we let $\widetilde{F}_t(s,x) = e^{-\lambda s} e^{-\lambda'(s-t)} |x|^2$, and apply Itô's formula to $\widetilde{F}_t(s, Y_s)$ from t to T to get

$$
e^{-\lambda t} E|\widehat{Y}_t|^2 + (\lambda' + \lambda) E \int_t^T e^{-\lambda \tau} e^{-\lambda'(\tau - t)} |\widehat{Y}_\tau|^2 d\tau
$$

$$
+ E \int_t^T e^{-\lambda \tau} e^{-\lambda'(\tau - t)} \| \widehat{Z}_\tau \|^2 d\tau
$$

$$
= e^{-\lambda T} e^{-\lambda'(T-t)} E|g(X_T) - g(X'_T)|^2
$$

$$
+ 2 \int_t^T e^{-\lambda'(\tau - t)} e^{-\lambda \tau} \langle \widehat{Y}_\tau, h(\tau, X_\tau, Y_\tau, Z_\tau) - h(\tau, X'_\tau, Y'_\tau, Z'_\tau) \rangle \, d\tau
$$

$$
+ 2E \int_t^T e^{-\lambda'(\tau - t)} e^{-\lambda \tau} \langle \widehat{Y}_\tau, d\widehat{\zeta}_\tau \rangle.
$$

Again, since $Y(t, \cdot), Y'(t, \cdot) \in \mathcal{O}_2(t, \cdot)$, P-a.s., by Definition 3.1-(4) we have $\langle \widehat{Y}_t(\omega), d\widehat{\zeta}_t(\omega) \rangle \le 0$, $dt \times dP$-a.s.. Thus, by using the similar argument as before, and using the definition of $\bar{\lambda}_2$, we obtain (3.5).

Now, letting $\lambda' = 0$ and $t = T$ in (3.4) yields (3.6); letting $\lambda' = \bar{\lambda}_1$ in (3.4) and then integrating both sides from 0 to T yields (3.7), since $B(\lambda_1, \cdot)$ is increasing; letting $\lambda' = \bar{\lambda}_2$ in (3.5) and integrating from 0 to T yields (3.8). Finally, note that if $\bar{\lambda}_2 \le 0$, then letting $\lambda' = \bar{\lambda}_2$ and $t = 0$ in (3.5) one has (remember $KC_4 = 1 - \alpha$)

$$
\|\widehat{Z}\|_\lambda^2 \le \int_0^T e^{-\lambda \tau} e^{-\bar{\lambda}_2(\tau - t)} \| \widehat{Z}_\tau \|^2 d\tau
$$

$$
\le \frac{e^{|\bar{\lambda}_2|T}}{\alpha} \Big\{ k_2^2 e^{-\lambda T} E|\widehat{X}_T|^2 + KC_3 \|\widehat{X}\|_\lambda^2 \Big\};
$$

while if $\bar{\lambda}_2 > 0$, then let $\lambda' = 0$ in (3.4) one has

$$\|\widehat{Z}\|_\lambda^2 \le \frac{1}{\alpha}\Big\{k_2^2 e^{-\lambda T} E|\widehat{X}_T|^2 + KC_3\|\widehat{X}\|_\lambda^2\Big\}.$$

Combining the above we obtain (3.9). $\qquad\qquad\qquad\qquad\qquad\qquad\qquad$ $\square$

We now present another set of useful a priori estimates for the adapted solution to FBSDER (3.2). Denote $\sigma^0(t, w) = \sigma(s, 0, 0, 0, w)$, $f^0(t, w) = f(s, 0, 0, 0, w)$, $h^0(t, w) = h(t, 0, 0, 0, w)$, and $g^0(w) = g(0, w)$.

Lemma 3.3. *Assume (A4). Let (X, Y, Z, η, ζ) be an adapted solution to the FBSDER (3.2). For any $\lambda, \lambda' \in \mathbb{R}, \varepsilon > 0, C_1, C_2, C_3, C_4 > 0$, we define $\bar{\lambda}_1^\varepsilon = \bar{\lambda}_1 - (1 + K^2)\varepsilon$ and $\bar{\lambda}_2^\varepsilon = \bar{\lambda}_2 - \varepsilon$, where $\bar{\lambda}_1$ and $\bar{\lambda}_2$ are those defined in Lemma 3.2. Then*

$$e^{-\lambda t} E|X_t|^2 + (\bar{\lambda}_1^\varepsilon - \lambda') \int_0^t e^{-\lambda'(t-\tau)} e^{-\lambda s} E|X_\tau|^2 d\tau \le e^{-\lambda' t}|x|^2$$

(3.10)
$$+ \int_0^t e^{-\lambda'(t-\tau)} e^{-\lambda \tau}\Big\{\frac{1}{\varepsilon}E|f(\tau, 0, 0, 0)|^2 + \Big(1 + \frac{1}{\varepsilon}\Big)|\sigma(\tau, 0, 0, 0)|^2$$
$$+ K(C_1 + K(1 + \varepsilon))E|Y_\tau|^2 + (KC_2 + k_1^2(1 + \varepsilon))E|Z_\tau|^2\Big\}d\tau.$$

and

$$e^{-\lambda t} E|Y_t|^2 + (\bar{\lambda}_2^\varepsilon - \lambda') \int_t^T e^{-\lambda'(\tau-t)} e^{-\lambda \tau} E|Y_\tau|^2 d\tau$$

$$+ (1 - k_4 C_4) \int_t^T e^{-\lambda'(\tau-t)} e^{-\lambda \tau} E|Z_\tau|^2 d\tau$$

(3.11)
$$\le k_2^2(1 + \varepsilon) e^{-\lambda'(T-t)} e^{-\lambda T} E|X_T|^2 + \Big(1 + \frac{1}{\varepsilon}\Big) e^{-\lambda'(T-t)} e^{-\lambda T} E|g(0)|^2$$

$$+ \int_t^T e^{-\lambda'(\tau-t)} e^{-\lambda \tau}\Big\{KC_3 E|X_\tau|^2 + \frac{1}{\varepsilon}E|h(\tau, 0, 0, 0)|^2\Big\}d\tau.$$

Consequently, if $C_4 = \frac{1-\alpha}{K}$, for some $\alpha \in (0, 1)$, we have

$$e^{-\lambda T} E|X_T|^2 + \bar{\lambda}_1^\varepsilon \|X\|_\lambda^2 \le \Big[|x|^2 + K(C_1 + K(1 + \varepsilon))\|Y\|_\lambda^2$$

(3.12)
$$+ (KC_2 + k_1^2(1 + \varepsilon))\|Z\|_\lambda^2 + \frac{1}{\varepsilon}\|f^0\|_\lambda^2 + \Big(1 + \frac{1}{\varepsilon}\Big)\|\sigma^0\|_\lambda^2\Big].$$

$$\|X\|_\lambda^2 \le B(\bar{\lambda}_1^\varepsilon, T)\Big[|x|^2 + K(C_1 + K(1 + \varepsilon))\|Y\|_\lambda^2$$

(3.13)
$$+ (KC_2 + k_1^2(1 + \varepsilon))\|Z\|_\lambda^2 + \frac{1}{\varepsilon}\|f^0\|_\lambda^2 + \Big(1 + \frac{1}{\varepsilon}\Big)\|\sigma^0\|_\lambda^2\Big].$$

$$\|Y\|_\lambda^2 \le B(\bar{\lambda}_2^\varepsilon, T)\left[k_2^2(1+\varepsilon)e^{-\lambda T}E|X_T|^2 + KC_3\|X\|_\lambda^2\right.$$

(3.14)

$$\left. + \left(1+\frac{1}{\varepsilon}\right)e^{-\lambda T}E|g^0|^2 + \frac{1}{\varepsilon}\|h^0\|_\lambda^2\right]$$

$$\|Z\|_\lambda^2 \le \frac{A(\bar{\lambda}_2^\varepsilon, T)}{\alpha}\left[k_2^2(1+\varepsilon)e^{-\lambda T}E|X_T|^2 + KC_3\|X\|_\lambda^2\right.$$

(3.15)

$$\left. + \left(1+\frac{1}{\varepsilon}\right)e^{-\lambda T}E|g^0|^2 + \frac{1}{\varepsilon}\|h^0\|_\lambda^2\right].$$

§3.2. Existence and uniqueness of the adapted solutions

We are now ready to study the well-posedness of the FBSDER (3.2). To begin with we introduce a mapping $\Gamma : \mathbf{H^c} \mapsto \mathbf{H^c}$ defined as follows: for fixed $x \in \mathbb{R}^n$, let $\overline{X} \stackrel{\Delta}{=} \Gamma(X)$ be the solution to the FSDER:

$$(3.16) \quad \overline{X}_t = x + \int_0^t b(s, \overline{X}_s, Y_s, Z_s)ds + \int_0^t \sigma(s, \overline{X}_s, Y_s, Z_s)dW_s + \bar{\eta}_t,$$

where the processes Y and Z are the solution to the following BSDER:

$$(3.17) \quad Y_t = g(X_T) + \int_t^T h(s, X_s, Y_s, Z_s)ds - \int_t^T Z_s dW_s + \zeta_T - \zeta_t.$$

Clearly, the assumption (A4) enables us to apply Theorem 2.5 to conclude that the BSDER (3.17) has a unique solution (Y, Z, ζ), which in turn guarantees the existence and uniqueness of the adapted solution $\overline{X}$ to the FSDER (3.16), thanks to Theorem 1.2. Furthermore, by definition of λ_1^ε (Lemma 3.3) we see that if λ is chosen so that $\bar{\lambda}_1 > 0$, then it is always possible to choose $\varepsilon > 0$ small enough so that $\bar{\lambda}_1^\varepsilon > 0$ as well; and (3.12) will lead to $\overline{X} \in \mathbf{H}_{\lambda, \bar{\lambda}_1}$ (since $\bar{\lambda}_1 > 0$ and $\bar{\lambda}_1^\varepsilon > 0$). Let us try to find a suitable $\bar{\lambda}_1 > 0$ so that Γ is a contraction on $\mathbf{H}_{\lambda, \bar{\lambda}_1}$, which will lead to the existence and uniqueness of the adapted solution to the FBSDER (3.2) immediately.

To this end, let $X^1, X^2 \in \mathbf{H^c}$; and let (Y^i, Z^i, ζ^i) and $(\overline{X}^i, \bar{\eta}^i)$, $i = 1, 2$, be the corresponding solutions to (3.17) and (3.16), respectively. Denote $\Delta\xi = \xi^1 - \xi^2$, for $\xi = X, Y, Z, \overline{X}$. Applying (3.6)–(3.9) (with $C_4 = \frac{1-\alpha}{K}$) we easily deduce that

(3.18)
$$e^{-\lambda T}E|\Delta\overline{X}_T|^2 + \bar{\lambda}_1\|\Delta\overline{X}\|_\lambda^2$$
$$\le \mu(\alpha, T)\{k_2^2 e^{-\lambda T}E|\Delta X_T|^2 + KC_3\|\Delta X\|_\lambda^2\}.$$

where

$$(3.19) \qquad \mu(\alpha, T) \stackrel{\Delta}{=} K(C_1 + K)B(\bar{\lambda}_2, T) + \frac{A(\bar{\lambda}_2, T)}{\alpha}(KC_2 + k_1^2);$$

and (recall Lemma 3.2)

$$(3.20) \quad \bar{\lambda}_1 = \lambda - K(2 + C_1^{-1} + C_2^{-1}) - K^2; \quad \bar{\lambda}_2 = -\lambda - 2\gamma - K(C_3^{-1} + C_4^{-1}).$$

Clearly, the function $\mu(\cdot,\cdot)$ depends on the constants K, k_1, k_2, γ, the duration $T > 0$, and the choice of C_1–C_4 as well as λ, α. To compensate the generality of the coefficients, we shall impose the following *compatibility conditions*.

(**C-1**) $0 \le k_1 k_2 < 1$;

(**C-2**) $k_2 = 0$; $\exists \alpha \in (0,1)$ such that $\mu(\alpha, T)KC_3 < \bar{\lambda}_1$,

(**C-3**) $k_2 > 0$; $\exists \alpha_0 \in (k_1 k_2, 1)$, such that $\mu(\alpha_0^2, T)k_2^2 < 1$ and $\bar{\lambda}_1 = \frac{KC_3}{k_2^2}$.

We remark here that the compatibility condition (C-1) is not a surprise. We already saw it in Chapter 1 (Theorem 1.5.1). In fact, in Example 1.5.2 we showed that such a condition is almost necessary for the solvability of an FBSDE with general coefficients, even in non-reflected cases with small duration. The first existence and uniqueness result for FBSDER (3.2) is the following.

Theorem 3.4. *Assume (A4) and fix $C_4 = \frac{1-\alpha_0^2}{K}$. Assume that the compatibility conditions (C-1), and either (C-2) or (C-3) hold for some choices of constants λ, α, and C_1–C_3. Then the FBSDER (3.2) has a unique adapted solution over $[0, T]$.*

Proof. Fix $C_4 = \frac{1-\alpha_0^2}{K}$. First assume that (C-1) and (C-2) hold. Since $k_2 = 0$, (3.18) leads to that

$$\|\Delta \overline{X}\|_\lambda^2 \le \frac{\mu(\alpha, T)KC_3}{\bar{\lambda}_1} \|\Delta X\|_\lambda^2,$$

Since we can find C_1–C_3 and $\alpha \in (0,1)$ so that $\mu(\alpha, T)KC_3 < 1$, Γ is a contraction mapping on $(H, \|\cdot\|_\lambda)$. The theorem follows.

Similarly, if (C-1) and (C-3) hold, then we can solve λ from (3.20) and $\bar{\lambda}_1 = KC_3/k_2^2$, and then derive from (3.18) that

$$|\Delta \overline{X}|_{\lambda^0, \bar{\lambda}_1}^2 \le \mu(\alpha_0^2, T)k_2^2 |\Delta X|_{\lambda^0, \bar{\lambda}_1}^2,$$

Let C_i, $i = 1, 2, 3$ and $\alpha_0 \in (k_1 k_2, 1)$ be such that $\mu(\alpha_0^2, T)k_2^2 < 1$, the mapping Γ is again a contraction, but on the space $\mathbf{H}_{\lambda, \bar{\lambda}_1}$, proving the theorem again. □

A direct consequence of Theorem 3.4 is the following.

Corollary 3.5. *Assume (A4) and the compatibility condition (C-1). Then there exists $T_0 > 0$ such that for all $T \in (0, T_0]$, the FBSDER (3.2) has a unique adapted solution.*

In particular, if either $k_1 = 0$ or $k_2 = 0$, then the FBSDER (3.2) is always uniquely solvable on $[0, T]$ for T small.

Proof. First assume $k_2 = 0$. In light of Theorem 3.4 we need only show that there exists $T_0 = T_0(C_1, C_2, C_3, \lambda, \alpha)$ such that (C-2) holds for some choices of C_1–C_3 and λ, α, for all $T \in (0, T_0]$.

For fixed C_1, C_2, C_3, λ, and $\alpha \in (0,1)$ we have from (3.19) that

$$\mu(\alpha, 0)KC_3 = \frac{(KC_2 + k_1^2)KC_3}{\alpha}.$$

Therefore, let C_1–C_3 and α be fixed we can choose λ large enough so that $\mu(\alpha, 0)KC_3 < \bar{\lambda}_1$ holds. Then, by the continuity of the functions $A(\alpha, \cdot)$ and $B(\alpha, \cdot)$, for this fixed λ we can find $T_0 > 0$ such that $\mu(\alpha, T)KC_3 < \bar{\lambda}_1$ for all $T \in (0, T_0]$. Thus (C-2) holds for all $T \in (0, T_0]$ and the conclusion follows from Theorem 3.4.

Now assume that $k_2 > 0$. In this case we pick an $\alpha_0 \in (k_1 k_2, 1)$, and define

(3.21)
$$\delta \stackrel{\triangle}{=} \frac{1}{k_2^2} - \frac{k_1^2}{\alpha_0^2} > 0.$$

Now let $C_2 = \frac{\alpha_0^2 \delta}{2K}$, $C_4 = \frac{1-\alpha_0^2}{K}$, and choose λ so that $\bar{\lambda}_1 = (k_3 C_3)/k_2^2 > 0$. Since in this case we have

$$\mu(\alpha_0^2, 0) = \frac{KC_2 + k_1^2}{\alpha_0^2} = \frac{1}{2k_2^2} + \frac{k_1^2}{2\alpha_0^2} < \frac{1}{k_2^2},$$

thanks to (3.21). Using the continuity of $\mu(\alpha_0^2, \cdot)$ again, for any $C_1, C_3 > 0$ we can find $T_0(C_1, C_3) > 0$ such that $\mu(\alpha_0^2, T)k_2^2 < 1$ for all $T \in (0, T_0]$. In other words, the compatibility condition (C-3) holds for all $T \in (0, T_0]$, proving our assertion again.

Finally if $k_1 = 0$, then (C-1) becomes trivial, thus the corollary always holds. □

From the proofs above we see that there is actually room for one to play with constant C_1–C_3 to improve the "maximum existence interval" $[0, T_0)$. A natural question is then *is there any possibility that $T_0 = \infty$ so that the FBSDER (3.2) is solvable over arbitrary duration $[0, T]$?* Unfortunately, so far we have not seen an affirmative answer for such a question, even in the non-reflecting case, under this general setting. Furthermore, in the reflecting case, even if we assume all the coefficients are deterministic and smooth, it is still far from clear that we can successfully apply the method of optimal control or Four Step Scheme (Chapters 3 and 4) to solve an FBSDER, because the corresponding PDE will become a quasilinear variational inequality, thus seeking its classical solution becomes a very difficult problem in general.

We nevertheless have the following result that more or less covers a class of FBSDERs that are solvable over arbitrary durations.

Theorem 3.6. *Assume (A4) and the compatibility condition (C-1). Then there exists a constant $\Lambda > 0$, depending only on the constants K, k_1, k_2, such that whenever $\gamma < -\Lambda$, the FBSDER (3.2) has a unique adapted solution for all $T > 0$.*

Proof. We shall prove that either (C-2) or (C-3) will hold for all $T > 0$ provided γ is negative enough, and we shall determine the constant Λ in each case, separately.

First assume $k_2 = 0$. In this case let us consider the following minimization problem with constraints:

(3.22)
$$\min_{\substack{C_i>0,\ i=1,2,3;\bar{\lambda}_1>0,0<\alpha<1,\\ \bar{\lambda}_1-2K(KC_2+k_1^2)C_3>0}} F(C_1,C_2,C_3,\bar{\lambda}_1,\alpha),$$

where

(3.23)
$$F(C_1,C_2,C_3,\bar{\lambda}_1,\alpha) \triangleq \frac{(C_1+K)K^2C_3}{\bar{\lambda}_1 - 2(KC_2+k_1^2)KC_3} + \bar{\lambda}_1$$
$$+ K(2 + C_1^{-1} + C_2^{-1} + C_3^{-1}) + K^2\left(\frac{2-\alpha}{1-\alpha}\right).$$

Let Λ be the value of the problem (3.22) and (3.23). We show that if $\gamma < -\Lambda/2$, then (C-2) holds for all $T > 0$.

Indeed, if $\gamma < -\Lambda/2$, then we can find $C_1, C_2, C_3, \bar{\lambda}_1 > 0$ and $\alpha \in (0,1)$, such that $\bar{\lambda}_1 - 2(KC_2 + k_1^2)KC_3 > 0$, and

(3.24)
$$-2\gamma > \frac{(C_1+K)K^2C_3}{\bar{\lambda}_1 - 2(KC_2+k_1^2)KC_3} + \bar{\lambda}_1$$
$$+ K(2 + C_1^{-1} + C_2^{-1} + C_3^{-1}) + K^2\left(\frac{2-\alpha}{1-\alpha}\right).$$

On the other hand, eliminating λ in the expressions of $\bar{\lambda}_1$ and $\bar{\lambda}_2$ in (3.20), and letting $C_4 = \frac{(1-\alpha)}{K}$ we have

$$\bar{\lambda}_2 = -\left(\bar{\lambda}_1 + K(2 + C_1^{-1} + C_2^{-1} + C_3^{-1}) + \frac{K^2}{1-\alpha} + K^2\right) - 2\gamma.$$

Thus (3.24) is equivalent to

(3.25)
$$\frac{1}{\bar{\lambda}_1}\left\{\frac{K(C_1+K)}{\bar{\lambda}_2} + \frac{(KC_2+k_1^2)}{\alpha}\right\}KC_3 < 1,$$

and $\bar{\lambda}_2 > 0$. Consequently, $A(\bar{\lambda}_2, T) = 1$ and $B(\bar{\lambda}_2, T) \leq \bar{\lambda}_2^{-1}$ (recall (3.3)); and (3.25) implies that $\mu(\alpha, T)KC_3 < \bar{\lambda}_1$, i.e., (C-2) holds for all $T > 0$.

Now assume $k_2 > 0$. Following the arguments in Corollary 3.5 we choose $\bar{\lambda}_1 = \frac{KC_3}{k_2^2} > 0$, $C_4 = \frac{1-\alpha_0^2}{K}$, and $\alpha_0 \in (k_1k_2, 1)$. Let $\delta > 0$ be that defined by (3.21), and consider the minimization problem:

(3.26)
$$\min_{\substack{C_i>0,\ i=1,2,3;\\ \delta\alpha_0^2 - KC_2>0}} \widetilde{F}(C_1,C_2,C_3),$$

where

(3.27)
$$\widetilde{F}(C_1,C_2,C_3) \triangleq \frac{\alpha_0^2 K(C_1+K)}{\delta\alpha_0^2 - KC_2} + K(2 + C_1^{-1} + C_2^{-1} + C_3^{-1})$$
$$+ \frac{KC_3}{k_2^2} + K^2\left(\frac{2-\alpha_0^2}{1-\alpha_0^2}\right).$$

Let Λ be the value of the problem (3.26) and (3.27), one can show as in the previous case that if $\gamma < -\Lambda/2$, then $\bar{\lambda}_2 > 0$ (hence $A(\bar{\lambda}_2, T) = 1$ and $B(\bar{\lambda}_2, T) \le \bar{\lambda}_2^{-1}$), and $\mu(\alpha_0^2, T)k_2^2 < 1$. Namely (C-3) holds for all $T > 0$. Combining the above we proved the theorem. $\square$

§3.3. A continuous dependence result

In many applications one would like to study the dependence of the adapted solution of an FBSDE on the initial data. For example, suppose that there exists a constant $T > 0$ such that the FBSDER (3.2) is uniquely solvable over any duration $[t, T] \subseteq [0, T]$, and denote its adapted solution by $(X^{t,x}, Y^{t,x}, Z^{t,x}, \eta^{t,x}, \zeta^{t,x})$. Then an interesting question would be how the random field $(t, x) \mapsto (X^{t,x}, Y^{t,x}, Z^{t,x}, \eta^{t,x}, \zeta^{t,x})$ behaves. Such a behavior is particularly useful when one wants to relate an FBSDE to a partial differential equation, as we shall see in the next chapter.

In what follows we consider only the case when $m = 1$, namely, the BSDER is one dimensional. We shall also make use of the following assumption:

(A5) (i) The coefficients b, h, σ, g are deterministic;
(ii) The domains $\{\mathcal{O}_2(\cdot, \cdot)\}$ are of the form $\mathcal{O}(s, \omega) = \mathcal{O}_2(s, X^{t,x}(s, \omega))$, $(s, \omega) \in [t, T] \times \mathbb{R}^n$, where $\mathcal{O}_2(t, x) = (L(t, x), U(t, x))$, where $L(\cdot, \cdot)$ and $U(\cdot, \cdot)$ are smooth deterministic functions of (t, x).

We note that the part (ii) of assumption (A5) does not cover, and is not covered by, the assumption (A4) with $m = 1$. This is because when $m = 1$ the domain $\mathcal{O}_2$ is simply an interval, and can be handled differently from the way we presented in §2 (see, e.g., Cvitanic & Karatzas [1]). Note also that if we can bypass §2 to derive the solvability of BSDERs, then the method we presented in the current section should always work for the solvability for FBSDERs. Therefore in what follows we shall discuss the continuous dependence in an a priori manner, without going into the details of existence and uniqueness again. Next, observe that under (A5) FBSDER (3.2) becomes "Markovian", we can apply the standard technique of "time shifting" to show that the process $\{Y^{t,x}(s)\}_{s \ge t}$ is $\mathcal{F}_t^s$-adapted, where $\mathcal{F}_s^t = \sigma\{W_r, t \le r \le s\}$. Consequently an application of the Blumenthal 0-1 law leads to that the function $u(t, x) = Y_t^{t,x}$ is always deterministic! In what follows we use the convention that $X^{t,x}(s) \equiv x$, $Y^{t,x}(s) \equiv Y^{t,x}(t)$, and $Z^{t,x}(s) \equiv 0$, for $s \in [0, t]$. Our main result of this subsection is the following.

Theorem 3.7. *Assume (A5) as well as (A4)-(iii)–(v). Assume also that the compatibility conditions (C-1) and either (C-2) or (C-3) hold. Let $u(t, x) \overset{\Delta}{=} Y_t^{t,x}$, $(t, x) \in [0, T] \times \mathcal{O}_1$. Then u is continuous on $[0, T] \times \mathcal{O}$ and there exists $C > 0$ depending only on T, b, h, g, and σ, such that the following estimate holds:*

$$(3.28) \quad |u(t_1, x_1) - u(t_2, x_2)|^2 \le C\Big(|x_1 - x_2|^2 + (1 + |x_1|^2 \vee |x_2|^2)|t_1 - t_2|\Big).$$

Proof. The proof is quite similar to that of Theorem 3.4, so we only sketch it.

Let (t_1, x_1) and (t_2, x_2) be given, and let $\widehat{X} = X^{t_1,x_1} - X^{t_2,x_2}$. Assume first $t_1 \geq t_2$, and recall the norms $\|\cdot\|_{t,\lambda}$ and $|\!|\!|\cdot|\!|\!|_{t,\lambda,\beta}$ at the beginning of §3.1. Repeating the arguments of Theorem 3.4 over the interval $[t_2, T]$, we see that (3.8) and (3.9) will look the same, with $\|\cdot\|_\lambda$ being replaced by $\|\cdot\|_{t_2,\lambda}$; but (3.6) and (3.7) become

$$(3.6)' \qquad \begin{aligned} & e^{-\lambda T} E|\widehat{X}_T|^2 + \bar{\lambda}_1 \|\widehat{X}\|_{t_1,\lambda}^2 \\ & \leq K(C_1 + K)\|\widehat{Y}\|_{t_2,\lambda}^2 + (KC_2 + k_1^2)\|\widehat{Z}\|_{t_2,\lambda}^2 + E|\widehat{X}(t_2)|^2. \end{aligned}$$

$$(3.7)' \qquad \begin{aligned} \|\widehat{X}\|_{t_2,\lambda}^2 \leq & \widetilde{B}(\bar{\lambda}_1, T)[K(C_1 + K)\|\widehat{Y}\|_{t_2,\lambda}^2 \\ & + (KC_2 + k_1^2)\|\widehat{Z}\|_{t_2,\lambda}^2 + E|\widehat{X}(t_2)|^2], \end{aligned}$$

where $\widetilde{B}(\lambda, T) \triangleq \frac{e^{-\lambda t_2} - e^{-\lambda T}}{\lambda}$. Now similar to (3.18), one shows that

$$(3.18)' \qquad \begin{aligned} & e^{-\lambda T} E|\widehat{X}_T|^2 + \bar{\lambda}_1 \|\widehat{X}\|_{t_2,\lambda}^2 \\ & \leq \mu(\alpha, T)\{k_2^2 e^{-\lambda T} E|\widehat{X}_T|^2 + KC_3\|\widehat{X}\|_{t_2,\lambda}^2\} + E|\widehat{X}(t_2)|^2. \end{aligned}$$

Arguing as in the proof of Theorem 3.4 and using compatibility conditions (C-1)–(C-3), we can find a constant $C > 0$ depending only on $T > 0$ and K, k_1, k_2 such that

$$(3.29) \qquad |\!|\!|\widehat{X}|\!|\!|_{t_2,\lambda,\beta}^2 \leq CE|\widehat{X}(t_2)|^2 = CE|x_2 - X^{t_1,x_1}(t_2)|^2,$$

where $\beta = \bar{\lambda}_1 - \mu(\alpha, T)KC_3$ if $k_2 = 0$; and $\beta = \mu(\alpha, T)k_2^2$ if $k_2 > 0$.

From now on by slightly abuse of notations we let $C > 0$ be a generic constant depending only on T, K, k_1 and k_2, and be allowed to vary from line to line. Applying standard arguments using Burkholder-Davis-Gundy inequality we obtain that

$$(3.30) \qquad E\sup_{t_2 \leq s \leq T}|X^1(s)|^2 + E\sup_{t_2 \leq s \leq T}|Y^1(s)|^2 \leq CE|\widehat{X}(t_2)|^2.$$

To estimate $E|\widehat{X}(t_2)|^2$ let us recall the parameters λ_1^ε and λ_2^ε defined in Lemma 3.3. For each $\varepsilon > 0$ define

$$\mu^\varepsilon(\alpha, T) \triangleq K(C_1 + K(1 + \varepsilon))B(\lambda_2^\varepsilon, T) + \frac{A(\lambda_2^\varepsilon, T)}{1 - KC_4}KC_2.$$

Since $\lambda_1^\varepsilon \to \lambda_1$, $\lambda_2^\varepsilon \to \lambda_2$, and $\mu^\varepsilon(\alpha, T) \to \mu(\alpha, T)$, as $\varepsilon \to 0$, if the compatibility condition (C-1) and either (C-2) or (C-3) hold, then we can choose $\varepsilon > 0$ such that $\mu^\varepsilon(\alpha, T)k_2^2(1 + \varepsilon) < 1$ when $k_2 = 0$ and $\mu^\varepsilon(\alpha, T)KC_3 < \lambda_1^\varepsilon$ when $k_2 \neq 0$. For this fixed $\varepsilon > 0$ we can then repeat the argument of Theorem 3.4 by using (3.12)—(3.15) to derive that

$$\left(1 - \frac{\mu^\varepsilon(\alpha, T)KC_3}{\bar{\lambda}_1^\varepsilon}\right)\|X^1\|_\lambda^2 \leq C(\varepsilon)\left[|x_1|^2 + \left(1 + \frac{1}{\varepsilon}\right)\right], \qquad k_2 = 0;$$

or

$$\left(1 - \mu^\varepsilon(\alpha, T)k_2^2\right)|X^1|_{\lambda,\beta}^2 \le C(\varepsilon)\left[|x_1|^2 + \left(1 + \frac{1}{\varepsilon}\right)\right], \quad k_2 \neq 0,$$

where $C(\varepsilon)$ is some constant depending on T, K, k_1, k_2, and ε. Since $\varepsilon > 0$ is now fixed, in either case we have, for a generic constant $C > 0$,

$$\|X^1\|_\lambda^2 \le C(1 + |x_1|^2),$$

which in turn shows that, in light of (3.12)–(3.15) $\|Y^1\|_\lambda^2 \le C(1 + |x_1|^2)$, and $\|Z\|_\lambda^2 \le C(1 + |x_1|^2)$. Again, applying the Burkholder and Hölder inequalities we can then derive

$$(3.31) \qquad E\left\{\sup_{t_1 \le s \le T} |X^1(t)|^2\right\} + E\left\{\sup_{t_1 \le s \le T} |Y^1(t)|^2\right\} \le C(1 + |x_1|^2).$$

Now, note that on the interval $[t_1, t_2]$ the process $(\widehat{X}, \widehat{Y}, \widehat{Z})$ satisfies the following SDE:

$$(3.32) \qquad \begin{cases} \widehat{X}(s) = (x_1 - x_2) + \displaystyle\int_{t_1}^s b^1(r)dr + \int_{t_1}^s \sigma^1(r)dW(r), \\[2mm] \widehat{Y}(s) = \widehat{Y}(t_2) + \displaystyle\int_s^{t_2} h^1(r)dr + \int_s^{t_2} Z^1(r)dW(r), \end{cases} \quad s \in [t_1, t_2],$$

where $b^1(r) = b(r, X^2(r), Y^1(r), Z^1(r))$, $\sigma^1(r) = \sigma(r, X^1(r), Y^1(r), Z^1(r))$, and $h^1(r) = h(r, X^1(r), Y^1(r), Z^1(r))$. Now from the first equation of (3.32) we derive easily that

$$E\{\sup_{t_1 \le s \le t_2} |\widehat{X}(s)|^2\} \le C\{|x_1 - x_2|^2 + (1 + |x_1|^2)|t_1 - t_2|\}.$$

Combining this with (3.30), (3.31), as well as the assumption (A4-iv), we derive from the second equation of (3.32) that

$$E|\widehat{Y}(t_1)|^2 \le E|\widehat{Y}(t_2)|^2 + C(1 + |x_1|^2 \vee |x_2|^2)|t_1 - t_2|$$
$$\le C\{|x_1 - x_2|^2 + (1 + |x_1|^2 \vee |x_2|^2)|t_1 - t_2|\}.$$

Since $\widehat{Y}(t_1) = u(t_1, x_1) - u(t_2, x_2)$ is deterministic, (3.28) follows. The case when $t_1 \le t_2$ can be proved by symmetry, the proof is complete. $\qquad\square$

Chapter 8
Applications of FBSDEs

In this chapter we collect some interesting applications of FBSDEs. These applications appear in various fields of both theoretical and applied probability problems, but our main interest will be those that related to the truly coupled FBSDEs and their applications in mathematical finance. Let us first recall the FBSDE in its general form: denote $\Theta = (X, Y, Z)$,

$$
(1.1) \quad
\begin{cases}
X(t) = x + \displaystyle\int_0^t b(s, \Theta(s))ds + \int_0^t \sigma(s, \Theta(s))dW(s), \\
Y(t) = g(X(T)) + \displaystyle\int_t^T \widehat{b}(s, \Theta(s))ds - \int_t^T Z(s)dW(s), \quad t \in [0, T],
\end{cases}
$$

In different applications we will make assumptions that are variations of what we have seen before, in order to suit the situation.

§1. An Integral Representation Formula

In this section we consider a special case: $\widehat{b} \equiv 0$, and σ is independent of z. Thus (1.1) takes the form:

$$
(1.2) \quad
\begin{cases}
X(t) = x + \displaystyle\int_0^t b(s, \Theta(s))ds + \int_0^t \sigma(s, X(s), Y(s))dW(s), \\
Y(t) = g(X(T)) - \displaystyle\int_t^T Z(s)dW(s), \quad t \in [0, T],
\end{cases}
$$

From the Four Step Scheme (see Chapter 4), we know that if we define $z(t, x, y, p) = p\sigma(t, x, y)$, and let $\theta(t, x)$ be the classical solution of the following system of PDEs:

$$
(1.3) \quad
\begin{cases}
\theta_t^k + \dfrac{1}{2}\mathrm{tr}\,[\theta_{xx}^k \sigma(t, x, \theta)\sigma(t, x, \theta)^T] + \langle\, b(t, x, \theta, z(t, x, \theta, \theta_x)), \theta_x^k \,\rangle = 0, \\
\hspace{6cm} k = 1, \cdots, m; \\
\theta(T, x) = g(x),
\end{cases}
$$

then the (unique) adapted solution of (1.2) is given by

$$
(1.4) \quad
\begin{cases}
X(t) = x + \displaystyle\int_0^t \widetilde{b}(s, X(s))ds + \int_0^t \widetilde{\sigma}(s, X(s))dW(s), \\
Y(t) = \theta(t, X(t)); \\
Z(t) = \theta_x(t, X(t))\sigma(t, X(t), \theta(t, X(t))).
\end{cases}
$$

where

$$
(1.5) \quad
\begin{cases}
\widetilde{b}(t, x) = b(t, x, \theta(t, x), \theta_x(t, x)\sigma(t, x, \theta(t, x))); \\
\widetilde{\sigma}(t, x) = \sigma(t, x, \theta(t, x)),
\end{cases}
$$

Now from the second (backward) equation in (1.2), and noting that Y_0 is non-random by Blumenthal 0-1 law, we have $Y_0 = EY_0 = Eg(X_T)$; and setting $t = 0$ in (1.2) we then have

$$(1.6) \quad g(X(T)) = Eg(X(T)) + \int_0^T \theta_x(s, X(s))\sigma(s, X(s), \theta(s, X(s)))dW(s).$$

Let us compare (1.6) with the Clark-Haussmann-Ocone formula in this special setting. For simplicity, we assume $m = n = 1$. Recall that the general form of the Clark-Haussmann-Ocone formula in this case is:

$$(1.7) \qquad g(X(T)) = Eg(X(T)) + \int_0^T E\{D_s g(X(T))|\mathcal{F}_s\}dW_s,$$

where D is the so-called "Malliavin derivative" operator. Note that by Malliavian calculus we have, for each $s \in [0, T]$, that $D_s g(X(T)) = g'(X(T))D_s X(T)$, and

$$D_s X(t) = \tilde{\sigma}(s, X(s)) + \int_s^t \tilde{b}_x(r, X(r))D_s X(r)dr$$
$$+ \int_s^t \tilde{\sigma}_x(r, X(r))D_s X(r)dW(r), \qquad t \in [s, T].$$

Denote

$$Z(t) = \int_s^t \tilde{b}_x(r, X(r))dr + \int_s^t \tilde{\sigma}_x(r, X(r))dW(r),$$

and let $\mathcal{E}(Z)_t$ be the Doléans-Dade stochastic exponential of Z, that is,

$$
(1.8) \quad
\begin{aligned}
\mathcal{E}(Z)_t &= \exp\{Z(t) - \frac{1}{2}[Z, Z](t)\} \\
&= \exp\left\{\int_s^t \tilde{\sigma}_x(r, X(r))dW(r) + \int_s^t [\tilde{b}_x(r, X(r)) - \frac{1}{2}\tilde{\sigma}_x^2(r, X(r))]dr\right\}.
\end{aligned}
$$

Then the process $u(t) \overset{\Delta}{=} D_s X(t)$, $t \in [s, T]$ can be written as $u(t) = \mathcal{E}(Z)_t \tilde{\sigma}(s, X(s))$. Therefore,

$$
(1.9) \quad
\begin{aligned}
E\{D_s g(X(T))|\mathcal{F}_s\} &= E\{g'(X(T))D_s X(T)|\mathcal{F}_s\} \\
&= E\{g'(X(T))\mathcal{E}(Z)_T|\mathcal{F}_s\}\tilde{\sigma}(s, X(s)).
\end{aligned}
$$

Putting this back into (1.7) and comparing it to (1.6) we obtain immediately that

$$\int_t^T \left\{E\{D_s g(X(T))|\mathcal{F}_s\} - \tilde{\sigma}(s, X(s))\theta_x(s, X(s))\right\}dW(s) = 0,$$

and consequently,

$$
(1.10) \quad
\begin{cases}
E\{D_s g(X(T))|\mathcal{F}_s\} = \tilde{\sigma}(s, X(s))\theta_x(s, X(s)); \\
E\{g'(X(T))\mathcal{E}(X)_T|\mathcal{F}_s\} = \theta_x(s, X(s)),
\end{cases}
\qquad d\mathbf{P} \otimes dt\text{-a.e.}
$$

Since the expressions on the right sides of (1.10) depend neither on the Malliavin derivatives, nor on the conditional expectations, they are more amenable in general. Also, since forward SDE in (1.4) depends actually on Y and Z, we thus obtained an *integral representation formula* (1.6) that is more general than the "classical" Clark-Haussmann-Ocone's formula, when the Brownian functional is of the form $g(X(T))$.

It is interesting to notice that the second equation in (1.10) does not contain the Malliavin derivative, and it leads to Haussmann's version of integral representation formula. Let us now prove it directly without using Malliavin calculus. To do this, we define a the process $p_t \overset{\triangle}{=} \theta_x(t, X(t))$ (such a process is often of independent interest in, e.g., stochastic control theory). For simplicity we assume $m = n = 1$ again and that the FBSDE is decoupled. That is

$$(1.11) \quad \begin{cases} X(t) = x + \displaystyle\int_0^t b(s, X(s))ds + \int_0^t \sigma(s, X(s))dW(s), \\ Y(t) = g(X(T)) - \displaystyle\int_t^T Z(s)dW(s), \quad t \in [0, T], \end{cases}$$

and the PDE (1.3) becomes

$$(1.12) \quad \begin{cases} \theta_t + \dfrac{1}{2}\theta_{xx}\sigma^2(t, x) + b(t, x)\theta_x = 0, \\ \theta(T, x) = g(x), \end{cases}$$

We should note that the following arguments are all valid for the coupled FBSDEs with $\hat{b} = 0$, in which case we should simply replace (1.11) by (1.4).

Proposition 1.1 *There exists an adapted process $\{K(t) : t \geq 0\}$ such that (p, K) is the unique adapted solution of the following backward SDE:*

$$(1.13) \quad \begin{aligned} p_t = g'(X(T)) + & \int_t^T [b_x(s, X(s))p_s + \sigma_x(s, X(s))K(s)]ds \\ & - \int_t^T K(s)dW(s). \end{aligned}$$

In particular, if the function θ is C^3, then $K(t) = \theta_{xx}(t, X(t))\sigma(t, X(t))$ for $t \geq 0$.

Proof. We first assume that θ is C^3. Taking one more derivative in the x variable to the equation (1.12) and denote $u = \theta_x$ we have

$$(1.14) \quad \begin{cases} u_t + \dfrac{1}{2}u_{xx}\sigma^2(t, x) + [b(t, x) + (\sigma\sigma_x)(t, x)]u_x + b(t, x)u = 0, \\ u(T, x) = g_x(x). \end{cases}$$

On the other hand, if we apply Itô's formula to u from t to τ $(0 \leq t \leq \tau)$, then we have

$$u(\tau, X(\tau)) = u(t, X(t)) + \int_t^\tau \{u_t(s, X(s)) + u_x(s, X(s))b(s, X(s))$$

(1.15)
$$+ \frac{1}{2}u_{xx}(s, X(s))\sigma^2(s, X(s))\}ds$$

$$+ \int_t^\tau u_x(s, X(s))\sigma(s, X(s))dW(s).$$

Using (1.14) and denoting $K(t) = u_x(t, X(t))\sigma(t, X(t))$, we obtain from (1.15) that

$$u(\tau, X(\tau)) = u(t, X(t)) - \int_t^\tau [ub_x + u_x(\sigma\sigma_x)](s, X(s))ds$$

$$+ \int_t^\tau u_x(s, X(s))\sigma(s, X(s))dW(s)$$

(1.16)
$$= u(t, X(t)) - \int_t^\tau [ub_x(s, X(s)) + K(s)\sigma_x(s, X(s))]ds$$

$$+ \int_t^\tau K(s)dW(s),$$

Now setting $p_t = u(t, X(t))$ and $\tau = T$, we obtain (1.13) immediately.

In the general case where θ is not necessarily C^3 we argue as follows. Let (p, K) be the adapted solution to the backward SDE (1.13), and we are to show that $p_t = \theta_x(t, X(t))$, that is, $\forall h \in \mathbb{R}$,

(1.17) $\theta(t, X(t) + h) - \theta(t, X(t)) = p_t h + o(h)$, $\forall t$, a.s.

To this end, fix $t \in [0, T]$ and consider the SDE

(1.18) $X^h(\tau) = X(t) + h + \int_t^\tau b(s, X^h(s))ds + \int_t^\tau \sigma(s, X^h(s))dW(s)$,

for $t \le \tau \le T$. Define $\zeta_\tau^h = X^h(\tau) - X(\tau)$, $\tau \in [t, T]$. Then it is easy to verify that ζ^h satisfies

(1.19) $d\zeta^h(\tau) = b_x(\tau, X(\tau))\zeta^h(\tau)d\tau + \sigma_x(\tau, X(\tau))\zeta^h(\tau)dW(\tau) + \varepsilon^h(\tau)$,

where

$$\varepsilon^h(\tau) = \int_t^\tau \left\{ \int_0^1 [b_x(s, X(s) + \beta\zeta^h(s)) - b_x(s, X(s))]d\beta \right\} \zeta^h(s)ds$$

$$+ \int_t^\tau \left\{ \int_0^1 [\sigma_x(s, X(s) + \beta\zeta^h(s)) - \sigma_x(s, X(s))]d\beta \right\} \zeta^h(s)dW(s).$$

Thus by the standard results in SDE we have $E\{\sup_{t \le \tau \le T} |\varepsilon^h(\tau)| \big| \mathcal{F}_t\} = o(h)$.

On the other hand, using Four Step Scheme one shows that

$$\theta(t, X(t)) = E\{g(X(T))|\mathcal{F}_t\}, \qquad \theta(t, X(t) + h) = E\{g(X(T))|\mathcal{F}_t\},$$

thus

(1.20)
$$\theta(t, X(t) + h) - \theta(t, X(t))$$
$$= E\{g(X^h(T)) - g(X(T))|\mathcal{F}_t\}$$
$$= E\{g'(X(T))\zeta_T^h|\mathcal{F}_t\} + E\left\{\int_0^1 [g'(X_T + \beta\zeta_T^h) - g'(X_T)]d\beta\zeta_T^h\Big|\mathcal{F}_t\right\}$$
$$= E\{g'(X(T))\zeta_T^h|\mathcal{F}_t\} + o(h).$$

Now applying Itô's formula to $p_\tau\zeta_\tau^h$ from $\tau = t$ to $\tau = T$ we have

$$\langle g'(X(T)\zeta^h(T) = p_t h + o(h) + m(T) - m(t),$$

where m stands for some $\{\mathcal{F}_t\}_{t\geq 0}$-martingale. Taking conditional expectation we obtain from (1.20) that

$$\theta(t, X(t) + h) - \theta(t, X(t)) = p_t h + o(h), \qquad \textbf{P-a.s.}, \ \forall t \in [0, T].$$

Using the continuity of both X and p we have $\theta_x(t, X(t)) = p_t$, $\forall t$, **P**-a.s., proving the proposition. $\qquad\qquad\qquad\qquad\qquad\qquad\qquad\qquad\qquad\quad\square$

§2. A Nonlinear Feynman-Kac Formula

In this section we establish a stochastic representation theorem for a class of quasilinear PDEs, via th route of FBSDEs. We note that following presentation will include the BSDEs as a special case. To begin with, let us rewrite (1.1) again, on an arbitrary time interval $[t, T]$, $t \in [0, T)$: for $t \leq s \leq T$,

(2.1)
$$\begin{cases} X(s) = x + \int_t^s b(r, \Theta(r))dr + \int_s^t \sigma(r, X(r), Y(r))dW(r), \\ Y(s) = g(X(T)) + \int_s^T h(r, \Theta(r))dr - \int_s^T Z(r)dW(r) \end{cases}$$

We would like to show that if the FBSDE (2.1) has unique adapted solutions on all subintervals $[t, T] \subseteq [0, T]$, denoted by $(X^{t,x}, Y^{t,x}, Z^{t,x})$, then the function $u(t, x) \overset{\triangle}{=} Y^{t,x}(t)$ would give a *viscosity solution* to a quasilinear PDE. Thus if we can prove the uniqueness of such viscosity solution (see Chapter 3, §3), then clearly we obtain a certain "probabilistic solution" to the corresponding PDE, in the spirit of the celebrated Feynman-Kac formula. For this purpose, in what follows we shall always assume the solvability of the the the FBSDE (2.1), under the following assumptions:

(A1) (i) $m = 1$; and the coefficients b, h, σ, g are deterministic.
   (ii) The functions b and h are differentiable in z.

Note that (A1)-(i) amounts to saying that coefficients of (1.2) are "Markovian". Thus the standard technique of "time shifting" can be used to show that the process $\{Y_s^{t,x}\}_{s\geq t}$ is $\mathcal{F}_s^t$-adapted, where $\mathcal{F}_s^t = \sigma\{W_r, t \leq r \leq s\}$. Consequently the function $u(t, x) = Y_t^{t,x}$ is deterministic, thanks again to the Blumenthal 0-1 law.

In order to describe the quasilinear PDE that an FBSDE is correspond-
ing to, let us denote $\mathcal{S}(n)$ to be the set of $n \times n$ symmetric non-negative
matrices, and for $p \in \mathbb{R}^n$, $Q \in \mathcal{S}(n)$, define

$$
\begin{aligned}
H(t, x, u, p, Q) &\stackrel{\Delta}{=} \frac{1}{2}\mathrm{tr}\left\{\sigma\sigma^T(t, x, u)Q\right\} + \langle b(t, x, u, \sigma(t, x, u)p), p \rangle \\
&\quad + h(t, x, u, \sigma(t, x, u)p),
\end{aligned}
$$

(2.2)

and denote $Du \stackrel{\Delta}{=} \nabla u = (\partial_{x_1}u, \cdots, \partial_{x_n}u)^T$, $D^2u = (\partial^2_{x_i x_j}u)_{i,j}$ (the Hessian
of u), and $u_t = \partial_t u$. The quasilinear PDE that we are interested in is of
the following form:

(2.3)
$$
\begin{cases}
u_t + H(t, x, u, Du, D^2u) = 0, \\
u(T, x) = g(x),
\end{cases}
$$

We have the following theorem.

Theorem 2.1. *Assume (A1). Suppose that for a given time duration*
$[t, T]$, *the FBSDE (2.1) has an adapted solution* $(X^{t,x}, Y^{t,x}, Z^{t,x})$. *Then*
the function $u(t, x) \stackrel{\Delta}{=} Y_t^{t,x}$, $(t, x) \in [0, T] \times \mathbb{R}^n$ *is a viscosity solution of the*
quasilinear PDE (2.3).

Proof. We shall prove only that u is a viscosity subsolution to (2.3).
The proof of the "supersolution" is left as an exercise. First note that
$u(t, x) = Y^{t,x}(t)$ is continuous on $[0, T] \times \mathbb{R}^n$, locally Lipschitz-continuous
in x, and locally Hölder-$\frac{1}{2}$ in t.
    Let $(t, x) \in [0, T) \times \mathbb{R}^n$ be given; and let $\varphi \in C^{1,2}([0, T] \times \mathbb{R}^n)$ be such
that (t, x) is a global maximum point of $u - \varphi$ such that $u(t, x) = \varphi(t, x)$.
We are to check that the inequality (3.27) of Chapter 3 holds.
    To simplify notations, in what follows we suppress the superscript " t,x "
for the processes X, Y, and Z. First note that by modifying φ slightly at
"infinite" if necessary we assume without loss of generality that and $D\varphi$
is uniformly bounded, thanks to the uniform Lipschitz property of u in x.
Next note that the pathwise uniqueness of the FBSDE leads to that for
any $0 \leq \tau \leq \tau \leq T$ one has $u(\tau, X(\tau)) = Y(\tau)$, hence we can rewrite the
backward SDE in (2.1) as

$$
\begin{aligned}
u(t, x) = u(\tau, X(\tau)) &+ \int_t^\tau h(s, X(s), Y(s), Z(s))ds \\
&- \int_t^\tau Z(s)dW(s).
\end{aligned}
$$

(2.8)

Now applying Itô's formula to $\varphi(\cdot, X(\cdot))$ from t to τ we have

(2.9)
$$
\begin{aligned}
\varphi(\tau, X(\tau)) = \varphi(t, x) &+ \int_t^\tau \varphi_t(s, X(s))ds \\
&+ \int_t^\tau \langle D\varphi(s, X(s)), b(s, X(s), u(s, X(s)), Z(s)) \rangle\, ds \\
&+ \int_t^\tau \frac{1}{2}\mathrm{tr}\{\sigma\sigma^T(s, X(s), u(s, X(s)))D^2\varphi(s, X(s))\}ds \\
&+ \int_t^\tau \langle D\varphi(s, X(s)), \sigma(s, X(s), u(s, X(s)))dW(s) \rangle\,.
\end{aligned}
$$

Write

(2.10)
$$
\begin{aligned}
h(s, X(s), Y(s), Z(s)) = {}& h(s, X(s), Y(s), [\sigma^T D\varphi](s, X(s), Y(s))) \\
&+ \langle \alpha(s), Z(s) - [\sigma^T D\varphi](s, X(s), Y(s)) \rangle; \\
b(s, X(s), Y(s), Z(s)) = {}& b(s, X(s), Y(s), [\sigma^T D\varphi](s, X(s), Y(s))) \\
&+ \beta(s)\{Z(s) - [\sigma^T D\varphi](s, X(s), Y(s)))\},
\end{aligned}
$$

where

(2.11)
$$
\begin{cases}
\alpha(s) = \displaystyle\int_0^1 \frac{\partial h}{\partial z}(s, X(s), Y(s), Z_\tau(s))d\theta; \\[2mm]
\beta(s) = \displaystyle\int_0^1 \frac{\partial b}{\partial z}(s, X(s), Y(s), Z_\theta(s))d\theta; \\[2mm]
Z_\theta(s) = \theta Z(s) + (1-\theta)\sigma^T(s, X(s), Y(s))D\varphi(s, X(s)).
\end{cases}
$$

By assumption (A1), we see that α and β are bounded, adapted processes. Therefore, subtracting (2.9) from (2.8), using (2.10) and (2.11), and noting the facts that $u(t, x) = \varphi(t, x)$ and $u(\tau, X(\tau)) \le \varphi(\tau, X(\tau))$, we obtain

(2.12)
$$
\begin{aligned}
0 \ge{}& u(\tau, X(\tau)) - \varphi(\tau, X(\tau)) \\
={}& \int_0^\tau \Big\{ -\frac{\partial \varphi}{\partial t}(s, X(s)) - F(s, X(s), Y(s), [\sigma^T D\varphi](s, X(s), Y(s)) \\
&- \langle Z(s) - [\sigma^T D\varphi](s, X(s), Y(s)), \alpha(s) - D\varphi(s, X(s))\beta(s) \rangle \Big\}ds \\
&+ \int_t^\tau \langle Z(s) - [\sigma^T D\varphi](s, X(s), Y(s)), dW(s) \rangle\,.
\end{aligned}
$$

Since $\theta(s) \overset{\Delta}{=} \alpha(s) + D\varphi(s, X(s))\beta(s)$, $s \in [t, T]$ is uniformly bounded, the following process is a $\mathbf{P}$-martingale on $[t, T]$:

$$
\Theta_s^t \overset{\Delta}{=} \exp\Big\{ -\int_t^s \langle \theta(r), dW(r) \rangle - \frac{1}{2}\int_t^s |\theta(r)|^2 dr \Big\}, \quad s \in [t, T].
$$

By Girsanov's Theorem, we can define a new probability measure $\widetilde{\mathbf{P}}$ via $\frac{d\widetilde{\mathbf{P}}}{d\mathbf{P}} = \Theta_T^t$, so that $\widetilde{W}^t(s) = W(s) - W(t) - \int_t^s \theta(r)dr$ is a $\widetilde{\mathbf{P}}$-Brownian motion on $[t, T]$. Furthermore, since the processes $(X, Y, Z))$ satisfies

$$E\Big\{ \sup_{t \leq s \leq T} |X(s)|^2 \Big\} + E\Big\{ \sup_{t \leq s \leq T} |Y(s)|^2 \Big\} + E \int_t^T |Z(s)|^2 ds < \infty,$$

the boundedness of $D\varphi$ and the uniform Lipschitz property of σ imply that, for some constant $C > 0$,

$$\widetilde{E}\Big\{ \int_t^T |Z(s) - [\sigma^T D\varphi](s, X(s), Y(s))|^2 ds \Big\}^{\frac{1}{2}}$$

$$\leq CE\Big\{ \Theta_T^t \Big(\int_t^T [1 + |Z(s)|^2 + |X(s)|^2 + |Y(s)|^2] ds \Big)^{\frac{1}{2}} \Big\}$$

$$\leq C\{E(\Theta_T^t)^2\}^{\frac{1}{2}} \Big\{ E \int_t^T [1 + |Z(s)|^2 + |X(s)|^2 + |Y(s)|^2] ds \Big\}^{\frac{1}{2}} < \infty.$$

In other words, the integral

$$M^t(u) \overset{\Delta}{=} \int_t^u \langle Z(s) - [\sigma^T D\varphi](s, X(s), Y(s)), d\widetilde{W}(s) \rangle, \qquad u \in [t, T]$$

is a $\widetilde{\mathbf{P}}$-local martingale on $[t, T]$ satisfying $E\langle M^t \rangle_T^{\frac{1}{2}} < \infty$, the by Burkholder-Davis-Gundy's inequality, one shows that it is a $\widetilde{\mathbf{P}}$-martingale on $[t, T]$. Hence, by taking expectation $\widetilde{E}\{\cdot\}$ on both sides of (2.12) we obtain that

·(2.13)
$$0 \geq \widetilde{E} \int_t^T \Big\{ -\frac{\partial \varphi}{\partial t}(s, X(s)) - H(s, X(s), Y(s), [\sigma^T D\varphi](s, X(s), Y(s))) \Big\} ds.$$

Dividing both sides by τ and then sending $\tau \to 0$ we obtain (3.27) of Chapter 3 immediately.   □

Remark 2.2. For a more complete theory, one should also prove that the viscosity solution to the quasilinear PDE (2.3) is unique. This is indeed the case when the coefficient σ is independent of y as well (i.e., $\sigma = \sigma(t, x)$); and when the solution class is restricted to, for example, bounded, continuous functions that are uniform Lipschitz in x and Hölder -$\frac{1}{2}$ in t. We note that due to the special quasilinearity, the function (2.2) is neither monotone, nor even one-sided uniform Lipschitz in the variable x, therefore H is not "proper" in the sense of Crandall-Ishii-Lions [1], or convertible to a proper function using the standard technique of "exponentiating" (see, e.g., Fleming-Soner [1]). Consequently, the uniqueness of the viscosity solution is by no means trivial. However, since this issue is more or less beyond the scope of this book, we will not include the proof here. We refer the interested readers to the works of Barles-Buckdahn-Pardoux [1], Pardoux-Tang [1], or Cvitanic-Ma [2].

§3. Black's Consol Rate Conjecture

One of the early applications of FBSDE is to confirm and explore a conjecture by Fischer Black regarding consol rate models for the term structure of interest rates. A consol is by definition a *perpetual annuity*, that is, a security that pays dividends continually and in perpetuity. A consol rate model is one in which the stochastic behavior of the short rate, taken as a non-negative progressively measurable process below, is influenced by the consol rate process. The relation between the two rate processes then yields a special term structure of interest rates.

In order to set up a mathematical model, let us consider the following simplest situation in which the short rate is a constant $r > 0$, then there should be no difference between the short rate and long term (consol) rate. In this case the consol price Y can be calculated as the simple actuarial present value of a perpetual annuity. Assuming, for instance, that the annuity is in a form of *annuity-immediate* in terms of actuarial mathematics, that is , it pays, say \$1, at the end of each year, then the price Y can be calculated easily as

$$(3.1) \qquad Y = \sum_{k=1}^{\infty} \frac{1}{(1+r)^k} = \frac{1}{(1+r)} \cdot \frac{1}{1 - \frac{1}{(1+r)}} = \frac{1}{r}.$$

In other words, the price for the (unit) consol is the reciprocal of the interest (consol) rate. In general, let us define the consol rate to be the reciprocal of the consol price, then instead of studying the original term structure of interest rates, it would be equivalent to study the relation between the consol price and the short rate.

Now let us generalize the above idea. For a given short rate process $r = \{r_t : t \geq 0\}$, we use the standard *expected discounted value formula* (an extension of the aforementioned actuarial present value formula) to evaluate the consol price process $Y = \{Y_t : t \geq 0\}$ †:

$$(3.2) \qquad Y(t) = E\left\{ \int_t^{\infty} e^{-\int_t^s r(u)du} ds \;\middle|\; \mathcal{F}_t \right\}, \quad t \geq 0.$$

(One can check that if $r(t) \equiv r$, then $Y(t) \equiv \frac{1}{r}$, as (3.1) shows!). The *Consol rate problem* can be formulated as follows. Assume that the short rate process depends on the consol price (whence consol rate) in a non-anticipating manner, via the following SDE:

$$(3.3) \qquad dr(t) = \mu(r(t), Y(t))dt + \alpha(r(t), Y(t))dW(t).$$

where W is a standard Brownian motion in $\mathbb{R}^2$, and μ, α are some appropriate functions. Then *is there actually a pair of adapted processes* (r, Y)

---

† Without getting into the associated definitions and related notions of arbitrage, it is not unusual in applications to work from the beginning with the so-called "equivalent martingale measure," in the sense of Harrison and Kreps [1], and we do so.

*that satisfies both (3.2) and (3.3)? If so, can Y also be described by an
SDE?* In an earlier work Brennan and Schwartz [1] proposed a model of
term structure of interest rates in which both short rate and long rate are
characterized by SDEs. However, it was shown later by Hogan [1] by examples that such a model may not be meaningful in practice. Sensing that
the controversy might be caused by the inappropriate specification of the
coefficients, together with a simple observations by using Ito's formula and
(3.1), the late economist/mathematician Fisher Black made the following
conjecture:

(Black's Conjecture). *Under at most technical conditions, for any* (μ, α)
there is always a function $A : (0, \infty) \times (0, \infty) \to (0, \infty)$ *depending on* μ *and*
α, *such that*

$$dY(t) = (r(t)Y(t) - 1) dt + A(r(t), Y(t))dW(t).$$

Black's conjecture essentially re-confirms the SDE model of Brennan
and Schwartz, but it was not clear at the time for how to determine the
function A, and how it should related to the coefficients μ and α in (3.3).

We now show how to confirm Black's conjecture by using the theory of
FBSDEs. To this end, let us assume first that the short rate r process is
"hidden Markovian". That is, there is a (Markovian) "state process" X in
$\mathbb{R}^n$ such that the short rate is given by $r_t = h(X_t)$, for some well behaved
function h. To be more specific, we will assume that X satisfies an SDE:

$$(3.4) \qquad \begin{cases} dX(t) = b(X(t), Y(t))dt + \sigma(X(t), Y(t))dW(t), \\ X(0) = x, \quad t \in [0, T], \end{cases}$$

where b, σ are some appropriate functions defined on $\mathbb{R}^n \times \mathbb{R}$. Since the
coefficients b and σ can be computed explicitly in terms of μ, α, and h using
Itô's formula, we can recast the consol rate problem as follows.

Infinite Horizon Consol Rate Problem (IHCR). *Find a pair of
adapted, locally square-integrable processes* (X, Y), *such that*

$$(3.5) \qquad \begin{cases} dX(t) = b(X(t), Y(t))dt + \sigma(X(t), Y(t))dW(t), \qquad t \in [0, \infty), \\ Y(t) = E\left\{ \int_t^\infty e^{-\int_t^s h(X(u))\, du} ds \;\Big|\; \mathcal{F}_t \right\}, \\ X_0 = x, \qquad t \in [0, \infty). \end{cases}$$

Any adapted process (X, Y) *satisfying (3.5) is called an adapted solution of
Problem IHCR. Moreover, an adapted solution* (X, Y) *of Problem IHCR is
called a nodal solution with representing function* θ *if there exists a bounded
C^2 function* θ *with* θ_x *being bounded, such that*

$$(3.6) \qquad Y(t) = \theta(X(t)), \qquad t \in [0, \infty).$$

Recall that the term "nodal solution" was first introduced in Chapter 4,
§3, where we studied the FBSDE in a infinite horizon $[0, \infty)$ of the following

type:

$$(3.7) \quad \begin{cases} dX(t) = b(X(t), Y(t))\, dt + \sigma(X(t), Y(t))dW(t), \\ dY(t) = (h(X(t))Y(t) - 1)dt - \langle Z(t), dW(t)\rangle, \quad t \in [0, \infty), \\ X(0) = x, \\ Y(t) \ \text{is bounded a.s., uniformly in } t \in [0, \infty). \end{cases}$$

The following theorem shows that (3.7) is exactly the system of SDEs that can characterize the process X and Y simultaneously, which will be the first step towards the resolution of Black's conjecture. First let us recall the technical assumptions (3.4)–(3.6) of Chapter 4:

(A2) The functions σ, b, h are C^1 with bounded partial derivatives and there exist constants $\lambda, \mu > 0$, and some continuous increasing function $\nu : [0, \infty) \to [0, \infty)$, such that

$$\begin{cases} \lambda I \le \sigma(x, y)\sigma(x, y)^\top \le \mu I, & (x, y) \in \mathbb{R}^n \times \mathbb{R}, \\ |b(x, y)| \le \nu(|y|), & (x, y) \in \mathbb{R}^n \times \mathbb{R}, \\ \inf_{x \in \mathbb{R}^n} h(x) = \delta > 0, & \sup_{x \in \mathbb{R}^n} h(x) = \gamma < \infty. \end{cases}$$

Theorem 3.1. *Assume (A2). If (X, Y, Z) is an adapted solution to (3.7), then (X, Y) is an adapted solution to Problem (IHCR).*

Conversely, if (X, Y) is an adapted solution to Problem IHCR, then there exists an adapted, $\mathbb{R}^d$-valued, locally square-integrable process Z, such that (X, Y, Z) is an adapted solution of (3.7).

Proof. To see the first assertion, let (X, Y, Z) be an adapted solution to (3.7). Let $\Gamma(t) = e^{-\int_0^t h(X(u))du}$, $t \in [0, T]$. Then using integration by parts (or Itô's formula) one shows easily that

$$\Gamma(T)Y(T) = \Gamma(t)Y(t) + \int_t^T \Gamma(s)dY(s) + \int_t^T Y(s)d\Gamma(s)$$

$$= -\int_t^T e^{-\int_t^s h(X(u))du}\, ds,$$

or

$$Y(t) = e^{-\int_t^T h(X(s))ds} + \int_t^T e^{-\int_t^T h(X(u))du}\, ds + m(T) - m(t).$$

where m denotes some $\{\mathcal{F}_t\}_{t\ge 0}$-martingale, as usual. Taking conditional expectation $E\{\cdot \,|\mathcal{F}_t\}$ on both sides and letting $T \to \infty$, we prove the first assertion.

Conversely, suppose that (X, Y) is an adapted solution to Problem IHCR. Define

$$(3.8) \qquad\qquad U(t) = \int_t^\infty e^{-\int_t^s h(X(u))du}\, ds.$$

Clearly, $U(t)$ is well-defined for each $t \geq 0$, thanks to (A2). We claim that U is the unique bounded solution of the following ordinary differential equation with random coefficients:

$$(3.9) \qquad \frac{dU(t)}{dt} = h(X(t))U(t) - 1, \qquad t \in [0, \infty).$$

Indeed, by a direct verification one shows that the function U defined by (3.8) is a bounded solution of (3.9). On the other hand, let U be any bounded solution to (3.9) defined on $[0, \infty)$. Then for any $0 \leq t \leq T$, we can apply the variation of constants formula to get

$$(3.10) \qquad U(t) = e^{-\int_t^T h(X(u))du} U(T) + \int_t^T e^{-\int_t^s h(X(u))du} ds.$$

Since $U(T)$ is bounded for all $T > 0$, and by (A2), $h(X(u)) \geq \delta > 0$ $\forall u \in [0, T]$, **P**-a.s., sending $T \to \infty$ on both sides of (3.10) we obtain (3.8), proving claim.

Next, define $Y(t) = E\{U(t)|\mathcal{F}_t\}$. Note that since the filtration $\{\mathcal{F}_t\}_{t \geq 0}$ is Brownian, the process Y is continuous and is indistinguishable from the optional (as well as predictable) projection of U. Hence, for any bounded, $\{\mathcal{F}_t\}_{t \geq 0}$-adapted process H, it holds that†

$$(3.11) \qquad E\left\{ \int_t^T H(s)U(s)ds \Big| \mathcal{F}_t \right\} = E\left\{ \int_t^T H(s)Y(s)ds \Big| \mathcal{F}_t \right\}.$$

Now for $0 \leq t < T < \infty$ we have from (3.9) and (3.11) that

$$(3.12) \qquad \begin{aligned} Y(t) &= E(U(t)|\mathcal{F}_t) = E\left\{ U(T) - \int_t^T [h(X(s))U(s) - 1]ds \Big| \mathcal{F}_t \right\} \\ &= E\left\{ Y(T) - \int_t^T [h(X(s))Y(s) - 1]ds \Big| \mathcal{F}_t \right\}. \end{aligned}$$

Thus, by using the martingale representation theorem one shows that there exists an adapted, square-integrable process $Z^{(T)}$ defined on $[0, T]$, such that for all $t \in [0, T]$,

$$(3.13) \quad Y(t) = Y(T) - \int_t^T [h(X(s))Y(s) - 1]ds + \int_t^T \langle Z^{(T)}(s), dW_s \rangle.$$

Since (3.13) holds for any $T > 0$, let $0 \leq T_1 < T_2 < \infty$, we have for $t \in [0, T_1]$ that

$$\begin{aligned} Y(t) &= Y(T_1) - \int_t^{T_1} [h(X(s))Y(s) - 1]ds + \int_t^{T_1} \langle Z^{(T_1)}(s)dW(s) \rangle \\ &= Y(T_2) - \int_t^{T_2} [h(X(s))Y(s) - 1]ds + \int_t^{T_2} \langle Z^{(T_2)}(s), dW(s) \rangle. \end{aligned}$$

---

† See, for example, Dellacherie and Meyer [1, Chapter VI].

From this one derives easily that

(3.14) $\displaystyle\int_t^{T_1} \langle\, Z^{(T_2)}(s) - Z^{(T_1)}(s), dW(s)\,\rangle = 0,$ for all

This leads to that $E\left\{\int_0^{T_1}[Z^{(T_2)}(s) - Z^{(T_1)}(s)]^2 ds\right\} = 0$. In other words, $Z^{(T_1)} = Z^{(T_2)}$, $dt \otimes d\mathbf{P}$-almost surely on $[0, T_1] \times \Omega$. Consequently, modulo a $dt \otimes d\mathbf{P}$-null set, we can define a process Z by $Z_t = Z^{(N)}(t)$, if $t \in [0, N]$, where $N = 1, 2, \cdots$. Clearly Z is locally square-integrable, and (3.13) can now be rewritten as

(3.15) $\displaystyle Y(t) = Y(T) - \int_t^T [h(X(s))Y(s) - 1]ds + \int_t^T \langle\, Z(s), dW(s)\,\rangle,$

for all $T > 0$, or equivalently, (X, Y) satisfies the SDE (3.7). Finally, the boundedness of Y follows easily from the definition of Y and the fact that $U_t \le \frac{1}{\delta}, \forall t \ge 0$, $\mathbf{P}$-a.s., proving the proposition. $\square$

We remark here that Theorem 3.1 shows that the Black's conjecture can be partially solved if the FBSDE (3.7) is solvable. However, in order to confirm Black's conjecture completely, we have to show that the process Z can actually be written as $Z(t) = \varphi(X(t), Y(t))$ for some function φ, which in turn will give $Z(t) = A(r(t), Y(t))$ for some function$A$, as the conjecture states. But this is exactly where the *nodal solution* comes into play, and the Chapter 4, Theorem 3.3 essentially solves the problem. We recast that theorem here in the new context.

Theorem 3.2. *Assume (A2). Then there exists at least one nodal solution (X, Y) of Problem IHCR. Moreover, the representing function θ satisfies*
    (i) $\gamma^{-1} \le \theta(x) \le \delta^{-1}$, *for all $x \in \mathbb{R}$.*
    (ii) θ *satisfies the following differential equation for $x \in \mathbb{R}^n$:*

(3.16) $\displaystyle \frac{1}{2}\mathrm{tr}\left(\theta_{xx}\sigma(x, \theta)\sigma^T(x, \theta)\right) + \langle\, b(t, \theta), \theta_x\,\rangle - h(x)\theta + 1 = 0,$

Consequently, The Black's conjecture is solved (in terms of Problem IHCR) with $A(x, y) = \sigma^T(x, y)\theta_x(x)$.

    Proof. This is the direct consequence of Chapter 4, Theorem 3.3; and the last statement if due to the fact that $Z(t) = A(X(t), Y(t))$ whenever the nodal solution exists. $\square$

Remark 3.3. We should point out here that although the bounded solution U of the random ODE (3.9) with infinite-horizon is unique, the uniqueness of the adapted solution to the FBSDE (3.7) over an infinite duration is still unknown. In fact, as we saw in Chapter 4 (§3), the uniqueness of the adapted solution, as well as that of the nodal solution, to FBSDE (3.7), is a more delicate issue, especially in the higher dimensional case. However, since Black's conjecture concerns only the existence of the function

A, Theorem 3.2 provides a sufficient answer. Interested readers could of course revisit Chapter 4, §3 for more details on various issues regarding uniqueness.

Finite-Horizon Valuation Problem and its limit.

In the standard theory of term structure of interest rates the time duration is often set to be finite. Namely, we content ourselves only in a finite time interval $[0, T]$. Let us now view the process Y as a long term interest rate (or the price of a long term bond to be comparable to the consol price). and view X as the state process for the short rate r, with $r(t) = h(X(t))$, and h satisfies (A2). In order to study the explicit relation between X and Y, let us assume that they have an explicit relation at terminal time T: $Y(T) = g(X(T))$. We consider the following *Finite-Horizon-Valuation Problem*. Note that Such a problem is a generalization of the well-known *finite horizon annuity valuation problem*, which corresponds to the case when $g \equiv 0$ below, by allowing the annuity price to influence the short rate.

Problem FHV. *Find an adapted process (X, Y) such that for*

(3.17)
$$
\begin{cases}
X(t) = x + \displaystyle\int_0^t b(X(s), Y(s))ds + \int_0^t \sigma(X(s), Y(s))dW(s), \\[2mm]
Y(t) = E\left\{ \Gamma_t^T g(X(T)) + \displaystyle\int_t^T \Gamma^s ds \,\Big|\, \mathcal{F}_t \right\}, \qquad t \in [0, T],
\end{cases}
$$

where $\Gamma_t^s \overset{\Delta}{=} e^{-\int_t^s h(X(u))du}$.

Any adapted process (X, Y) satisfying (3.17) is called an *adapted solution* of Problem FHV. Further, an adapted solution (X, Y) of Problem FHV is called a *nodal solution* of Problem FHV if there exists a function $\theta : [0, T] \times \mathbb{R}^n \to \mathbb{R}$, which is C^1 in t and C^2 in x, such that

(3.18)
$$
Y(t) = \theta(t, X(t)), \qquad t \in [0, T].
$$

Conceivably the Problem FHV will associate to an FBSDE as well, as was seen in the IHCR case. In fact, some similar arguments as those in Theorem 3.1 shows that if (X, Y) is an adapted solution to the Problem FHV, then there exist a progressively measurable, square integrable process Z such that (X, Y, Z) is an adapted solution to the following FBSDE:

(3.19)
$$
\begin{cases}
dX(t) = b(X(t), Y(t))\, dt + \sigma(X(t), Y(t))dW(t), \\
dY(t) = (h(X(t))Y(t) - 1)dt - \langle Z(t), dW(t) \rangle, \qquad t \in [0, T], \\
X(0) = x, \quad Y(T) = g(X(T)).
\end{cases}
$$

Conversely, if (X, Y, Z) is an adapted solution to (3.19), then a variation of constant formula applied to the backward SDE in (3.19) would lead immediately to that Y satisfies (3.17). Furthermore, using the results in

Chapter 4 (Four Step Scheme) we see that if g is regular enough, then any adapted solution of (3.19) must be a nodal solution. These facts, together with Chapter 4, Theorem 3.10, give us the following theorem, which slightly goes beyond the Black Conjecture.

Theorem 3.4. *In addition to (A2), assume further that the function g belongs boundedly to $C^{2+\alpha}(\mathbb{R}^n)$ for some $\alpha \in (0,1)$. Then, Problem FHV admits a unique adapted solution (X, Y). Moreover, this solution is in fact a nodal solution.*

Furthermore, if the Problem IHCR has a unique nodal solution, denoted by (X, Y), where $Y = \theta(X)$ and θ satisfies the differential equation (3.16); and if we denote (X^K, Y^K) to be the nodal solution of Problem FHV on the interval $[0, K]$, then it holds that

$$(3.20) \qquad \lim_{K \to \infty} E|Y_t^K - Y_t|^2 + E|X_t^K - X_t|^2 = 0,$$

uniformly in $t \in [0, \infty)$ on compacts.

§4. Hedging Options for a Large Investor

In this section we apply the theory of FBSDEs to another problem in finance: hedging contingent claims for a large investor. We recall that the problem of hedging a contingent claim was discussed briefly in Chapter 1, §1.3. In this section we shall remove one of the fundamental assumptions on which the Black-Scholes theory is built, that is, the "small investor" assumption. Roughly speaking, the "small investor" assumption says that no individual investor is influential enough so that his/her investment strategy, or wealth, once exposed, could affect the market prices. Mathematically, under such an assumption the coefficients of the stochastic differential equation that characterizes the price of underlying security should be independent of the portfolio of any investor. Although such an assumption has long been deemed as common sense, it has been also noted recently that the investors that are "not-so-small" could really make disastrous effect to a financial market. A probably indisputable evidence, for example, is the "Hedge Fund" crisis of 1998 in the global financial market, in which the "large investors" obviously played some important roles. In this section, we try to attack the problem of hedging a contingent claim involving "large investors". We should point out here that the model that we will be studying is still quite "ad hoc", and we shall only concentrate on the mathematical side of the problem.

Recall from Chapter 1, §1.3 the mathematical model of a continuous-time financial market. There are $d+1$ assets traded continuously: a *money market account* and d stocks, whose prices at each time t are denoted by $P_0(t)$, $P_i(t)$, $i = 1, \cdots, d$, respectively. An investor is allowed to trade continuously and frictionlessly. The "wealth" of the investor at time t is denoted by $X(t)$; and the amount of money that the investor puts into the i-th stock at time t is denoted by $\pi_i(t)$, $1 = 1, \cdots, d$ (thus the amount of money that the investor puts into the money market at time t is $X(t) -$

$\sum_{i=1}^{d} \pi_i(t)$). We assume that the investor is "large" in the sense that his wealth and strategy, once exposed, might influence the prices of the financial instruments. More precisely, let us assume that the prices $(P_0, P_1, \cdots, P_d)$ evolves according to the following (stochastic) differential equations on a given finite time horizon $[0, T]$ (comparing to Chapter 1, (1.26)):

(4.1)
$$\begin{cases} dP_0(t) = P_0(t)r(t, X(t), \pi(t))dt, \qquad 0 \le t \le T; \\ dP_i(t) = P_i(t)\{b_i(t, P(t), X(t), \pi(t))dt \\ \qquad\qquad + \displaystyle\sum_{j=1}^{d} \sigma_{ij}(t, P(t), X(t), \pi(t))dW_j(t)\}, \\ P_0(0) = 1, \quad P_i(0) = p_i > 0, \qquad i = 1, \cdots, d, \end{cases}$$

where $W = (W_1, \cdots, W_d)$ is a d-dimensional standard Brownian motion defined on a complete probability space $(\Omega, \mathcal{F}, \mathbf{P})$, and we assume as usual that $\{\mathcal{F}_t\}_{t \ge 0}$ is the $\mathbf{P}$-augmentation of the natural filtration generated by W. To be consistent with the classical model, we call b the *appreciation rate* and σ the *volatility matrix* of the stock market.

Further, we assume that the investor is provided an initial endowment $x \ge 0$, and is allowed to consume, and denote $C(t)$ to be the cumulative consumption time t. It is not hard to argue that the change of the wealth "$dX(t)$" should now follow the dynamics:

(4.2)
$$\begin{cases} dX(t) = \displaystyle\sum_{i=1}^{d} \frac{\pi_i(t)}{P_i(t)} dP_i(t) + \frac{(X(t) - \sum_{i=1}^{d} \pi_i(t))}{P_0(t)} dP_0(t) - dC(t) \\ \\ X(0) = x > 0. \end{cases}$$

To simplify presentation, from now on we assume that $d = 1$ and that the interest rate r is independent of π and X, i.e., $r \equiv r(t)$, $t \ge 0$. Denote

(4.3)
$$\begin{aligned} \widehat{b}(t, p, x, \pi) &\stackrel{\Delta}{=} (x - \pi)r(t) + \pi b(t, p, x, \pi); \\ \widehat{\sigma}(t, p, x, \pi) &\stackrel{\Delta}{=} \pi\sigma(t, p, x, \pi), \end{aligned}$$

for $(t, p, x, \pi) \in [0, T] \times \mathbb{R}^3$. We can rewrite (4.1) and (4.2) as

(4.4)
$$\begin{cases} P_0(t) = \exp\left\{ \displaystyle\int_0^t r(s)ds \right\}, \\ P(t) = p + \displaystyle\int_0^t P(s)\{b(s, P(s), X(s), \pi(s))ds \\ \qquad\qquad + \sigma(s, P(s), X(s), \pi(s))dW(s)\}, \end{cases} \qquad t \in [0, T];$$

(4.5)
$$\begin{aligned} X(t) = x &+ \int_0^t \widehat{b}(s, P(s), X(s), \pi(s))ds \\ &+ \int_0^t \widehat{\sigma}(s, P(s), X(s), \pi(s))dW(s) - C(t); \end{aligned}$$

Before we proceed, we need to make some technical observations: first, we say a pair of $\{\mathcal{F}_t\}_{t\geq 0}$-adapted processes (π, C) is a *hedging strategy* (or simply *strategy*) if $C(\cdot)$ has nondecreasing and RCLL paths, such that $C(0) = 0$ and $C(T) < \infty$, a.s.-**P**; and $E \int_0^T |\pi(s)|^2 ds < \infty$. Clearly, under suitable conditions, for a given strategy (π, C) and the initial values $p > 0$ and $x \geq 0$ the SDEs (4.4) and (4.5) have unique strong solutions, which will be denoted by $P = P^{p,x,\pi,C}$ and $X = X^{p,x,\pi,C}$, whenever the dependence of the solution on p, x, π, C needs to be specified.

Next, for a given $x \geq 0$, we say that a hedging strategy (π, C) is *admissible w.r.t.* x, if for any $p > 0$, it holds that $P^{p,x,\pi,C}(t) > 0$ and $X^{p,x,\pi,C}(t) \geq 0$, $\forall t \in [0, T]$, a.s. **P**. We denote the set of strategies that are admissible w.r.t. x by $\mathcal{A}(x)$. It is not hard to show that $\mathcal{A}(x) \neq \emptyset$ for all x. Indeed, for any $x > 0$, and $p > 0$, consider the pair $\pi \equiv 0$ and $C \equiv 0$. Therefore, under very mild conditions on the coefficients (e.g., the standing assumptions below) we see that both P and X can be written as "exponential" functions:

$$
(4.6) \quad
\begin{cases}
P(t) = p \exp\left\{ \int_0^t [b(s) - \frac{1}{2}|\sigma(s)|^2] ds + \int_0^t \sigma(s) dW(s) \right\} > 0, \\
X(t) = x \exp\{ \int_0^t r(s) ds \} \geq 0,
\end{cases}
$$

where $b(s) = b(s, P(s), 0, 0)$ and $\sigma(s) = \sigma(s, P(s), 0, 0)$. Thus $(0, 0) \in \mathcal{A}(x)$.

Recall from Chapter 1, §1.3 that an *option* is an $\mathcal{F}_T$-measurable random variable $B = g(P(T))$, where g is a real function; and that the *hedging price* of the option is

$$
(4.7) \quad h(B) \overset{\Delta}{=} \inf\{x \in \mathbb{R} : \exists(\pi, C) \in \mathcal{A}(x), \text{s.t. } X^{x,\pi,C}(T) \geq B \text{ a.s. }\}.
$$

In light of the discussion in Chapter 1, §1.3, we will be interested in the forward-backward version of the SDEs (4.4) and (4.5):

$$
(4.8) \quad
\begin{cases}
P(t) = p + \int_0^t P(s)\{b(s, P(s), X(s), \pi(s)) ds \\
\qquad\qquad\qquad + \sigma(s, P(s), X(s), \pi(s)) dW(s)\}, \\
X(t) = g(P(T)) - \int_t^T \widehat{b}(s, P(s), X(s), \pi(s)) ds \\
\qquad\qquad\qquad - \int_t^T \widehat{\sigma}(s, P(s), X(s), \pi(s)) dW(s);
\end{cases}
$$

We first observe that under the standard assumptions on the coefficients and that $g \geq 0$, if (P, X, π) is a solution to FBSDE (4.8), then the pair $(\pi, 0)$ must be admissible w.r.t. $X(0)$ (a deterministic quantity by Blumenthal $0 - 1$ law). Indeed, let (P, X, π) be an adapted solution to (4.8). Then a similar representation as that in (4.6) shows that $P(t) > 0$, $\forall t$, a.s. Further, define a (random) function

$$
f(t, x, z) = r(t)x + z\sigma^{-1}(t, P(t), X(t), \pi(t))[b(t, P(t), X(t), \pi(t)) - r(t)],
$$

then $x(t) = X(t)$, $z(t) = \sigma(t, P(t), X(t), \pi(t))\pi(t)$ solves the following backward SDE

$$x(t) = g(P(T)) + \int_t^T f(s, x(s), z(s))ds + \int_t^T z(s)dW(s).$$

Applying the Comparison theorem (Chapter 1, Theorem 6.1), we conclude that $X(t) = x(t) \geq 0$, $\forall t$, **P**-a.s., since $g(P(T)) \geq 0$, **P**-a.s. The assertion follows.

§4.1. Hedging without constraint

We first seek the solution to the hedging problem (4.7) under the following assumptions.

(H3) The functions b, $\sigma : [0, T] \times \mathbb{R}^3 \mapsto \mathbb{R}$ are twice continuously differentiable, with first order partial derivatives in p, x and π being uniformly bounded. Further, we assume that there exists a $K > 0$, such that for all (t, p, x, π),

$$\left| p \frac{\partial b}{\partial p} \right| + \left| p \frac{\partial \sigma}{\partial p} \right| + \left| x \frac{\partial \sigma}{\partial x} \right| + \left| x \frac{\partial \sigma}{\partial \pi} \right| \leq K.$$

(H4) There exist constants $K > 0$ and $\mu > 0$, such that for all (t, p, x, π) with $p > 0$, it holds that

$$\mu < \sigma^2(t, p, x, \pi) \leq K.$$

(H5) $g \in C_b^{2+\alpha}(\mathbb{R})$ for some $\alpha \in (0, 1)$; and $g \geq 0$.

Remark 4.1. Assumption (H4) amounts to saying that the market is complete. Assumption (H5) is inherited from Chapter 4, for the purpose of applying the Four step scheme. However, since the boundedness of g excludes the simplest, say, European call option case, it is desirable to remove the boundedness of g. One alternative is to replace (H5) by the following condition.

(H5)' $\lim_{|p| \to \infty} g(p) = \infty$; but $g \in C^3(\mathbb{R})$ and $g' \in C_b^2(\mathbb{R})$. Further, there exists $K > 0$ such that for all $p > 0$,

(4.9) $|pg'(p)| \leq K(1 + g(p))$; $|p^2 g''(p)| \leq K$.

The point will be revisited after the proof of our main theorem. Finally, all the technical conditions in (H3)–(H5) are verified by the classical models. An example of a non-trivial function σ that satisfies (H3) and (H4) could be $\sigma(t, p, x, \pi) = \sigma(t) + \arctan(x^2 + |\pi|^2)$.

We shall follow the "Four Step Scheme" developed in Chapter 4 to solve the problem. Assuming $C = 0$ and consider the FBSDE (4.8). Since we have seen that the solution to (4.8), whenever exists, will satisfy $P(t) > 0$, we shall restrict ourselves to the region $(t, p, x, \pi) \in [0, T] \times (0, \infty) \times \mathbb{R}^2$

without further specification. The Four Step Scheme in the current case is the following:

Step 1: Find $z : [0, T] \times (0, \infty) \times \mathbb{R}^2 \to \mathbb{R}$ such that

(4.10) $qp\sigma(t, p, x, z(t, p, x, q)) - z(t, p, x, q)\sigma(t, p, x, z(t, p, x, q)) = 0,$

In other words, $z(t, p, x, q) = pq$ since $\sigma > 0$ by (H4).

Step 2: Using the definition of $\hat{b}$ and $\hat{\sigma}$ in (4.3), we deduce the following extension of Black-Scholes PDE:

(4.11) $\begin{cases} 0 = \theta_t + \dfrac{1}{2}\sigma^2(t, p, \theta, p\theta_p)p^2\theta_{pp} + p\theta_p - r(t)\theta, \\ \theta(T, p) = g(p), \qquad p > 0. \end{cases}$

Step 3: Let θ be the (classical) solution of (4.11), set

(4.12) $\begin{cases} \tilde{b}(t, p) = b(t, p, \theta(t, p), p\theta_p(t, p)) \\ \tilde{\sigma}(t, p) = \sigma(t, p, \theta(t, p), p\theta_p(t, p)), \end{cases}$

and solve the following SDE:

(4.13) $P(t) = p + \displaystyle\int_0^t P(s)\tilde{b}(s, P(s))ds + \int_0^t P(s)\tilde{\sigma}(s, P(s))dW(s).$

Step 4: Setting

(4.14) $\begin{cases} X(t) = \theta(t, P(t)) \\ \pi(t) = P(t)\theta_p(t, P(t)), \end{cases}$

show that (P, X, π) is an adapted solution to (4.8) with $C \equiv 0$.

The resolution of the Four Step Scheme depends heavily on the existence of the classical solution to the quasilinear PDE (4.11). Note that in this case the PDE is "degenerate" near $p = 0$, the result of Chapter 4 does not apply directly. We nevertheless have the following result that is of interest in its own right:

Theorem 4.2. *Assume (H3)–(H5). There exists a unique classical solution $\theta(\cdot, \cdot)$ to the PDE (4.11), defined on $(t, p) \in [0, T] \times (0, \infty)$, which enjoys the following properties:*

   (i) $\theta - g$ *is uniformly bounded for* $(t, p) \in [0, T] \times (0, \infty)$;

   (ii) *The partial derivatives of θ satisfy: for some constant $K > 0$,*

(4.15) $|p\theta_p(t, p)| \le K(1 + |p|); \qquad |p^2\theta_{pp}(t, p)| \le K.$

Proof. First consider the function $\widehat{\theta} \stackrel{\Delta}{=} \theta - g$. It is obvious that $\widehat{\theta}_t = \theta_t$, $\widehat{\theta}_p = \theta_p - g_p$ and $\widehat{\theta}_{pp} = \theta_{pp} - g_{pp}$; and $\widehat{\theta}$ satisfies the following PDE:

(4.16)
$$
\begin{cases}
0 = \widehat{\theta}_t + \dfrac{1}{2}\sigma^2(t,p,\widehat{\theta}+g(p),p(\widehat{\theta}_p+g'(p)))p^2(\widehat{\theta}_{pp}+g'') \\
\qquad + r(t)[p(\widehat{\theta}_p+g') - (\widehat{\theta}+g)], \\
\widehat{\theta}(T,p) = 0, \qquad p > 0.
\end{cases}
$$

To simplify notations, let us set $\bar{\sigma}(t,p,x,\pi) = \sigma(t,p,x+g(p),\pi+pg'(p))$, then we can rewrite (4.16) as

(4.17)
$$
\begin{cases}
0 = \widehat{\theta}_t + \dfrac{1}{2}\bar{\sigma}^2(t,p,\widehat{\theta},p\widehat{\theta}_p)p^2\widehat{\theta}_{pp} + r(t)p\widehat{\theta}_p + \widehat{r}(t,p,\widehat{\theta},p\widehat{\theta}_p), \\
\widehat{\theta}(T,p) = 0, \qquad p > 0,
\end{cases}
$$

where

(4.18) $\quad \widehat{r}(t,p,x,\pi) = \dfrac{1}{2}\bar{\sigma}^2(t,p,x,\pi)p^2 g''(p) + r(t)pg'(p) - r(t)(x+g(p)).$

Next, we apply the standard Euler transformation: $p = e^\xi$, and denote $\widetilde{\theta}(t,\xi) \stackrel{\Delta}{=} \widehat{\theta}(t,e^\xi)$. Since $\widetilde{\theta}_t(t,\xi) = \widehat{\theta}_t(t,e^\xi)$, $\widetilde{\theta}_\xi(t,\xi) = e^\xi\widehat{\theta}_p(t,e^\xi)$, and $\widetilde{\theta}_{\xi\xi}(t,\xi) = e^{2\xi}\widehat{\theta}_{pp}(t,e^\xi) + e^\xi\widehat{\theta}_p(t,e^\xi)$, we we derive from (4.17) a quasilinear parabolic PDE for $\widetilde{\theta}$:

(4.19)
$$
\begin{cases}
0 = \widetilde{\theta}_t + \dfrac{1}{2}\bar{\sigma}^2(t,e^\xi,\widetilde{\theta},\widetilde{\theta}_\xi)(\widetilde{\theta}_{\xi\xi} - e^\xi\widetilde{\theta}_\xi) + r(t)\widetilde{\theta}_\xi + \widehat{r}(t,e^\xi,\widetilde{\theta},\widetilde{\theta}_\xi), \\
\quad = \widetilde{\theta}_t + \dfrac{1}{2}\bar{\sigma}_0^2(t,\xi,\widetilde{\theta},\widetilde{\theta}_\xi)\widetilde{\theta}_{\xi\xi} + b_0(t,\xi,\widetilde{\theta},\widetilde{\theta}_\xi)\widetilde{\theta}_\xi + \widehat{b}_0(t,\xi,\widetilde{\theta},\widetilde{\theta}_\xi), \\
\widetilde{\theta}(T,\xi) = 0, \qquad \xi \in \mathbb{R},
\end{cases}
$$

where

(4.20)
$$
\begin{aligned}
\bar{\sigma}_0(t,\xi,x,\pi) &= \bar{\sigma}(t,e^\xi,x,\pi); \\
b_0(t,\xi,x,\pi) &= r(t) - \dfrac{1}{2}[\bar{\sigma}_0^2(t,\xi,x,\pi)]; \\
\widehat{b}_0(t,\xi,x,\pi) &= \widehat{r}(t,e^\xi,x,\pi).
\end{aligned}
$$

Now by (H3) and (H4) we see that $\bar{\sigma}_0(t,\xi,x,\pi) \geq \mu > 0$, for all $(t,\xi,x,\pi) \in [0,T] \times \mathbb{R}^3$ and for all (t,ξ,x,π), it holds (suppressing the variables) that

$$
\frac{\partial\bar{\sigma}_0}{\partial\xi} = \frac{\partial\bar{\sigma}}{\partial p}e^\xi + \frac{\partial\bar{\sigma}}{\partial x}g'(e^\xi)e^\xi + \frac{\partial\bar{\sigma}}{\partial\pi}\left[g''(e^\xi)e^{2\xi} + e^\xi g'(e^\xi)\right].
$$

Thus, either (H5) or (H5)', together with (H3), will imply the boundedness of $\frac{\partial \bar{\sigma}_0}{\partial \xi}$. Similarly, we have

$$\sup_{(t,\xi,x,\pi)} \left| \frac{\partial \sigma}{\partial p}(t, e^\xi, x + g(e^\xi), \pi + e^\xi g'(e^\xi)) e^\xi \right| < \infty;$$

$$\sup_{(t,\xi,x,\pi)} \left| \frac{\partial \sigma}{\partial x}(t, x + g(e^\xi), \pi + e^\xi g'(e^\xi)) g'(e^\xi) e^\xi \right|$$

$$\leq K \sup_{(t,\xi,x,\pi)} \left| \frac{\partial \sigma}{\partial x}(t, x + g(e^\xi), \pi + e^\xi g'(e^\xi)) \right| [1 + (x + g(e^\xi))] < \infty;$$

$$\sup_{(t,\xi,x,\pi)} \left| \frac{\partial \sigma}{\partial \pi}(t, x + g(e^\xi), \pi + e^\xi g'(e^\xi)) g'(e^\xi) e^\xi \right|$$

$$\leq K \sup_{(t,\xi,x,\pi)} \left| \frac{\partial \sigma}{\partial \pi}(t, x + g(e^\xi), \pi + e^\xi g'(e^\xi)) \right| [1 + (x + g(e^\xi))] < \infty,$$

Consequently, we conclude that the function $\bar{\sigma}_0$ has bounded first order partial (thus uniform Lipschitz) in the variables ξ, x and π, and thus so is b_0. Moreover, note that for any

$$\frac{1}{2}\bar{\sigma}^2(t, e^\xi, x, \pi)g''(e^\xi) = \frac{1}{2}\bar{\sigma}_0^2(t, \xi, x, \pi)e^{2\xi}2g''(e^\xi)$$

is uniformly bounded and Lipschitz in ξ, x and π by either (H5) or (H5)$'$, we see that $\hat{b}_0$ is also uniform bounded and uniform Lipschitz in (x, ξ, π). Now we can apply Chapter 4, Theorem 2.2 to conclude that the PDE (4.11) has a unique classical solution $\tilde{\theta}$ in $C^{1+\frac{\alpha}{2}, 2+\alpha}$ (for any $\alpha \in (0,1)$). Furthermore, $\tilde{\theta}$, together with its first and second partial derivatives in ξ, is uniformly bounded throughout $[0, T] \times \mathbb{R}$. If we go back to the original variable, then we obtain that the function $\hat{\theta}$ is uniformly bounded and its partial derivatives satisfy:

$$\sup_{(t,p)} |p\hat{\theta}_p(t, p)| < \infty; \qquad \sup_{(t,p)} |p^2\hat{\theta}_{pp}(t, p)| < \infty.$$

This, together with the definition of $\hat{\theta}$ and condition (H5) (or (H5)$'$), leads to the estimates (4.15), proving the proposition.   □

A direct consequence of Theorem 4.2 is the following

Theorem 4.3. *Assume (H3), (H4), and either (H5) or (H5)$'$. Then for any given $p > 0$, the FBSDE (4.8) admits an adapted solution (P, X, π).*

Proof. We follow the Four Step Scheme. Step 1 is obvious. Step 2 is the consequence of Theorem 4.2. For step 3, we note that since θ_p and θ_{pp} may blow up when $p \downarrow 0$, a little bit more careful consideration is needed here. However, observe that $\tilde{b}$ and $\tilde{\sigma}$ are locally Lipschitz in $[0, T] \times (0, \infty) \times \mathbb{R}^2$, thus one can show that for ant $p > 0$, the SDE (4.13) always has a "local solution" for t sufficiently small. It is then standard to show (or simply note the exponential form (4.6)) that the solution, whenever exists, will neither go across the boundary $p = 0$ nor explode before T. Hence step 3 is complete. Since step 4 is trivial, we proved the theorem.   □

Our next goal is to show that the adapted solution of FBSDE (4.8) does give us the optimal strategy. Also, we would like to study the uniqueness of the adapted solution to the FBSDE (4.8), which cannot be easily deduced from Chapter 4, since in this case the function σ depends on π (see Chapter 4, Remark 1.2). It turns out, however, under the special setting of this section, we can in fact establish some comparison theorems which will resolve all these issues simultaneously. We should note that given the counterexample in Chapter 1, §6 (Example 6.2 of Chapter 1), these comparison theorems should be interesting in their own rights.

Theorem 4.4. *(Comparison Theorem): Suppose that the assumptions of the Theorem 4.3 are in force. For given $p \in \mathbb{R}_+^d$, let (π, C) be any admissible pair such that the corresponding price/wealth process (P, X) satisfies $X(T) \geq g(P(T))$, a.s. Then $X(\cdot) \geq \theta(\cdot, P(\cdot))$, where θ is the solution to (4.11).*

Consequently, if (P', X') is an adapted solution to FBSDE (4.8) starting from $p \in \mathbb{R}_+^d$, constructed by the Four-Step scheme. Then it holds that $X(0) \geq \theta(0, p) = X'(0)$.

Proof. We only consider the case when condition (H5)' holds, since the other case is much easier. Let (P, X, π, C) be given such that $(\pi, C) \in \mathcal{A}(Y(0))$ and $X(T) \geq g(P(T))$, a.s. We first define a change of probability measure as follows: let

$$(4.21) \quad \begin{cases} \theta_0(t) = [\sigma^{-1}[b - r\mathbf{1}](t, P(t), X(t), \pi(t)). \\[2mm] Z_0(t) = \exp\left\{ -\int_0^t \theta_0(s)dW(s) - \frac{1}{2}\int_0^t |\theta_0(s)|^2 ds\right\}; \\[2mm] \dfrac{d\mathbf{P}_0}{d\mathbf{P}} = Z_0(T), \end{cases}$$

so that the process $W_0(t) \stackrel{\Delta}{=} W(t) + \int_0^t \theta_0(s)ds$ is a Brownian motion on the new probability space $(\Omega, \mathcal{F}, \mathbf{P}_0)$. Then, the price/wealth FBSDE (4.4) and (4.5) become

$$(4.22) \quad \begin{cases} P(t) = p + \int_0^t P(s)\{r(s, X(s), \pi(s))ds \\[2mm] \qquad\qquad + \int_0^t \sigma(s, P(s), X(s), \pi(s))dW_0(s)\}, \\[2mm] X(t) = g(P(T)) - \int_t^T r(s, X(t), \pi(s))X(s)ds \\[2mm] \qquad\qquad - \int_t^T \pi(s)\sigma(s, P(s), X(s), \pi(s))dW_0(s) + C(T) - C(t), \end{cases}$$

Since in the present case the PDE (4.11) is degenerate, and the function g is not bounded, the solution θ to (4.11) and its partial derivatives could blow up as p approaches to $\partial\mathbb{R}_+^d$ and infinity. Therefore some modification of the method in Chapter 4 are needed here. First, we apply Itô's formula to the process $g(P(\cdot))$ from t to T to get

$$g(P(t)) = g(P(T)) - \int_t^T \{g_p(P)r(s, X, \pi)P - \frac{1}{2}\sigma^2(s, P, X, \pi)g_{pp}(P)\}ds$$
$$- \int_t^T g_p(P)\sigma(s, P, X, \pi)dW_0(s),$$

here and in what follows we write (P, X, π) instead of $(P(s), X(s), \pi(s))$ in all the integrals for notational convenience.

Next, we define a process $\widehat{X} = X - g(P)$, then $\widehat{X}$ satisfies the following (backward) SDE:

$$\widehat{X}(t) = \widehat{X}(T) - \int_t^T \{r(s, X, \pi)[X - g_p(P)P] - \frac{1}{2}\sigma^2(s, P, X, \pi)g_{pp}(P)\}ds$$
$$- \int_t^T (\pi(s) - Pg_p(P))\sigma(s, P, X, \pi)dW_0(s) + C(T) - C(t)$$

We now use the notation $\widehat{\theta} = \theta - g$ as that in the proof of Theorem 4.2; then it suffices to show that $\widehat{X}(t) \geq \widehat{\theta}(t, P(t))$ for all $t \in [0, T]$, a.s. $\mathbf{P_0}$. To this end, let us denote $\tilde{X}(t) = \widehat{\theta}(t, P(t))$, $\tilde{\pi}(t) = P(t)[\widehat{\theta}_p(t, P(t)) + g_p(P(t))]$; and $\Delta_X(t) = \widehat{X}(t) - \tilde{X}(t)$, $\Delta_\pi(t) = \pi(t) - \tilde{\pi}(t)$. Applying Itô's formula to the process $\Delta_X(t)$, we obtain

$$\Delta_X(t) = \widehat{X}(T) - \int_t^T \{r(s, X, \pi)[Y - (g_p(P) + \widehat{\theta}_p(s, P))P]$$
$$- \widehat{\theta}_s(s, P) - \frac{1}{2}\sigma^2(s, P, X, \pi)[\widehat{\theta}_{pp}(s, P) + g_{pp}(P)]\}ds$$

(4.23)
$$- \int_t^T (\pi - P[g_p(P) + \widehat{\theta}_p(s, P)]\sigma(s, P, X, \pi)dW_0(s)$$
$$= \widehat{X}(T) - \int_t^T A(s)ds - \int_t^T \Delta_\pi\sigma(s, P, X, \pi)dW_0(s)$$
$$+ C(T) - C(t),$$

where the process $A(\cdot)$ in the last term above is defined in the obvious way. Recall that the function $\widehat{\theta}$ satisfies PDE (4.16), that $\widehat{\theta}(t, P(t)) + g(P(t)) = X(t) - \Delta_X(t)$, and the definition of $\tilde{\pi}$, we can easily rewrite $A(\cdot)$ as follows:

$$A(s) = r(s, X, \pi)X(s) - r(s, X - \Delta_X, \tilde{\pi})[X(s) - \Delta_X(s)]$$
$$- r(s, X, \pi)\pi(s) - r(s, \widehat{\theta}(s, P), \tilde{\pi})\tilde{\pi}(s)$$
$$+ \frac{1}{2}\{\Delta\sigma(s, P, X, \pi, \tilde{\pi}, \widehat{\theta}(s, P))\Theta(s, P) = I_1(s) + I_2(s) + I_3(s),$$

where

$$\begin{cases} \Theta(t, p) \stackrel{\Delta}{=} p^2(\widehat{\theta}_{pp}(t, p) + g_{pp}(p)); \\ \Delta\sigma(t, p, x, \pi, \tilde{\pi}, q) \stackrel{\Delta}{=} \sigma^2(t, p, q + g(p), \tilde{\pi}) - \sigma^2(t, p, x, \pi)), \end{cases}$$

and I_i's are defined in the obvious way. Now noticing that

$$I_1(s) = [r(s, X, \pi)X(s) - r(s, X - \Delta_X, \pi)(X - \Delta_X)]$$
$$+ [r(s, X - \Delta_X, \pi) - r(s, X - \Delta_X, \widetilde{\pi})][X(s) - \Delta_X(s)]$$
$$= \Big\{ \int_0^1 \frac{\partial}{\partial x} \{r(s, x, \pi)x\} \Big|_{x=(X(s)-\lambda\Delta_X(s))} d\lambda \Big\} \Delta_X(s)$$
$$+ \int_0^1 \frac{\partial r}{\partial \pi}(s, X - \Delta_X, \pi + \lambda\Delta_\pi)[X - \Delta_X]d\lambda\Delta_\pi(s)$$
$$= \alpha_1(s)\Delta_X(s) + \beta_1(s)\Delta_\pi(s) \rangle,$$

we have from condition (A3) that both α_1 and β_2 are adapted processes and are uniformly bounded in (t, ω). Similarly, by conditions (H1)–(H3) and (H5'), we see that the process $\Theta(\cdot, P(\cdot))$ is uniformly bounded and that there exist uniformly bounded, adapted processes α_2, α_3 and β_2, β_3 such that

$$I_2(s) = r(s, X, \pi)\pi(s) - r(s, \widehat{\theta}(s, P), \pi)\pi(s)$$
$$+ [r(s, \widehat{\theta}(s, P), \pi)\pi(s) - r(s, \widehat{\theta}(s, P), \widetilde{\pi})\widetilde{\pi}(s)]$$
$$= \alpha_2(s)\Delta_X(s) + \beta_2(s)\Delta_\pi(s);$$
$$I_3(s) = \alpha_3(s)\Delta_X(s) + \beta_3(s)\Delta_\pi(s).$$

Therefore, letting $\alpha = \sum_{i=1}^3 \alpha_i$, $\beta = \sum_{i=1}^3 \beta_i$, we obtain that

$$A(t) = \alpha(t)\Delta_X(t) + \beta(t)\Delta_\pi(t),$$

where α and β are both adapted, uniformly bounded processes. In other words, we have from (4.23) that

(4.24)
$$\Delta_X(t) = \widehat{X}(T) - \int_t^T \{\alpha(s)\Delta_X(s) + \beta(s)\Delta_\pi(s)\}ds$$
$$- \int_t^T \Delta_\pi(s)\sigma(s, P, X, \pi)dW_0(s) \rangle + C(T) - C(t).$$

Now following the same argument as that in Chapter 1, Theorem 6.1 for BSDE's, one shows that (4.24) leads to that

(4.25)
$$\exp\Big(-\int_0^t \alpha(s)ds\Big)\Delta_X(t) = E\Big\{ \exp\Big(-\int_0^T \alpha(s)ds\Big)\Delta_X(T)$$
$$+ \int_t^T \exp\Big(-\int_0^s \alpha(u)du\Big)dC(s)\Big|\mathcal{F}_t\Big\}.$$

Therefore $\Delta_X(T) = X(T) - g(P(T)) \geq 0$ implies that $\Delta_X(t) \geq 0$, $\forall t \in [0, T]$, **P**-a.s. We leave the details to the reader.

Finally, note that if (P', X') is an adapted solution of (4.8) starting from p and constructed by Four Step Scheme, then it must satisfy that $X'(0) = \theta(0, p)$, hence $X(0) \geq X'(0)$ by the first part, completing the proof.   □

Note that if (P, X, π) is any adapted solution of FBSDE (4.8) starting from p, then (4.25) leads to that $X(t) = \theta(t, P(t))$, $\forall t \in [0, T]$, P-a.s., since $C \equiv 0$ and $\Delta(T) = X(T) - g(P(T)) = 0$. We derived the following uniqueness result of the FBSDE (4.8).

Corollary 4.5. *Suppose that assumptions of Theorem 4.4 are in force. Let (P, X, π) be an adapted solution to FBSDE (4.8), then it must be the same as the one constructed from the Four Step Scheme. In other words, the FBSDE (4.8) has a unique adapted solution and it can be constructed via (4.13) and (4.14).*

Reinterpreting Theorem 4.4 and Corollary 4.5 in the option pricing terms we derive the following optimality result.

Corollary 4.6. *Under the assumptions of Theorem 4.4, it holds that $h(g(P(T))) = X(0)$, where P, X are the first two components of the adapted solution to the FBSDE (4.8). Furthermore, the optimal hedging strategy is given by $(\pi, 0)$, where π is the third component of the adapted solution to FBSDE (4.8). Furthermore, the optimal hedging prince for (4.7) is given by $X(0)$, and the optimal hedging strategy is given by $(\pi, 0)$.*

Proof. We need only show that $(\pi, 0)$ is the optimal Strategy. Let $(\pi', C) \in H(B)$. Denote P' and X' be the corresponding price/wealth pair, then it holds that $X'(T) \geq g(P'(T))$ by definition. Theorem 4.4 then tells us that $X'(0) \geq X(0)$, where X is the backward component of the solution to the FBSDE (4.8), namely the initial endowment with respect to the strategy $(\pi, 0)$. This shows that $h(g(P(T))) = X(0)$, and therefore $(\pi, 0)$ is the optimal strategy. $\qquad\square$

To conclude this section, we present another comparison result that compares the adapted solutions of FBSDE (4.8) with different terminal condition. Again, such a comparison result takes advantage of the special form of the FBSDE considered in this section, which may not be true for general FBSDEs.

Theorem 4.7. (*Monotonicity in terminal condition*) *Suppose that the conditions of Theorem 4.3 are in force. Let (P^i, X^i, π^i), $i = 1, 2$ be the unique adapted solutions to (4.8), with the same initial prices $p > 0$ but different terminal conditions $X^i(T) = g^i(P^i(T))$, $i = 1, 2$ respectively. If g^1, g^2 all satisfy the condition (H5) or (H5)', and $g^1(p) \geq g^2(p)$ for all $p > 0$, then it holds that $X^1(0) \geq X^2(0)$.*

Proof. By Corollary 4.5 we know that X^1 and X^2 must have the form

$$X^1(t) = \theta^1(t, P^1(t)); \qquad X^2(t) = \theta^2(t, P^2(t)),$$

where θ^1 and θ^2 are the classical solutions to the PDE (4.11) with terminal conditions g^1 and g^2, respectively. We claim that the inequality $\theta^1(t, p) \geq \theta^2(t, p)$ must hold for all $(t, p) \in [0, T] \times \mathbb{R}^d_+$.

To see this, let us use the Euler transformation $p = e^\xi$ again, and define $u^i(t, \xi) = \theta^i(T - t, e^\xi)$. It follows from the proof of Theorem 4.2 that u^1

and u^2 satisfy the following PDE:

(4.26)
$$\begin{cases} 0 = u_t - \dfrac{1}{2}\bar{\sigma}^2(t,\xi,u,u_\xi)u_{\xi\xi} - b_0(t,\xi,u,u_\xi)u_\xi + u\bar{r}(t,u,u_\xi), \\ u(0,\xi) = g^i(e^\xi), \qquad \xi \in \mathbb{R}^d, \end{cases}$$

respectively, where

$$\bar{\sigma}(t,\xi,x,\pi) = e^{-\xi}\sigma(T-t,e^\xi,x,\pi);$$
$$b_0(t,\xi,x,\pi) = r(T-t,x,\pi) - \frac{1}{2}\bar{\sigma}^2(T-t,\xi,x,\pi);$$
$$\bar{r}(t,x,\pi) = r(T-t,x,\pi).$$

Recall from Chapter 4 that u^i's are in fact the (local) uniform limits of the solutions of following initial-boundary value problems:

(4.27)
$$\begin{cases} 0 = u_t - \dfrac{1}{2}\bar{\sigma}_i^2(t,\zeta,u,u_\xi)u_{\xi\xi} - b_0(t,\xi,u,u_\xi)u_\xi + ur(t,u,u_\xi), \\ u\big|_{\partial B_R}(t,\xi) = g^i(e^\xi), \qquad |\xi| = R; \\ u(0,\xi) = g^i(e^\xi), \qquad \xi \in B_R, \end{cases}$$

$i = 1,2$, respectively, where $B_R \overset{\Delta}{=} \{\xi; |\xi| \le R\}$. Therefore, we need only show that $u_R^1(t,\xi) \ge u_R^2(t,\xi)$ for all $(t,\xi) \in [0,T] \times B_R$ and $R > 0$.

For any $\varepsilon > 0$, consider the PDE:

(4.27ε)
$$\begin{cases} u_t = \dfrac{1}{2}\bar{\sigma}^2(t,\xi,u,u_\xi)u_{\xi\xi} + b_0(t,\xi,u,u_\xi)u_\xi - ur(t,u,u_\xi) + \varepsilon, \\ u\big|_{\partial B_R}(t,\xi) = g^1(e^\xi) + \varepsilon, \qquad |\xi| = R; \\ u(0,\xi) = g^1(e^\xi) + \varepsilon, \qquad \xi \in B_R, \end{cases}$$

and denote its solution by $u_{R,\varepsilon}^1$. It is not hard to check, using a standard technique of PDEs (see, e.g., Friedman [1]), that $u_{R,\varepsilon}^1$ converges to u_R^1, uniformly in $[0,T] \times \mathbb{R}^d$. Next, We define a function

$$F(t,\xi,x,q,\widehat{q}) = \frac{1}{2}\bar{\sigma}^2(t,\xi,x,q)\widehat{q} + b_0(t,\xi,q,\widehat{q})\widehat{q} - x\bar{r}(t,x,q).$$

Clearly F is continuously differentiable in all variables, and $u_{R,\varepsilon}^1$ and u_R^2 satisfies

$$\begin{cases} \dfrac{\partial u_{R,\varepsilon}^1}{\partial t} > F(t,\xi,u_{R,\varepsilon}^1,(u_{R,\varepsilon}^1)_\xi,(u_{R,\varepsilon}^1)_{\xi\xi}); \\ \dfrac{\partial u_R^2}{\partial t} = F(t,\xi,u_R^2,(u_R^2)_\xi,(u_R^2)_{\xi\xi}); \\ u_{R,\varepsilon}^1(t,\xi) > u_R^2(t,\xi), \qquad (t,\xi) \in [0,T] \times B_T \bigcup \{0\} \times \partial B_R, \end{cases}$$

Therefore by Theorem II.16 of Friedman [1], we have $u_{R,\varepsilon}^1 > u_R^2$ in B_R. By sending $\varepsilon \to 0$ and then $R \to \infty$, we obtain that $u^1(t,\xi) \ge u^2(t,\xi)$

for all $(t, \xi) \in [0, T] \times \mathbb{R}^d$, whence $\theta^1(\cdot, \cdot) \geq \theta^2(\cdot, \cdot)$. In particular, we have $X^1(0) = \theta^1(0, p) \geq \theta^2(0, p) = X^2(0)$, proving the theorem. $\qquad\square$

Remark 4.8. We should note that from $\theta^1(t, p) \geq \theta^2(t, p)$ we *cannot* conclude that $X^1(t) \geq X^2(t)$ for all t, since in general there is no comparison between $\theta^1(t, P^1(t))$ and $\theta^2(t, P^2(t))$, as was shown in Chapter 1, Example 6.2!

§4.2. Hedging with constraint

In this section we try to solve the hedging problem (4.7) with an extra condition that the portfolio of an investor is subject to a certain constraint, namely, we assume that

(Portfolio Constraint) *There exists a constant $C_0 > 0$ such that $|\pi(t)| \leq C_0$, for all $t \in [0, T]$, a.s.*

Recall that $\pi(t)$ denotes the amount of money the investor puts in the stock, an equivalent condition is that the total number of shares of the stock available to the investor is limited, which is quite natural in the practice.

In what follows we shall consider the log-price/wealth pair instead of price/wealth pair like we did in the last subsection. We note that these two formulations are not always equivalent, we do this for the simplicity of the presentation. Let P be the price process that evolves according to the SDE (4.1). We assume the following

(H6) b and σ are independent of π and are time-homogeneous; $g \geq 0$ and belongs boundedly to $C^{2+\alpha}$ for some $\alpha \in (0, 1)$; and r is uniformly bounded.

Define $\chi(t) = \ln P(t)$. Then by Itô's formula we see that χ satisfies the SDE:

$$\chi(t) = \chi_0 + \int_0^t [b(e^{\chi(s)}, X(s)) - \frac{1}{2}\sigma^2(e^{\chi(s)}, X(s))]ds$$

(4.21)
$$+ \sigma(e^{\chi(s)}, X(s))dW(s)$$

$$= \chi_0 + \int_0^t \tilde{b}(\chi(s), X(s))ds + \int_0^t \tilde{\sigma}(\chi(s), X(s))dW(s),$$

where $\chi_0 = \ln p$; $\tilde{b}(\chi, x) = b(e^\chi, x) - \frac{1}{2}\sigma^2(e^\chi, x)$; and $\tilde{\sigma}(\chi, x) = \sigma(e^\chi, x)$. Next, we rewrite the wealth equation (4.5) as follows.

$$X(t) = x + \int_0^t [r(s)X(s) + \pi(s)(b(P(s), X(s)) - r(s))]ds$$

(4.22)
$$+ \int_0^t \pi(s)\sigma(P(s), X(s))dW(s) - C(t)$$

$$= x - \int_0^t f(s, \chi(s), X(s), \pi(s))ds + \int_0^t \pi(s)d\chi(s) - C(t).$$

where

(4.23)
$$f(t, \chi, x, \pi) = -r(s)x - \pi[\frac{1}{2}\tilde{\sigma}^2(\chi, x) - r(s)].$$

In light of the discussion in the previous subsection, we see that in order to solve a hedging problem (4.7) with portfolio constraint, one has to solve the following FBSDE

$$
(4.24) \quad
\begin{cases}
\chi(t) = \chi_0 + \displaystyle\int_0^t \tilde{b}(\chi(s), X(s))ds + \int_0^t \tilde{\sigma}(\chi(s), X(s))dW(s), \\[2mm]
X(t) = g(\chi(T)) + \displaystyle\int_t^T f(s, \chi(s), X(s), \pi(s))ds - \int_t^T \pi(s)d\chi(s) \\[2mm]
\qquad\quad + C(T) - C(t), \\[2mm]
|\pi(\cdot)| \le C_0, \qquad dt \times d\mathbf{P}\text{-a.e. } (t, \omega) \in [0, T] \times \Omega.
\end{cases}
$$

In the sequel we call the set of all adapted solutions (χ, X, π, C) to the FBSDE (4.24) the set of *admissible solutions*. We will be interested in the nonemptyness of this set and the existence of the *minimal solution*, which will give us the solution to the hedging problem (4.7). To simplify discussion let us make the following assumption:

(H7) $\tilde{b}$ and $\tilde{\sigma}$ are uniformly bounded in (χ, x) and both have bounded first order partial derivatives in χ and x.

We shall apply a *Penalization procedure* similar to the one used in Chapter 7 to prove the existence of the admissible solution. Namely, we let φ be a smooth function defined on $\mathbb{R}$ such that

$$
(4.25) \quad
\begin{aligned}
\varphi(x) &= \begin{cases}
0 & |x| \le C_0; \\
x - (C_0 + 1) & x \ge C_0 + 2; \\
-x - (C_0 + 1) & x \le -C_0 - 2;
\end{cases} \\
|\varphi'(x)| &\le 1, \qquad \forall x \in \mathbb{R};
\end{aligned}
$$

and consider the *penalized* FBSDEs corresponding to (4.24) with $C = 0$: for each $n > 0$, and $0 \le t \le s \le T$,

(4.26)
$$
\begin{cases}
\chi^n(s) = \chi_0 + \displaystyle\int_t^s \tilde{b}(\chi^n(r), X^n(r))dr + \int_t^s \tilde{\sigma}(\chi^n(r), X^n(r))dW(r), \\[2mm]
X^n(s) = g(\chi^n(T)) + \displaystyle\int_s^T [f(r, \chi^n(r), X^n(r), \pi^n(r)) + n\varphi(\pi^n(r))]dr \\[2mm]
\qquad\quad - \displaystyle\int_s^T \pi^n(r)d\chi^n(r).
\end{cases}
$$

Applying the Four Step Scheme in Chapter 4 (in the case $m = 1$), we see that (4.26) has a unique adapted solution that can be written explicitly as

$$
(4.27) \quad
\begin{cases}
X^n(s) = \theta^n(s, \chi^n(s)); \\
\pi^n(s) = \theta^n_\chi(s, \chi^n(s)),
\end{cases}
\qquad s \in [t, T],
$$

where θ^n is the classical solution to the following parabolic PDE:

$$(4.28) \quad \begin{cases} \theta_t^n + \dfrac{1}{2}\tilde{\sigma}^2(\chi,\theta^n)\theta_{\chi\chi}^n + f(t,\chi,\theta^n,\theta_\chi^n) + n\varphi(\theta_\chi^n) = 0, \\ \theta^n(T,\chi) = g(\chi). \end{cases}$$

Further, the solution θ^n, along with its partial derivatives θ_t^n, θ_χ^n and $\theta_{\chi\chi}^n$ are all bounded (with the bound depending possibly on n). The following lemma shows that the bound for θ^n and θ_χ^n can actually be made independent of n.

Lemma 4.9. *Assume (H6) and (H7). Then there exists and constant $C > 0$ such that*

$$0 \le \theta^n(\chi,x) \le C; \qquad |\theta_\chi^n(\chi,x)| \le C, \qquad \forall(\chi,x) \in \mathbb{R}^2.$$

Proof. By (H6) and (H7), definitions (4.23) and (4.25), we see that there exist adapted process α^n and β^n such that $|\alpha^n(s)| \le L$, $|\beta^n(s)| \le L_n$, $\forall s \in [t,T]$, $\forall n > 0$, P-a.s., for some $L > 0$, $L_n > 0$; and that

$$f(s,\chi(s),X^n(s),\pi^n(s)) + n\varphi(\pi^n(s)) = \alpha^n(s)X^n(s) + \beta^n(s)\pi^n(s).$$

Define $R^n(s) = \exp\{\int_t^s \alpha_r^n dr\}$, $s \in [t,T]$. Then by Itô's formula one has

$$R^n(s)X^n(s) = R^n(T)g(\chi^n(T)) + \int_s^T R^n(r)\beta^n(r)\pi^n(r)dr$$

$$(4.29) \qquad\qquad - \int_s^T R^n(s)\pi^n(r)dW(r)$$

$$= R^n(T)g(\chi^n(T)) - \int_s^T R^n(r)\pi^n(r)dW^n(r).$$

where $W^n(s) = W(s) - W(t) - \int_t^s \beta^n(r)dr$. Since β^n is bounded for each n, there exists probability measure $Q^n \ll P$ such that W^n is a Q^n Brownian motion on $[t,T]$, thanks to Girsanov's Theorem. We derive from (4.29) and (H6) that

$$0 \le R^n(s)X^n(s) = E^{Q^n}\{R^n(T)g(\chi^n(T))|\mathcal{F}_s\} \le C, \qquad Q^n\text{-a.s.}$$

where the constant $C > 0$ is independent of n. Consequently X^n, is uniformly bounded, uniformly in n, almost surely. In particular, there exists $C > 0$ such that $0 \le X_t^n = \theta^n(t,\chi) \le C$, proving the first part of the lemma.

To see the second part, denote $Z^n(s) = \theta_{\chi\chi}^n(s,\chi(s))\tilde{\sigma}(s,\chi(s))$. Since we can always assume that θ^n is actually C^3 by the smoothness assumptions in (H6) and (H7), we can use the similar argument as that in Proposition 1.1 to show that the pair (π^n, Z^n) is an adapted solution to the BSDE:

$$\pi^n(s) = g'(\chi(T)) + \int_s^T [A^n(r)Z^n(r) + B^n(r)\pi^n(r) + C^n(r)]dr$$

$$- \int_s^T Z^n(r)dW(r),$$

222

Chapter 8. Applications of FBSDEs

where

$$
(4.30)
\begin{cases}
A^n(s) = \Big\{ \tilde{\sigma}(\chi, x)[\tilde{\sigma}_\chi(\chi, x) + \tilde{\sigma}_x(\chi, x)\theta_\chi^n(t, \chi) + n\varphi'(\pi) \\
\qquad\qquad + f_\pi(s, \chi, x, \pi)] \Big\}\Big|_{(\chi,x,\pi)=(\chi^n(s),\theta^n(s,\chi^n(s)),\theta_\chi^n(s,\chi^n(s)))} \\
B^n(s) = f_x(s, \chi, x, \pi)\Big|_{(\chi,x,\pi)=(\chi^n(s),\theta^n(s,\chi^n(s)),\theta_\chi^n(s,\chi^n(s)))} \\
C^n(s) = f_\chi(s, \chi, x, \pi)\Big|_{(\chi,x,\pi)=(\chi^n(s),\theta^n(s,\chi^n(s)),\theta_\chi^n(s,\chi^n(s)))}
\end{cases}
$$

Since B^n and C^n are uniformly bounded, uniformly in n, by (H6) and (H7), and A^n is bounded for each n, a similar argument as that of part 1 will lead to the uniform boundedness of π^n, with the bounded independent of n. The proof of the lemma is now complete. $\qquad\square$

Next, we prove a comparison theorem that is not covered by those in Chapter 1, §6.

Lemma 4.10. *Assume (H6) and (H7). For any $n \geq 1$ it holds that $\theta^{n+1}(t, \chi) \geq \theta^n(t, \chi)$, $\forall (t, \chi) \in [0, T] \times \mathbb{R}$.*

Proof. For each n, let (χ^n, X^n, π^n) be the adapted solution to (4.26), defined on $[t, T]$. Define $\tilde{X}^n(s) = \theta^n(s, \chi^{n+1}(s))$ and $\tilde{\pi}^n(s) = \theta_\chi^n(s, \chi^{n+1}(s))$. Applying Itô's formula and using the definition of $\tilde{X}^n$, $\tilde{\pi}^n$, and θ^n one shows that

$$
d\tilde{X}^n(s) = \Big\{ -f(s, \chi^{n+1}(s), \tilde{X}^{n+1}(s), \tilde{\pi}^{n+1}(s)) \\
\qquad - (n+1)\varphi(\tilde{\pi}^{n+1}(s)) + \tilde{\pi}^n(s)\tilde{b}(\chi^{n+1}(s), X^{n+1}(s)) \Big\} ds \\
\qquad + \tilde{\pi}^n(s)\tilde{\sigma}(\chi^{n+1}(s), X^{n+1}(s)) dW(s).
$$

On the other hand, by definition we have

$$
dX^{n+1}(s) = \Big\{ -f(s, \chi^{n+1}(s), \tilde{X}^{n+1}(s), \pi^{n+1}(s)) \\
\qquad - (n+1)\varphi(\pi^{n+1}(s)) + \pi^n(s)\tilde{b}(\chi^{n+1}(s), X^{n+1}(s)) \Big\} ds \\
\qquad + \pi^n(s)\tilde{\sigma}(\chi^{n+1}(s), X^{n+1}(s)) dW(s).
$$

Now denote $\hat{X}^n = X^{n+1} - \tilde{X}^n$ and $\hat{\pi}^n = \pi^{n+1} - \tilde{\pi}^n$, and note that φ is uniform Lipschitz with Lipschitz constant 1, b and θ^n are uniformly bound, we see that for some some bounded processes α^n and β^n it holds that

$$
d\hat{X}^n(s) = \Big\{ -\alpha^n(s)\hat{X}^n(s) - \beta^n(s)\hat{\pi}^n(s) - \varphi(\pi^{n+1}(s)) \\
\qquad + \hat{\pi}^n(s)\tilde{\sigma}(\chi^{n+1}(s), X^{n+1}(s)) dW(s).
$$

Since $\hat{X}^n(T) = 0$ and $\varphi \geq 0$, the same technique of Theorem 4.4 than shows that under some probability measure $\tilde{Q}$ which is equivalent to $\mathbf{P}$ one has

$$
\hat{X}^n(s) = E^{\tilde{Q}} \Big\{ \int_s^T R^n(r)\varphi(\pi^{n+1}(r)) dr \Big| \mathcal{F}_s \Big\} \geq 0,
$$

where $\Gamma(s) = \exp\{\int_t^s \alpha^n(r)dr\}$. Setting $s = t$ we derive that $\theta^{n+1}(t, \chi) \geq \theta^n(t, \chi)$. □

Combining Lemmas 4.9 and 4.10 we see that there exists function $\theta(t, x)$ such that $\theta^n(t, \chi) \nearrow \theta(t, \chi)$, as $n \to \infty$. Clearly θ is jointly measurable, uniformly bounded, and uniform Lipschitz in χ, thanks to Lemma 4.9. Thus the following SDE is well-posed:

$$(4.31) \qquad \chi(s) = \int_t^s \tilde{b}(\chi(r), \theta(r, \chi(r)))dr + \int_t^s \tilde{\sigma}(\chi(r), \theta(r, \chi(r)))dW(r);$$

Now define $X(s) = \theta(s, \chi(s))$. It is easy to show, using the uniform Lipschitz property of θ (in x) and some standard argument for the stability of SDEs, that

$$(4.32) \qquad \lim_{n \to \infty} E\Big\{ \sup_{t \leq s \leq T} |\chi^n(s) - \chi(s)| \Big\} = 0,$$

and, together with a simple application of Dominated Convergence Theorem, that

$$(4.33) \quad \begin{aligned} E\{|X^n(s) - X(s)|\} &= E\{|\theta^n(s, \chi^n(s)) - \theta(s, \chi(s))|\} \\ &\leq 2C_2^2 E\{|\chi^n(s) - \chi(s)|\} + 2E\{|\theta^n(s, \chi(s)) - \theta(s, \chi(s))|\} \to 0, \end{aligned}$$

as $n \to \infty$. We should note that at this point we do not have any information about the regularity of the paths of process X, and neither do we know that it is even a semimartingale. Let us now take a closer look.

First notice that Lemma 4.9 and the boundedness of $r(\cdot)$ and $\tilde{\sigma}$

$$E\int_t^T |\pi^n(s)|^2 ds \leq C; \qquad E\int_t^T |f(s, \chi^n(s), X^n(s), \pi^n(s))|^2 ds \leq C.$$

Therefore for some processes $\pi, f^0 \in L_{\mathcal{F}}^2(t, T; \mathbb{R})$ such that, possibly along a subsequence, one has

$$(4.34) \qquad (\pi^n, f(s, \chi^n(s), X^n(s), \pi^n(s))) \xrightarrow{w} (\pi, f^0), \text{ in } (L_{\mathcal{F}}^2(t, T; \mathbb{R}))^2.$$

Next, let us define

$$(4.35) \qquad \begin{cases} A^n(s) = \int_t^s n\varphi(\pi^n(r))dr, & 0 \leq t \leq s \leq T, \\ A(s) = \theta(t, \chi) - X(s) - \int_t^s f^0(r)dr - \int_t^s \pi(r)d\chi(r). \end{cases}$$

Since

$$(4.36) \quad \begin{aligned} A^n(s) &= \theta^n(t, \chi) - X^n(s) - \int_t^s [f(r, \chi^n(r), X^n(r), \pi^n(r))dr \\ &\quad - \int_t^s \pi^n(r)d\chi^n(r), \end{aligned}$$

Combining (4.32)–(4.34), one shows easily that, A^n converges weakly in $L^2(0, T; \mathbb{R})$ to $A(s)$. Therefore it is not hard to see that for any fixed $t \le s_1 < s_2 \le T$ it holds that

$$(4.37) \qquad\qquad P\{A_{s_1} \le A_{s_2}\} = 0,$$

since A^n's are all continuous, monotone increasing processes. Thus one shows that both $A(s-)$ and $A(s+)$ exist for all $s \in [t, T]$. Denote $\widetilde{A}(s) = A(s+)$, then $\widetilde{A}$ is càdlàg, and for fixed s, $A(s) \le \widetilde{A}(s)$, P-a.s.. We claim that the equality actually holds. Indeed, from (4.35) we see that $X(\cdot) + A(\cdot)$ is continuous. Let $\mathbf{Q}$ be the rationals in $\mathbb{R}$, then for each $s \in [t, T]$, it holds almost surely that

$$(4.38) \qquad \lim_{\substack{r \downarrow s \\ r \in \mathbf{Q}}} X(r) = \lim_{\substack{r \downarrow s \\ r \in \mathbf{Q}}} [X(r) + A(r) - A(r)] = X(s) + A(s) - \widetilde{A}(s).$$

On the other hand, since for each $r \in [t, T]$ one has $X(r) = \theta(r, \chi(r)) \ge \theta^n(r, \chi(r))$, using the continuity of the functions θ^n's and the process $\chi(\cdot)$ we have

$$\lim_{r \downarrow s} X(r) \ge \lim_{r \downarrow s} \theta^n(r, \chi(r)) = \theta^n(s, \chi(s)).$$

Letting $n \to \infty$ and using (4.38) we derive

$$X(s) + A(s) = \lim_{\substack{r \downarrow s \\ r \in \mathbf{Q}}} X(r) + \widetilde{A}(s) \ge \lim_{n \to \infty} \theta^n(s, \chi(s)) + \widetilde{A}(s)$$

$$= \theta(s, \chi(s)) + \widetilde{A}(s) = X(s) + \widetilde{A}(s).$$

Consequently, $A(s) \ge \widetilde{A}(s)$, P-a.s., whence $A(s) = \widetilde{A}(s)$, P-a.s.. In other words, $\widetilde{A}(s)$ is a càdlàg version of A.

From now on we replace A by its càdlàg version in (4.35) without further specification. Namely the process X $(\overset{\Delta}{=} \theta(\cdot, \chi(\cdot)))$ is a semimartingale with the decomposition:

$$(4.39) \quad X(s) = \theta(t, \chi) - \left(\int_s^T f^0(r)dr + A(s) \right) - \int_s^T \pi(r)d\chi(r), \quad t \le s \le T,$$

and is càdlàg as well. We have the following theorem.

Theorem 4.11. *Assume (H6) and (H7). Let χ, f^0, π be defined by (4.31) and (4.34), respectively; and let $X(s) = \theta(s, \chi(s))$, where θ is the (monotone) limit of the solutions of PDEs (4.28), $\{\theta^n\}$. Define*

$$(4.40) \quad C(s) \overset{\Delta}{=} \int_t^s \{f^0(s) - f(\chi(r), X(r), \pi(r))\}dr + A(s), \quad t \le s \le T.$$

Then (χ, X, π, C) is an adapted solution to the FBSDE with constraint (4.24).

Furthermore, if $(\tilde{\chi}, \widetilde{X}, \tilde{\pi}, \widetilde{C})$ is any adapted solution to (4.24) on $[t, T]$, then it must hold that $X(t) \le \widetilde{X}(t)$. Consequently, $x^* \overset{\Delta}{=} X(0)$ is the minimum hedging price to the problem (4.7) with the portfolio constraint $|\pi(t)| \leq C_0$.

Proof. We first show that $f^0(r) - f(\chi(r), X(r), Z(r))\} \geq 0$, $dt \times d\mathbf{P}$-a.e. In fact, using the convexity of f in the variable z and that $f(t, x, 0, 0) = 0$ we have, for each n,

$$
\begin{aligned}
&f(s, \chi^n(s), X^n(s), \pi^n(s)) - f(s, \chi(s), X(s), \pi(s)) \\
&\geq -L(|\chi^n(s) - \chi(s)| + |X^n(s) - X(s)|) \\
&\qquad + [f(s, \chi(s), X(s), \pi^n(s)) - f(s, \chi(s), X(s), \pi(s))] \\
&\geq -L(|\chi^n(s) - \chi(s)| + |X^n(s) - X(s)|) \\
&\qquad + (\pi^n(s) - \pi(s))f_\pi(\chi(s), X(s), \pi(r)).
\end{aligned}
$$

Using the boundedness of f_π, we see that for any $\eta \in L^2_{\mathcal{F}}(0, T; \mathbb{R})$ such that $\eta \geq 0$, $dt \times d\mathbf{P}$-a.e., it holds that, as $n \to \infty$,

$$
\begin{aligned}
&E \int_t^T [f^0(r) - f(r, \chi(r), X(r), \pi(r))]\eta_r \, dr \\
&= \lim_{n \to \infty} E \int_t^T [f(r, \chi^n(r), X^n(r), \pi^n(r)) - f(r, \chi(r), X(r), \pi(r))]\eta_r \, dr \\
&\geq -LE \int_t^T [|\chi^n(r) - \chi(r)| + |X^n(r) - X(r)|]\eta(r) dr \\
&\quad + E \int_t^T (\pi^n(r) - \pi(r))f_\pi(\chi(s), X(s), \pi(r))\eta(r)dr \to 0.
\end{aligned}
$$

Therefore $f^0(s) - f(s, \chi(s), X(s), \pi(s)) \geq 0$, $dt \times dP$-a.e., namely $C(\cdot)$ is a càdlàg, nondecreasing process. Now rewriting (4.39) as

$$
X(s) = g(\chi(T)) + \int_s^T f(r, \chi(r), X(r), \pi(r))dr + \int_s^T \pi(r)d\chi(r) + C(T) - C(s),
$$

for $t \leq s \leq T$, we see that (χ, X, π, C) solves the FBSDE in (4.24). It remains to check that $\pi(t) \in \Gamma$, $dt \times d\mathbf{P}$-a.e. But since $\varphi_\Gamma(0) = 0$ and $|\varphi_\Gamma'| \leq 1$, we have

$$
E \int_0^T |\varphi_\Gamma(\pi^n(s))|^2 ds \leq E \int_0^T |\pi^n(s)|^2 ds \leq C.
$$

Thus, possibly along a subsequence, we have $\varphi_\Gamma(\pi^n(\cdot)) \overset{w}{\to} \varphi_\Gamma^0$ for some $\varphi_\Gamma^0 \in L^2(0, T; \mathbb{R})$. Since φ_G is convex and C^1 by construction, we can repeat the argument as before to conclude that $\varphi_\Gamma^0(s) \geq \varphi_\Gamma(\pi(s)) \geq 0$, $dt \times d\mathbf{P}$-a.e.. But on the other hand,

$$
\begin{aligned}
E \int_t^T \varphi_\Gamma(\pi(r))dr &\leq E \int_t^T \varphi_\Gamma^0(r)dr = \lim_{n \to \infty} E \int_t^T \varphi_\Gamma(\pi^n(r))dr \\
&= \lim_{n \to \infty} \frac{1}{n} EA^n(T) = 0,
\end{aligned}
$$

we have that $\varphi_\Gamma(\pi(s)) = 0$, $dt \times d\mathbf{P}$-a.e.

To prove the last statement of the theorem let $(\widetilde{\chi}, \widetilde{X}, \widetilde{\pi}, \widetilde{C})$ be any other solutions of the FBSDE (4.24). Denote for each n, $\widetilde{X}^n(s) = \theta^n(s, \chi(s))$ and $\widetilde{\pi}^n(s) = \theta^n_\chi(s, \chi(s))$. Applying Itô's formula and Using (4.28) one can show that $\widetilde{X}^n$ is a solution to the BSDE

(4.41)
$$\widetilde{X}^n(s) = g(\widetilde{\chi}(T)) + \int_s^T [f(r, \widetilde{\chi}(r), \widetilde{X}^n(r), \widetilde{\pi}^n(r)) + n\varphi(\widetilde{\pi}^n(r))]$$
$$-\frac{1}{2}(\sigma^2(\widetilde{\chi}(r), \widetilde{X}(r)) - \sigma^2(\widetilde{\chi}(r), \widetilde{X}^n(r)))]dr - \int_s^T \widetilde{\pi}^n(r)d\chi(r).$$

It then follows, with $\hat{X} \overset{\Delta}{=} \widetilde{X} - \widetilde{X}^n$, $\hat{\pi} \overset{\Delta}{=} \widetilde{\pi} - \widetilde{\pi}^n$, that

$$\hat{X}(s) = \int_s^T [\alpha^n(r)\hat{X}(r) + \beta^n(r)\hat{\pi}(r)]dr - \int_s^T \hat{\pi}(r)d\chi(r) + \widetilde{C}(T) - \widetilde{C}(s),$$

where α^n and β^n are some bounded, adapted processes, thanks to the assumptions on the coefficients. Thus some similar arguments as those in Lemma 4.10 shows that $\hat{X}(s) \geq 0$, $\forall s \in [t, T]$, $\mathbf{P}$-a.s. In particular, one has $\widetilde{X}(t) \geq \widetilde{X}^n(t) = \theta^n(t, \chi)$, for all n. Letting $n \to \infty$ we obtain that $\widetilde{X}(t) \geq \theta(t, \chi) = X(t)$. Thus (χ, X, π, C) is the minimum solution of (4.24) on $[t, T]$. Finally, if $t = 0$, then we conclude that $x^* = X(0)$ is the minimum hedging price to (4.7) with portfolio constraint, proving the theorem. $\qquad\square$

§5. A Stochastic Black-Scholes Formula

In this section we present another application of the theory established in the previous chapters to the theory of option pricing. First recall that in the last section we essentially assumed that the market is "Markovian", that is, we assumed that all the coefficients in the price equation are deterministic so that the Four Step Scheme could be applied. We now try to explore the possibility of considering more general market models in which the market parameters can be random. To compensate this relaxation, we return to a standard "small investor" world. Namely, we assume that the price equations are (compared to (4.4)):

(5.1)
$$\begin{cases} dP^0(t) = r(t)P^0(t)dt; & \text{(bond)} \\ dP(t) = P(t)[b(t)dt + \sigma(t)dW(t)], & \text{(stock)} \end{cases}$$

where r, b, and σ are now assumed to be bounded, progressively measurable stochastic processes. We also assume that σ is bounded away from zero. To simplify discussion, we shall assume that both P and W are one dimensional. Thus the wealth equation (4.5) now becomes (replacing X by Y in this section)

(5.2) $dY(t) = [Y(t)r(t) + \pi(t)(b(t) - r(t))]dt + \pi(t)\sigma(t)dW(t) - dC(t).$

In the case where $r(\cdot) \equiv r$, $b(\cdot) \equiv b$, and $\sigma(\cdot) \equiv \sigma$ are all constants, the standard Black-Scholes theory tells us that the fair price of an option of the form $g(P(T))$ at any time $t \in [0, T]$ is given by

$$(5.3) \qquad Y(t) = \tilde{E}\{e^{-r(T-t)}g(P(T))|\mathcal{F}_t\},$$

Here $\tilde{E}$ is the expectation with respect to some *risk-neutral* probability measure (or "equivalent martingale measure"). Furthermore, if we denote $u(t, x)$ to be the (classical) solution to the backward PDE:

$$(5.4) \qquad \begin{cases} u_t + \dfrac{1}{2}\sigma^2 x^2 u_{xx} + rxu_x - ru = 0, & (t, x) \in [0, T) \times (0, \infty); \\ u(T, x) = g(x), \end{cases}$$

then it holds that $Y(t) = u(t, P(t))$, $\forall t \in [0, T]$, a.s.. Further, using the theory of BSDE, it is not hard to show that if (Y, Z) is the unique adapted solution of the backward SDE:

$$Y(t) = g(P(T)) - \int_t^T [rY(s) + \sigma^{-1}(b - r)Z(s)]ds - \int_t^T Z(s)dW(s),$$

then Y coincides with that in (5.3); and the optimal hedging strategy is given by $\pi(t) = \sigma^{-1}Z(t) = v_x(t, P(t))$.

In light of the result of §4, we see that the valuation formula (5.3) is not hard to prove even in the general cases when r, b, σ, and $g(\cdot)$ are allowed to be random. But a more subtle problem is to find a proper replacement, if possible, of the "Black-Scholes PDE" (5.4). We note that since the coefficients are now random, a "PDE" would no longer be appropriate. It turns out that the BSPDE established in Chapter 5 will serve for this purpose.

§5.1. Stochastic Black-Scholes formula

Let us consider the price equation (5.1) with random coefficients r, b, σ; and we consider the general terminal value g as described at the beginning of the section. We allow further that r and b may depend on the stock price in a nonanticipating way. In other words, we assume that $r(t, \omega) = r(t, P(t, \omega), \omega)$; $b(t, \omega) = b(t, P(t, \omega), \omega)$, and $\sigma(t, \omega) = \sigma(t, P(t, \omega), \omega)$ where for each fixed $p \in \mathbb{R}$, $r(\cdot, p, \cdot)$, $b(\cdot, p, \cdot)$, and $\sigma(\cdot, p, \cdot)$ are predictable processes. Thus we can write (5.1) and (5.5) as an (decoupled) FBSDE:

$$(5.6) \qquad \begin{cases} P(t) = p + \int_0^t P(s)b(s, P(s))ds + \int_0^t P(s)\sigma(s)dW_s, \\ Y(t) = g(P(T)) - \int_t^T [Y(s)r(s, P(s)) + Z(s)\theta(s, P(s))]ds \\ \qquad\qquad - \int_t^T Z(s)dW(s), \end{cases}$$

where θ is the so-called *risk premium* process defined by

$$\theta(t, P(t)) = \sigma^{-1}(t, P(t))[b(t, P(t)) - r(t, P(t))], \qquad \forall t \in [0, T];$$

and $Z(t) \overset{\Delta}{=} \pi(t)\sigma(t)$. We shall again make use of the Euler transformation $x = \log p$ introduced in the last section. By Itô's formula we see that the *log-price* process $X \overset{\Delta}{=} \log P$ and the wealth process Y will satisfy

(5.7)
$$
\begin{cases}
X(t) = \xi + \int_0^t \bar{b}(s, X(s))ds + \int_0^t \bar{\sigma}(s, X(s))dW(s), \\[2mm]
Y(t) = \bar{g}(X(T)) - \int_t^T [Y(s)\bar{r}(s, X(s)) + Z(s)\bar{\theta}(s, X(s))]ds \\[2mm]
\qquad\quad - \int_t^T Z(s)dW(s),
\end{cases}
$$

where

(5.8)
$$
\begin{aligned}
&\bar{b}(t, x, \omega) = b(t, e^x, \omega) - \frac{1}{2}\sigma^2(t, e^x\omega); \\[2mm]
&r(t, x, \omega) = r(t, e^x, \omega); \quad \sigma(t, x, \omega) = \sigma(t, e^x, \omega); \\[2mm]
&\bar{\theta}(t, x, \omega) = \theta(t, e^x, \omega), \quad \bar{g}(p, \omega) = g(e^x, \omega).
\end{aligned}
$$

We have the following result.

Theorem 5.1. (*Stochastic Black-Scholes Formula*) *Suppose that the random fields $\bar{b}$, $\bar{r}$, $\bar{\theta}$ and $\bar{g}$ defined in (5.8) are progressively measurable in (t, ω), and are m-th continuously differentiable in the variable x, with all partial derivatives being uniformly bounded, for some $m > 2$. Let the unique adapted solution of (5.7) be (X, Y, Z). Then the hedging price against the contingent claim $g(P(T), \cdot)$ at any $t \in [0, T]$ is given by*

(5.9)
$$
\begin{aligned}
Y(t) &= \tilde{E}\left\{ e^{-\int_t^T \bar{r}(s, X(s))ds} \bar{g}(X(T), \cdot) \Big| \mathcal{F}_t \right\} \\[2mm]
&= \tilde{E}\left\{ e^{-\int_t^T r(s, P(s))ds} g(P(T), \cdot) \Big| \mathcal{F}_t \right\},
\end{aligned}
$$

where $\tilde{E}\{\cdot|\mathcal{F}_t\}$ is the conditional expectation with respect to the equivalent martingale measure $\tilde{P}$ defined by

$$
\frac{d\tilde{P}}{dP} = \exp\left\{ -\int_0^T \bar{\theta}(t, X(t))dW(t) - \frac{1}{2}\int_0^T |\bar{\theta}(t, X(t))|^2 dt \right\}.
$$

Furthermore, the backward SPDE

(5.10)
$$
\begin{aligned}
u(t, x) &= \bar{g}(x) + \int_t^T \left\{ \frac{1}{2}\bar{\sigma}^2 u_{xx} + (\bar{b} - \bar{\sigma}\bar{\theta})u_x \right. \\[2mm]
&\qquad\qquad \left. - \bar{r}u + \bar{\sigma}q_x - q\bar{\theta} \right\}ds - \int_t^T q(s, x)dW_s
\end{aligned}
$$

has a unique adapted solution (u, q), such that the log-price X and the wealth process Y are related by

(5.11)
$$
Y(t) = u(t, X(t), \cdot), \qquad \forall t \in [0, T], \quad \text{a.s.}
$$

Finally, the optimal hedging strategy π is given by, for all $t \in [0, T]$,

(5.12)
$$\pi(t) = \bar{\sigma}^{-1}(t, X(t))Z(t)$$
$$= \nabla u(t, X(t), \cdot) + \bar{\sigma}(t, X(t))^{-1}q(t, X(t), \cdot), \quad \text{a.s.}$$

Proof. First, since the FBSDE (5.7) is decoupled, it must have unique adapted solution. Next, under the assumption, the backward SPDE (5.10) admits a (classical) adapted solution, thanks to Chapter 5, Theorems 2.1–2.3. Applying the generalized Itô's formula, and the following the Four Step Scheme one shows that the adapted solution (X, Y, Z) to (5.7) satisfies

(5.13) $Y(t) = u(t, X(t)), \qquad Z(t) = q(t, X(t)) + \sigma(t, X(t))\nabla u(t, X(t)).$

On the other hand, using the comparison theorem for BSDE (Chapter 1, Theorem 6.1), and following the same argument of Corollary 4.6, one shows that the hedging price at any time t is $Y(t)$, and the hedging strategy is given by (5.12). Finally, since the Y satisfies a BSDE in (5.7), an argument as that in Theorem 4.4 gives the expression (5.9). □

Remark 5.2. In the case when all the coefficients are constants, by uniqueness we see that the adapted solution to the BSPDE (5.10) is simply $(u, 0)$, where u is the classical solution to a backward PDE which, after a change of variable $x = \log x'$ and by setting $v(t, x') = u(t, \log x')$, becomes exactly the Black-Scholes PDE (5.4). Thus Theorem 5.1 recovers the classical Black-Scholes formula.

§5.2. Convexity of the European contingent claims

In this and the following subsection we apply the comparison theorems for backward SPDEs derived in Chapter 5 to obtain some interesting consequences in the option pricing theory, in a general setting that allows random coefficients in the market models. Our discussion follows the lines of those of El Karoui-Jeanblanc-Picqué-Shreve [1].

The first result concerns the convexity of the European contingent claims. In the Markovian case such a property was discussed by Bergman-Grundy-Wiener [1] and El Karoui-Jeanblanc-Picqué-Shreve [1]. Let us now assume that r and σ are stochastic processes, independent of the current stock price. From Theorem 5.1 we know that the option price at time t with stock price x is given by $\bar{u}(t, x) \overset{\Delta}{=} u(t, \log x)$, where u is the adapted solution to the BSPDE (5.10). (Note, here we slightly abuse the notations x and p!). The convexity of the European option states that the function $\bar{u}(t, \cdot)$ is a convex function, provided g is convex. To prove this we first note that by using the inverse Euler transformation one can show that $\bar{u}$ is the (classical) adapted solution to the BSPDE:

(5.14)
$$\begin{cases} d\bar{u} = \{-\frac{1}{2}p^2\sigma^2\bar{u}_{xx} - xr\bar{u}_x + r\bar{u} - x\sigma q_x - \theta q\}dt - qdW(t), \\ \qquad\qquad\qquad [0, T) \times (0, \infty) \\ \bar{u}(T, x) = g(x), \qquad x \geq 0. \end{cases}$$

Differentiating (5.14) with respect to x twice and denote $v = u_{xx}$, $p = q_{xx}$, then we see that (v, p) satisfies the following (linear) BSPDE:

(5.15)
$$\begin{cases} dv = \Big\{ -\dfrac{1}{2}x^2\sigma^2 v_{xx} - (2x\sigma^2 + xr)v_x - (\sigma^2 + r)v \\ \qquad\quad - x\sigma p_x - (2\sigma - \theta)p \Big\} dt - p\,dW(t); \quad (t, y) \in [0, T) \times \mathbb{R}; \\ v(T, x) = g''(x). \end{cases}$$

Here again the well-posedness of (5.15) can be obtained by considering its equivalent form after the Euler transformation (since r and σ are independent of x!). Now we can applying Chapter 5, Corollary 6.3 to conclude that $v \geq 0$, whenever $g'' \geq 0$, and hence $\bar{u}$ is convex provided g is.

We can discuss more complicated situation by using the comparison theorems in Chapter 5. For example, let us assume that both r and σ are *deterministic* functions of (t, x), and we assume that they are both C^2 for simplicity. Then (5.10) coincides with (5.4). Now differentiating (5.4) twice and denoting $v = u_{xx}$, we see that v satisfies the following PDE:

(5.16)
$$\begin{cases} 0 = v_t + \dfrac{1}{2}x^2\sigma^2 v_{xx} + \hat{a}x v_x + \hat{b}v + r_{xx}(x u_x - u), \\ v(T, x) = g''(x), \qquad x \geq 0, \end{cases}$$

where

$$\hat{a} = 2\sigma^2 + 2x\sigma\sigma_x + r;$$
$$\hat{b} = \sigma^2 + 4x\sigma\sigma_x + (x\sigma_x)^2 + x^2\sigma\sigma_{xx} + 2xr_x + r.$$

Now let us denote $V = x u_x - u$, then some computation shows that V satisfies the equation:

(5.17)
$$\begin{cases} 0 = V_t + \dfrac{1}{2}x^2\sigma^2 V_{xx} + \hat{c}x V_x + (x r_x - r)V, \quad \text{on } [0, T) \times (0, \infty), \\ V(T, x) = x g'(x) - g(x), \qquad x \geq 0, \end{cases}$$

for some function $\hat{c}$ depending on $\hat{a}$ and $\hat{b}$ (whence r and σ). Therefore applying the comparison theorems of Chapter 5 (use Euler transformation if necessary) we can derive the following results: assume that g is convex, then

(i) if r is convex and $x g'(x) - g(x) \geq 0$, then u is convex.
(ii) if r is concave and $x g'(x) - g(x) \leq 0$, then u is convex.
(iii) if r is independent of x, then u is convex.

Indeed, if $x g'(x) - g(x) \geq 0$, then $V \geq 0$ by Chapter 5, Corollary 6.3. This, together with the convexity of r and g, in turn shows that the solution v of (5.16) is non-negative, proving (i). Part (ii) can be argued similarly. To see (iii), note that when r is independent of x, (5.16) is homogeneous, thus the convexity of h implies that of $\bar{u}$, thanks to Chapter 5, Corollary 6.3 again.

§5.3. Robustness of Black-Scholes formula

The robustness of the Black-Scholes formula concerns the following problem: suppose a practitioner's information leads him to a misspecified value of, say, volatility σ, and he calculates the option price according to this misspecified parameter and equation (5.4), and then tries to hedge the contingent claim, what will be the consequence?

Let us first assume that the only misspecified parameter is the volatility, and denote it by $\sigma = \sigma(t, x)$, which is C^2 in x; and assume that the interest rate is deterministic and independent of the stock price. By the conclusion (iii) in the previous part we know that u is convex in x. Now let us assume that the true volatility is an $\{\mathcal{F}_t\}_{t\geq 0}$-adapted process, denoted by $\hat{\sigma}$, satisfying

$$(5.18) \qquad\qquad \hat{\sigma}(t) \geq \sigma(t, x), \qquad \forall (t, x), \text{a.s.}$$

Since in this case we have proved that u is convex, it is easy to check that in this case (6.16) of Chapter 5 reads

$$(5.19) \qquad (\hat{\mathcal{L}} - \mathcal{L})u + (\hat{\mathcal{M}} - \mathcal{M})q + \hat{f} - f = \frac{1}{2}x^2[\hat{\sigma}^2 - \sigma^2]u_{xx} \geq 0,$$

where $(\hat{\mathcal{L}}, \hat{\mathcal{M}})$ is the differential operator corresponding to the misspecified coefficients $(r, \hat{\sigma})$. Thus we conclude from Chapter 5, Theorem 6.2 that $\hat{u}(t, x) \geq u(t, x)$, $\forall (t, x)$, a.s. Namely the misspecified price dominates the true price.

Now let us assume that the inequality in (5.18) is reversed. Since both (5.4) and (5.14) are linear and homogeneous, $(-\hat{u}, -\hat{q})$ and $(-u, 0)$ are both solutions to (5.14) and (5.4) as well, with the terminal condition being replaced by $-g(x)$. But in this case (5.19) becomes

$$(\hat{\mathcal{L}} - \mathcal{L})(-u) = \frac{1}{2}x^2[\hat{\sigma}^2 - \sigma^2](-u_{xx}) \geq 0,$$

because u is convex, and $\hat{\sigma}^2 \leq \sigma^2$. Thus $-\hat{u} \geq -u$, namely $\hat{u} \leq u$.

Using the similar technique we can again discuss some more complicated situations. For example, let us allow the interest rate r to be misspecified as well, but in the form that it is convex in x, say. Assume that the payoff function h satisfies $xh'(x) - h(x) \geq 0$, and that $\hat{r}$ and $\hat{\sigma}$ are true interest rate and volatility such that they are $\{\mathcal{F}_t\}_{t\geq 0}$-adapted random fields satisfying $\hat{r}(t, x) \geq r(t, x)$, and $\hat{\sigma}(t, x) \geq \sigma(t, x)$, $\forall (t, x)$. Then, using the notation as before, one shows that

$$(\hat{\mathcal{L}} - \mathcal{L})u = \frac{1}{2}x^2[\hat{\sigma}^2 - \sigma^2]u_{xx} + (\hat{r} - r)[xu_x - u] \geq 0,$$

because u is convex, and $xu_x - u = V \geq 0$, thanks to the arguments in the previous part. Consequently one has $\hat{u}(t, x) \geq u(t, x)$, $\forall (t, x)$, a.s. Namely, we also derive a one-sided domination of the true values and misspecified values.

We remark that if the misspecified volatility is not the deterministic function of the stock price, the comparison may fail. We refer the interested readers to El Karoui-Jeanblanc-Picqué-Shreve [1] for an interesting counterexample.

§6. An American Game Option

In this section we apply the result of Chapter 7 to derive an ad hoc option pricing problem which we call the *American Game Option*.

To begin with let us consider the following FBSDE with reflections (compare to Chapter 7, (3.2))

$$(6.1) \quad \begin{cases} X_t = x + \displaystyle\int_0^t b(s, X_s, Y_s, Z_s)ds + \int_0^t \sigma(s, X_s, Y_s, Z_s)dW_s; \\[2mm] Y_t = g(X_T) + \displaystyle\int_t^T h(s, X_s, Y_s, Z_s)ds - \int_t^T Z_s dW_s + \zeta_T - \zeta_t. \end{cases}$$

Note that the forward equation does not have reflection; and we assume that $m = 1$ and $\mathcal{O}_2(t, x, \omega) = (L(t, x, \omega), U(t, x, \omega))$, where L and U are two random fields such that $L(t, x, \omega) \leq U(t, x, \omega)$, for all $(t, x, \omega) \in [0, T] \times \mathbb{R}^n \times \Omega$. We assume further that both L and U are continuous functions in x for all (t, ω), and are $\{\mathcal{F}_t\}_{t \geq 0}$-progressively measurable, continuous processes for all x.

In light of the result of the previous section, we can think of X in (6.1) as a price process of financial assets, and of Y as a wealth process of an (large) investor in the market. However, we should use the latter interpretation only up until the first time we have $d\zeta < 0$. In other words, no external funds are allowed to be added to the investor's wealth, although he is allowed to consume.

The *American game option* can be described as follows. Unlike the usual American option where only the buyer has the right to choose the exercise time, in a game option we allow the seller to have the same right as well, namely, the seller can force the exercise time if he wishes. However, in order to get a nontrivial option (i.e., to avoid immediate exercise to be optimal), it is required that the payoff be higher if the seller opts to force the exercise. Of course the seller may choose not to do anything, then the game option becomes the usual American option.

To be more precise, let us denote by $\mathcal{M}_{t,T}$ the set of $\{\mathcal{F}_t\}_{t \geq 0}$-stopping times taking values in $[t, T]$, and $t \in [0, T)$ be the time when the "game" starts. Let $\tau \in \mathcal{M}_{t,T}$ be the time the buyer chooses to exercise the option; and $\sigma \in \mathcal{M}_{t,T}$ be that of the seller. If $\tau \leq \sigma$, then the seller pays $L(\tau, X_\tau)$; if $\sigma < \tau$, then the seller pays $U(\sigma, X_\sigma)$. If neither exercises the option by the maturity date T, then the seller pays $B = g(X_T)$. We define the *minimal hedging price* of this contract to be the infimum of initial wealth amounts Y_0, such that the seller can deliver the payoff, a.s., without having to use additional outside funds. In other words, his wealth process has to follow the dynamics of Y (with $d\zeta \geq 0$), up to the exercise time $\sigma \wedge \tau \wedge T$, and at the exercise time we have to have

(6.2) $Y_{\sigma \wedge \tau \wedge T} \geq g(X_T)\mathbf{1}_{\{\sigma \wedge \tau = T\}} + L(\tau, X_\tau)\mathbf{1}_{\{\tau < T, \tau \leq \sigma\}} + U(\sigma, X_\sigma)\mathbf{1}_{\{\sigma < \tau\}}.$

Our purpose is to determine the minimal hedging price, as well as the corresponding minimal hedging process.

To solve this option pricing problem, let us first study the following stochastic game (Dynkin game): there are two players, each can choose a (stopping) time to stop the game over an given horizon $[t, T]$. Let $\sigma \in \mathcal{M}_{t,T}$ be the time that player I chooses, and $\tau \in \mathcal{M}_{t,T}$ be that of player II's. If $\sigma < \tau$, the player I pays $U(\sigma)(= U(\sigma, X_\sigma))$ to player II; whereas if $\tau \leq \sigma < T$, player I pays $L(\tau)(= L(\tau, X_\tau))$ (yes, in both cases the player I pays!). If no one stops by time T, player I pays B. There is also a *running cost* $h(t)(= h(t, X_t, Y_t, Z_t))$. In other words the payoff player I has to pay is given by

(6.3)
$$R_t^B(\sigma, \tau) \triangleq \int_t^{\sigma \wedge \tau} h(u)du + B\mathbf{1}_{\{\sigma \wedge \tau = T\}}$$
$$+ L(\tau)\mathbf{1}_{\{\tau < T, \ \tau \leq \sigma\}} + U(\sigma)\mathbf{1}_{\{\sigma < \tau\}},$$

where $B \in L^2(\Omega)$ is a given $\mathcal{F}_T$–measurable random variable satisfying $L(T) \leq B \leq U(T)$. Suppose that player II is trying to maximize the payoff, while player I attempts to minimize it. Define the *upper* and *lower* value s of the game by

(6.4)
$$\overline{V}(t) \triangleq \operatorname*{essinf}_{\sigma \in \mathcal{M}_{t,T}} \operatorname*{esssup}_{\tau \in \mathcal{M}_{t,T}} E\{R_t^B(\sigma, \tau)|\mathcal{F}_t\},$$
$$\underline{V}(t) \triangleq \operatorname*{esssup}_{\tau \in \mathcal{M}_{t,T}} \operatorname*{essinf}_{\sigma \in \mathcal{M}_{t,T}} E\{R_t^B(\sigma, \tau)|\mathcal{F}_t\}$$

respectively; and we say that the game has a value if $\overline{V}(t) = \underline{V}(t) \triangleq V(t)$.

The solution to the Dynkin game is given by the following theorem, which can be obtained by a line by line analogue of Theorem 4.1 in Cvitanić and Karatzas [2]. Here we give only the statement.

Theorem 6.1. *Suppose that there exists a solution (X, Y, Z, ζ) to FB-SDER (6.1) (with $\mathcal{O}_2(t, x) = (L(t, x), U(t, x))$. Then the game (6.3) with $B = g(X_T)$, $h(t) = h(t, X_t, Y_t, Z_t)$, and $L(t, \omega) = L(t, X_t(\omega))$, $U(t, \omega) = U(t, X_t(\omega))$ has value $V(t)$, given by the backward component Y of the solution to the FBSDER, i.e. $V(t) = \overline{V}(t) = \underline{V}(t) = Y_t$, a.s., for all $0 \leq t \leq T$. Moreover, there exists a saddle-point $(\widehat{\sigma}_t, \widehat{\tau}_t) \in \mathcal{M}_{t,T} \times \mathcal{M}_{t,T}$, given by*

$$\widehat{\sigma}_t \triangleq \inf\{s \in [t, T) : Y_s = U(s, X_s)\} \wedge T,$$
$$\widehat{\tau}_t \triangleq \inf\{s \in [t, T) : Y_s = L(s, X_s)\} \wedge T,$$

namely, we have

$$E\{R_t^{g(X_T)}(\hat{\sigma}_t, \tau)|\mathcal{F}_t\} \le E\{R_t^{g(X_T)}(\hat{\sigma}_t, \hat{\tau}_t)|\mathcal{F}_t\}$$
$$=Y_t \le E\{R_t^{g(X_T)}(\sigma, \hat{\tau}_t)|\mathcal{F}_t\}, \text{ a.s.}$$

for every $(\sigma, \tau) \in \mathcal{M}_{t,T} \times \mathcal{M}_{t,T}$. $\square$

In what follows when we mention FBSDER, we mean (6.1) specified as that in Theorem 6.1.

Theorem 6.2. *The minimal hedging price of the American Game Option is greater or equal to $\bar{V}(0)$, the upper value of the game (at $t = 0$) of Theorem 6.1. If the corresponding FBSDER has a solution $(\tilde{X}, \tilde{Y}, \tilde{Z}, \zeta)$, then the minimal hedging price is equal to $\tilde{Y}_0$.*

Proof: Fix the exercise times σ, τ of the seller and the buyer, respectively. If Y is the seller's hedging process, it satisfies the following dynamics for $t \le \tau \wedge \sigma \wedge T$:

$$Y_t + \int_0^t h(s, X_s, Y_s, Z_s)ds = \int_0^t Z_s dW_s - \zeta_t,$$

with ζ non-decreasing. Hence, the left-hand side is a supermartingale. From this and the requirement that Y be a hedging process, we get $Y_t \ge E\{R_t^{g(X_T)}(\sigma, \tau)|\mathcal{F}_t\}$, $\forall t$, a.s. in the notation of Theorem 4.1. Since the buyer is trying to maximize the payoff, and the seller to minimize it, we get $Y_t \ge \bar{V}_t$. $\forall t$, a.s.. Consequently, the minimal hedging price is no less than $\bar{V}(0)$.

Conversely, if the FBSDER has a solution with $\tilde{Y}$ as the backward component, then by Theorem 6.1, process $\tilde{Y}$ is equal to the value process of the game, and by (4.4) (with $t = 0$) and (2.10), up until the optimal exercise time $\hat{\sigma} := \hat{\sigma}_0$ for the seller, it obeys the dynamics of a wealth process, since ζ_t is nondecreasing for $t \le \hat{\sigma}_0$. So, the seller can start with $\tilde{Y}_0$, follow the dynamics of $\tilde{Y}$ until $t = \hat{\sigma}$ and then exercise, if the buyer has not exercised first. In general, from the saddle-point property we know that, for any $\tau \in \mathcal{M}_{0,T}$,

$$\tilde{Y}_{\hat{\sigma} \wedge \tau} \ge g(X_T)\mathbf{1}_{\{\hat{\sigma} \wedge \tau = T\}} + L(\tau, X_\tau)\mathbf{1}_{\{\tau < T, \tau \le \hat{\sigma}\}} + U(\hat{\sigma}, X_{\hat{\sigma}})\mathbf{1}_{\{\hat{\sigma} < \tau\}}.$$

This implies that that the seller can deliver the required payoff if he uses $\hat{\sigma}$ as his exercise time, no matter what the buyer's exercise time τ is. Consequently, $\tilde{Y}_0 = V(0)$ is no less than the minimal hedging price. $\square$

Chapter 9

Numerical Methods for FBSDEs

In the previous chapter we have seen various applications of FBSDEs in theoretical and applied fields. In many cases a satisfactory numerical simulation is highly desirable. In this chapter we present a complete numerical algorithm for a fairly large class of FBSDEs, and analyze its consistency as well as its rate of convergence. We note that in the standard forward SDEs case two types of approximations are often considered: a strong scheme which typically converges pathwisely at a rate $\mathcal{O}(\frac{1}{\sqrt{n}})$, and a weak scheme which approximates only approximates $E\{f(X(T))\}$, with a possible faster rate of convergence. However, as we shall see later, in our case the weak convergence is a simple consequence of the pathwise convergence, and the rate of convergence of our scheme is the same as the strong scheme for pure forward SDEs, which is a little surprising because a FBSDE is much more complicated than a forward SDE in nature.

§1. Formulation of the Problem

In this chapter we consider the following FBSDE: for $t \in [0, T]$,

(1.1)
$$\begin{cases} X(t) = x + \int_0^t b(s, \Theta(s))ds + \int_0^t \sigma(s, X(s), Y(s))dW(s); \\ Y(t) = g(X(T)) + \int_t^T \widehat{b}(s, \Theta(s))ds - \int_t^T Z(s)dW(s), \end{cases}$$

where $\Theta = (X, Y, Z)$. We note that in some applications (e.g., in Chapter 8, §3, Black's Consol Rate Conjecture), the FBSDE (1.1) takes a slightly simpler form:

(1.2)
$$\begin{cases} X(t) = x + \int_0^t b(s, X(s), Y(s))ds + \int_0^t \sigma(s, X(s), Y(s))dW_s; \\ Y(t) = g(X(T)) + \int_t^T \widehat{b}(s, X(s), Y(s))ds - \int_t^T Z(s)dW(s). \end{cases}$$

That is, the coefficients b and $\widehat{b}$ do not depend on Z explicitly, and often in these cases only the components (X, Y) are of significant interest. In what follows we shall call (1.2) the "special case" when only the approximation of (X, Y) are considered; and we call (1.1) the "general case" if the approximation of (X, Y, Z) is required. We note that in what follows we restrict ourselves to the case where all processes involved are one dimensional. The higher dimensional case can be discussed under the same idea, but technically much more complicated. Furthermore, we shall impose the following standing assumptions:

(**A1**) The functions b, $\widehat{b}$ and σ are continuously differentiable in t and twice continuously differentiable in x, y, z. Moreover, if we denote any one of these functions generically by ψ, then there exists a constant $\alpha \in (0,1)$, such that for fixed y and z, $\psi(\cdot, \cdot, y, z) \in C^{1+\frac{\alpha}{2}, 2+\alpha}$. Furthermore, for some $L > 0$,

$$\|\psi(\cdot, \cdot, y, z)\|_{1,2,\alpha} \le L, \qquad \forall(y, z) \in \mathbb{R}^2.$$

(**A2**) The function σ satisfies

(1.3) $\qquad \mu \le \sigma(t, x, y) \le C, \qquad \forall(t, x, y) \in [0, T] \times \mathbb{R}^2,$

where $0 < \mu \le C$ are two constants.

(**A3**) The function g belongs boundedly to $C^{4+\alpha}$ for some $\alpha \in (0, 1)$ (one may assume that α is the same as that in (A1)).

It is clear that the assumptions (A1)–(A3) are stronger that those in Chapter 4, therefore applying Theorem 2.2 of Chapter 4, we see that the FBSDE (1.1) has a unique adapted solution which can be constructed via the Four Step Scheme. That is, the adapted solution (X, Y, Z) of (1.1) can be obtained in the following way:

(1.4)
$$\begin{cases} X(t) = x + \int_0^t \tilde{b}(s, X(s)) ds + \int_0^t \tilde{\sigma}(s, X(s)) dW(s), \\ Y(t) = \theta(t, X(t)), \qquad Z(t) = \sigma(t, X(t), \theta(t, X(t))) \theta_x(t, X(t)), \end{cases}$$

where

$$\tilde{b}(t, x) = b(t, x, \theta(t, x), \sigma(t, x, \theta(t, x)) \theta_x(t, x))),$$
$$\tilde{\sigma}(t, x) = \sigma(t, x, \theta(t, x));$$

and $\theta \in C^{1+\frac{\alpha}{2}, 2+\alpha}$ for some $0 < \alpha < 1$ is the unique classical solution to the quasilinear parabolic PDE:

(1.5)
$$\begin{cases} \theta_t + \dfrac{1}{2}\sigma(t, x, \theta)^2 \theta_{xx} + b(t, x, \theta, \sigma(t, x, \theta)\theta_x)\theta_x \\ \qquad + \widehat{b}(t, x, \theta, \sigma(t, x, \theta)\theta_x) = 0, \qquad (t, x) \in (0, T) \times \mathbb{R}, \\ \theta(T, x) = g(x), \qquad x \in \mathbb{R}. \end{cases}$$

We should point out that, by using standard techniques for gradient estimates, that is, applying parabolic Schauder interior estimates to the difference quotients repeatedly (cf. Gilbarg & Trudinger [1]), it can be shown that under the assumptions (A1)–(A3) the solution θ to the quasilinear PDE (1.5) actually belongs to the space $C^{2+\frac{\alpha}{2}, 4+\alpha}$. Consequently, there exists a constant $K > 0$ such that

(1.6) $\|\theta\|_\infty + \|\theta_t\|_\infty + \|\theta_{tt}\|_\infty + \|\theta_x\|_\infty + \|\theta_{xx}\|_\infty + \|\theta_{xxx}\|_\infty + \|\theta_{xxxx}\|_\infty \le K.$

Our line of attack is now clear: we shall first find a numerical scheme for the quasilinear PDE (1.5), and then find a numerical scheme for the (forward) SDE (1.4). We should point out that although the numerical analysis for the quasilinear PDE is not new, but the special form of (1.5) has not been covered by existing results. In the next Section 2 we shall study the numerical scheme of the quasilinear PDE (1.5) in full details, and then in Section 3 we study the (strong) numerical scheme for the forward SDE in (1.4).

§2. Numerical Approximations of the Quasilinear PDE

In this section we study the numerical approximation scheme and its convergence analysis for the quasilinear parabolic PDE (1.5). We will first carry out the discussion for the special case completely, upon which the study of the general case will be built.

§2.1. A special case

In this case the coefficients b and $\widehat{b}$ are independent of Z, we only approximate (X, Y). Note that in this case the PDE (1.5), although still quasilinear, takes a much simpler form:

$$(2.1) \quad \begin{cases} \theta_t + \dfrac{1}{2}\sigma(t, x, \theta)^2 \theta_{xx} + b(t, x, \theta)\theta_x + \widehat{b}(t, x, \theta) = 0, & t \in (0, T), \\ \theta(T, x) = g(x), & x \in \mathbb{R}. \end{cases}$$

Let us first standardize the PDE (2.1). Define $u(t, x) = \theta(T - t, x)$, and for $\varphi = \sigma$, b, and $\widehat{b}$, respectively, we define

$$\bar{\varphi}(t, x, y) = \varphi(T - t, x, y), \qquad \forall(t, x, y).$$

Then u satisfies the PDE

$$(2.2) \quad \begin{cases} u_t - \dfrac{1}{2}\bar{\sigma}^2(t, x, u)u_{xx} - \bar{b}(t, x, u)u_x - \widehat{\bar{b}}(t, x, u) = 0; \\ u(0, x) = g(x). \end{cases}$$

To simplify notation we replace $\bar{\sigma}$, $\bar{b}$ and $\widehat{\bar{b}}$ by σ, b and $\widehat{b}$ themselves in the rest of this section. We first determine the characteristics of the first order nonlinear PDE

$$(2.3) \qquad\qquad u_t - b(t, x, u)u_x = 0.$$

Elementary theory of PDEs (see, e.g., John [1]) tells us that the characteristic equation of (2.3) is

$$\det|a_{ij}t'(s) - \delta_{ij}x'(s)| = 0, \quad s \geq 0,$$

where s is the parameter of the characteristic and (a_{ij}) is the matrix

$$\begin{bmatrix} 0 & 0 & 0 \\ 0 & -b(t, x, u) & 0 \\ 0 & -1 & 0 \end{bmatrix}.$$

In other words, if we let parameter $s = t$, then the characteristic curve $\mathcal{C}$ is given by the ODE:

$$(2.4) \qquad\qquad x'(t) = -b(t, x(t), u(t, x(t))).$$

Further, if we let τ be the arclength of $\mathcal{C}$, then along $\mathcal{C}$ we have

$$d\tau = \left[1 + b^2(t, x, u(t, x))\right]^{\frac{1}{2}} dt,$$

and

$$\frac{\partial}{\partial \tau} = \frac{1}{\psi} \left\{ \frac{\partial}{\partial t} - b \frac{\partial}{\partial x} \right\},$$

where $\psi(t, x) = \left[1 + b^2(t, x, u(t, x))\right]^{\frac{1}{2}}$. Thus, along $\mathcal{C}$, equation (2.2) is simplified to

$$(2.5) \qquad\qquad \begin{cases} \psi \dfrac{\partial u}{\partial \tau} = \dfrac{1}{2}\sigma^2(t, x, u)u_{xx} + \widehat{b}(t, x, u); \\ u(0, x) = g(x). \end{cases}$$

We shall design our numerical scheme based on (2.5).

§2.1.1. Numerical scheme

Let $h > 0$ and $\Delta t > 0$ be fixed numbers. Let $x_i = ih$, $i = 0, \pm 1, \cdots$, and $t^k = k\Delta t$, $k = 0, 1, \cdots, N$, where $t^N = T$. For a function $f(t, x)$, let $f^k(\cdot) = f(t^k, \cdot)$; and let $f_i^k = f(t^k, x_i)$ denote the grid value of the function f. Define for each k the approximate solution w^k by the following recursive steps:

Step 0: Set $w_i^0 = g(x_i)$, $i = \cdots, -1, 0, 1, \cdots$; use linear interpolation to obtain a function $w^0(x)$ defined on $x \in \mathbb{R}$.

Suppose that $w^{k-1}(x)$ is defined for $x \in \mathbb{R}$, let $w_i^{k-1} = w^{k-1}(x_i)$ and

$$(2.6) \qquad \begin{cases} b_i^k = b(t^k, x_i, w_i^{k-1}); \quad \sigma_i^k = \sigma(t^k, x_i, w_i^{k-1}); \quad \widehat{b}_i^k = \widehat{b}(t^k, x_i, w_i^{k-1}); \\ \bar{x}_i^k = x_i - b_i^k \Delta t, \quad \bar{w}_i^{k-1} = w^{k-1}(\bar{x}_i^k); \\ \delta_x^2(w)_i^k = h^{-2}[w_{i+1}^k - 2w_i^k + w_{i-1}^k]. \end{cases}$$

Step k: Obtain the grid values for the k-th step approximate solution, denoted by $\{w_i^k\}$, via the following difference equation:

$$(2.7) \qquad \frac{w_i^k - \bar{w}_i^{k-1}}{\Delta t} = \frac{1}{2}(\sigma_i^k)^2 \delta_x^2(w)_i^k + (\widehat{b})_i^k; \qquad -\infty < i < \infty,$$

Since by our assumption σ is bounded below positively and $\widehat{b}$ and g are bounded, there exists a unique *bounded* solution of (2.7) as soon as an evaluation is specified for $w^{k-1}(x)$.

Finally, we use *linear interpolation* to extend the grid values of $\{w_i^k\}_{i=-\infty}^{\infty}$ to all $x \in \mathbb{R}$ to obtain the k-th step approximate solution $w^k(\cdot)$.

Before we do the convergence analysis for this numerical scheme, let us point out a standard *localization* idea which is essential in our future discussion, both theoretically and computationally. We first recall from Chapter 4 that the (unique) classical solution of the Cauchy problem (2.2) (therefore (2.5)) is in fact the uniform limit of the solutions $\{u^R\}$ $(R \to \infty)$ to the *initial-boundary* problems:

$$(2.2)_R \qquad \begin{cases} u_t - \dfrac{1}{2}\bar\sigma(t, x, u)^2 u_{xx} - \bar b(t, x, u)u_x - \widetilde{\bar b}(t, x, u) = 0, \\ u(0, x) = g(x), \qquad x \in \mathbb{R}; \\ u(t, x) = g(x), \qquad |x| = R, \quad 0 < t \le T. \end{cases}$$

It is conceivable that we can also restrict the corresponding difference equation (2.7) so that $-i_0 \le i \le i_0$, for some $i_0 < \infty$. Indeed, if we denote $w^{i_0, k}$ to be the following *localized* difference equation

$$(2.7)_{i_0} \qquad \begin{cases} \dfrac{w_i^k - \bar w_i^{k-1}}{\Delta t} = \dfrac{1}{2}(\sigma_i^k)^2 \delta_x^2(w)_i^k + (\widehat b)_i^k; \qquad -i_0 \le i \le i_0, \\ w_i^0 = g(x_i), \qquad -i_0 \le i \le i_0; \\ w_{\pm i_0}^k = g(x_{\pm i_0}), \qquad k = 0, 1, 2, \cdots, \end{cases}$$

then by (A1) and (A2), one can show that w_i^k is the uniform limit of $\{w_i^{i_0, k}\}$, as $i_0 \to \infty$, uniformly in i and k. In particular, if we fix the mesh size $h > 0$, and let $R = i_0 h$, then the quantities

$$(2.8) \qquad \max_i |u(t^k, x_i) - w_i^k| \quad \text{and} \quad \max_{-i_0 \le i \le i_0} |u^R(t^k, x_i) - w_i^{i_0, k}|$$

differ only by a error that is uniform in k, and can be taken to be arbitrarily small as i_0 (or $i_0 h = R$) is sufficiently large. Consequently, as we shall see later, if for fixed h and Δt we choose R (or i_0) so large that the error between the two quantities in (2.8) differ by $\mathcal{O}(h + |\Delta t|)$, then we can replace (2.2) by $(2.2)_R$, and (2.7) by $(2.7)_{i_0}$ without changing the desired results on the rate of convergence. But on the other hand, since for the localized solutions the error $|u^R(t^k, x_{\pm i_0}) - w_{\pm i_0}^{i_0, k}| \equiv 0$ for all $k = 0, 1, 2, \cdots$, the maximum absolute value of the error $|u^R(t^k, x_i) - w_i^{i_0, k}|$, $i = -i_0, \cdots, i_0$, will always occur in an "interior" point of $(-R, R)$. Such an observation will be particularly useful when a maximum-principle argument is applies (see, e.g., Theorem 2.3 below). Based on the discussion above, from now on we will use the localized version of the solutions to (2.2) and (2.7) whenever necessary, without further specifications.

To conclude this subsection we note that the approximate solutions $\{w^k(\cdot)\}$ are defined only on the times $t = t^k$, $k = 0, 1, \cdots, N$. An approximate solution defined on $[0, T] \times \mathbb{R}$ is defined as follows: for given $h > 0$

and $\Delta t > 0$,

$$(2.9) \qquad w^{h,\Delta t}(t,x) = \begin{cases} \displaystyle\sum_{k=1}^{N} w^k(x) 1_{(t^{k-1},t^k]}(t), & t \in (0,T]; \\ w^0(x), & t = 0. \end{cases}$$

Clearly, for each k and i, $w^{h,\Delta t}(t^k, x_i) = w_i^k$, where $\{w_i^k\}$ is the solution to (2.7).

§2.1.2. Error analysis

We first analyze the approximate solution $\{w^k(\cdot)\}$. To begin with, let us introduce some notations: for each k and i, let

$$(2.10) \qquad \bar{x}_i^k \stackrel{\Delta}{=} x_i + b(t^k, x_i, u_i^{k-1})\Delta t, \qquad \bar{u}_i^{k-1} \stackrel{\Delta}{=} u(t^{k-1}, \bar{x}_i^k).$$

Let $\{x(t) : t^{k-1} \le t \le t^k\}$ be the characteristic such that $x(t^k) = x_i$. That is, by (2.4),

$$x(t) = x_i + \int_t^{t^k} b(s, x(s), u(s, x(s))) ds, \qquad t^{k-1} \le t \le t^k.$$

Denote $\bar{x} = x(t^{k-1})$. It is then easily seen that

$$\begin{cases} \displaystyle\sup_{t^{k-1} \le t \le t^k} |x(t) - x_i| \le \|b\|_\infty \Delta t; \\ \displaystyle|\bar{x}_i^k - \bar{x}| \le \int_{t^{k-1}}^{t^k} |b(t^k, x_i, u_i^{k-1}) - b(t, x(t), u(t, x(t)))| dt \\ \qquad \le \{\|b_t\|_\infty + \|b_x\|_\infty \|b\|_\infty + \|b_u\|_\infty (\|u_t\|_\infty + \|u_x\|_\infty \|b\|_\infty)\}\Delta t^2 \end{cases}$$

To simplify notations from now on we let $C > 0$ to be a generic constant depending only on b, $\hat{b}$, σ, T, and the constant K in (1.6), which may vary from line to line. Thus the above becomes

$$(2.11) \qquad \sup_{t^{k-1} \le t \le t^k} |x(t) - x_i| \le C\Delta t; \qquad |\bar{x}_i^k - \bar{x}| \le C\Delta t^2.$$

We now derive an equation for the approximation error. To this end, recall $\bar{x}$ and $\bar{x}_i^k$ defined by (2.10); and note that along the characteristic curve $\mathcal{C}$,

$$\psi \frac{\partial u}{\partial \tau} \approx \psi \frac{u(t^k, x) - u(t^{k-1}, \bar{x})}{\Delta \tau} \approx \psi(x) \frac{u(t^k, x) - u(t^{k-1}, \bar{x})}{[(x - \bar{x})^2 + (\Delta t)^2]^{\frac{1}{2}}}$$

$$= \frac{u(t^k, x) - u(t^{k-1}, \bar{x})}{\Delta t}.$$

The solution of (2.5) thus satisfies a difference equation of the following form: for $-\infty < i < \infty$ and $k = 1, \cdots, N$,

$$(2.12) \qquad \frac{u_i^k - \bar{u}_i^{k-1}}{\Delta t} = \frac{1}{2}(\sigma(u)_i^k)^2 \delta_x^2(u)_i^k + \hat{b}(u)_i^k + e_i^k,$$

where $\bar{u}_i^{k-1} = u^{k-1}(\bar{\mathbf{x}}_i^k)$ and $\widehat{b}(u)_i^k$ and $\sigma(u)_i^k$ correspond to $\widehat{b}_i^k$ and σ_i^k defined in (2.6), except that the values $\{w_i^{k-1}\}$ are replaced by $\{u_i^{k-1}\}$; e_i^k is the error term to be estimated. We have the following lemma.

Lemma 2.1. *There exists a constant $C > 0$, depending only on b, $\widehat{b}$, σ, T, and the constant K in (1.6), such that for all $k = 0, \cdots, N$ and $-\infty < i < \infty$,*

$$|e_i^k| \leq C(h + \Delta t).$$

Proof. First observe that at each grid point (t^k, x_i)

$$\psi(t^k, x_i)\frac{\partial u}{\partial \tau}\bigg|_{(t^k, x_i)} = \frac{1}{2}\sigma^2(t^k, x_i, u_i^k)u_{xx}\bigg|_{(t^k, x_i)} + \widehat{b}(t^k, x_i, u_i^k).$$

Therefore, for $-\infty < i < \infty$, $k = 1, \cdots, N$,

$$
\begin{aligned}
e_i^k = &\left\{\frac{u_i^k - \bar{u}_i^{k-1}}{\Delta t} - \psi(t^k, x_i)\frac{\partial u}{\partial \tau}\bigg|_{(t^k, x_i)}\right\} \\
&+ \left\{\frac{1}{2}\sigma^2(t^k, x_i, u_i^k)u_{xx}\bigg|_{(t^k, x_i)} - \frac{1}{2}(\sigma(u)_i^k)^2\delta^2(u)_i^k\right\} \\
&+ \left\{\widehat{b}(t^k, x_i, u_i^k) - \widehat{b}(u)_i^k\right\} = I_i^{1,k} + I_i^{2,k} + I_i^{3,k}.
\end{aligned}
$$

We estimate $I_i^{1,k}$, $I_i^{2,k}$ and $I_i^{3,k}$ separately. Recall that C will denote a generic constant that might vary from line to line. Using the uniform boundedness of $\widehat{b}_y$ and u_t we have

(2.13) $\qquad |I_i^{3,k}| = |\widehat{b}(t^k, x_i, u_i^k) - \widehat{b}(t^k, x_i, u_i^{k-1})| \leq C\Delta t,$

Similarly,

(2.14)
$$
\begin{aligned}
|I_i^{2,k}| \leq &\frac{1}{2}\bigg\{|\sigma^2(t^k, x_i, u_i^k) - \sigma^2(t^k, x_i, u_i^{k-1})|\,|u_{xx}(t^i, x_i)| \\
&+ |\sigma^2(t^k, x_i, u_i^{k-1})|\bigg|u_{xx}(t^k, x_i) - \frac{u_{i+1}^k - 2u_i^k + u_{i-1}^k}{h^2}\bigg|\bigg\} \\
&\leq C(\|u_t\|_\infty\Delta t + \|u_{xxx}\|_\infty h) \leq C(h + \Delta t).
\end{aligned}
$$

To estimate $I_i^{1,k}$ we note from (2.11) that

(2.15) $\qquad \left|\frac{u(t^{k-1}, \bar{x}) - u(t^{k-1}, \bar{\mathbf{x}}_i^k)}{\Delta t}\right| \leq \frac{\|u_x\|_\infty|\bar{x} - \bar{\mathbf{x}}_i^k|}{\Delta t} \leq C\Delta t,$

On the other hand, integrating along the characteristic from $(t^{k-1}, \bar{x})$ to

(t^k, x_i), we have

$$
\frac{u(t^k, x_i) - u(t^{k-1}, \bar{x})}{\Delta t} = \frac{1}{\Delta t} \int_{t^{k-1}}^{t^k} \frac{d}{dt} u(t, x(t)) dt
$$

(2.16)
$$
= \frac{1}{\Delta t} \int_{t^{k-1}}^{t^k} [u_t - b(\cdot, \cdot, u) u_x](t, x(t)) dt
$$

$$
= \frac{1}{\Delta t} \int_{t^{k-1}}^{t^k} \left[\psi \frac{\partial u}{\partial \tau} \right](t, x(t)) dt
$$

$$
= \left[\psi \frac{\partial u}{\partial \tau} \right](t^k, x_i) + \frac{1}{\Delta t} \int_{t^{k-1}}^{t^k} \left\{ \left[\psi \frac{\partial u}{\partial \tau} \right](t, x(t)) - \left[\psi \frac{\partial u}{\partial \tau} \right](t^k, x_i) \right\} dt.
$$

Since along the characteristics $\dfrac{\partial^2 u}{\partial \tau^2}$ depends on u_{tt}, u_{tx} and u_{xx} and b, which are all bounded, one can easily deduce that

(2.17) $$
\left| \frac{1}{\Delta t} \int_{t^{k-1}}^{t^k} \left\{ \left[\psi \frac{\partial u}{\partial \tau} \right](t, x(t)) - \left[\psi \frac{\partial u}{\partial \tau} \right](t^k, x_i) \right\} dt \right| \leq C(h + \Delta t),
$$

Combining (2.11)–(2.17), we have

$$
|I_i^{1,k}| \leq \left| \frac{u(t^k, x_i) - u(t^{k-1}, \bar{x})}{\Delta t} - \left[\psi \frac{\partial u}{\partial \tau} \right](t^k, x_i) \right| + \left| \frac{u(t^k, x_i) - \bar{u}_i^{k-1}}{\Delta t} \right|
$$
$$
\leq C(h + \Delta t),
$$

proving the lemma. $\square$

We are now ready to analyze the error between the approximate solution $w^{h, \Delta t}(t, x)$ and the true solution $u(t, x)$. To do this we define the error function $\zeta(t, x) = u(t, x) - w^{h, \Delta t}(t, x)$ for $(t, x) \in [0, T] \times \mathbb{R}$; as before, let $\zeta_i^k = \zeta(t^k, x_i) = u_i^k - w_i^k$. We have the following theorem.

Theorem 2.2. *Assume (A1)—(A3). Then*

$$
\sup_{k, i} |\zeta_i^k| = \mathcal{O}(h + \Delta t).
$$

Proof. First, by subtracting (2.7) from (2.12), we see that $\{\zeta_i^k\}$ satisfies the difference equation

(2.18)
$$
\begin{cases}
\dfrac{\zeta_i^k - (\bar{u}_i^{k-1} - \bar{w}_i^{k-1})}{\Delta t} = \dfrac{1}{2} \left\{ (\sigma(u)_i^k)^2 \delta_x^2 (u)_i^k - (\sigma_i^k)^2 \delta^2 (w)_i^k \right\} \\
\qquad\qquad\qquad\qquad + [\hat{b}(u)_i^k - \hat{b}_i^k] + e_i^k; \\
\zeta_i^0 = 0.
\end{cases}
$$

Since

$$
\bar{u}_i^{k-1} - \bar{w}_i^{k-1} = [u(t^{k-1}, \bar{\mathbf{x}}_i^k) - u(t^{k-1}, \bar{x}_i^k)] + [u(t^{k-1}, \bar{x}_i^k) - w^{k-1}(\bar{x}_i^k)]
$$
$$
= \bar{\zeta}_i^{k-1} + [u(t^{k-1}, \bar{\mathbf{x}}_i^k) - u(t^{k-1}, \bar{x}_i^k)],
$$

where $\bar{\zeta}_i^{k-1} = u(t^{k-1}, \bar{x}_i^k) - w^{k-1}(\bar{x}_i^k)$, and

$$(\sigma(u)_i^k)^2 \delta_x^2(u)_i^k - (\sigma_i^k)^2 \delta_x^2(w)_i^k$$
$$= (\sigma_i^k)^2 \delta_x^2(\zeta)_i^k + [\sigma^2(t^k, x_i, u_i^{k-1}) - \sigma^2(t^k, x_i, w_i^{k-1})]\delta_x^2(u)_i^k,$$

we can rewrite (2.18) as

(2.19)
$$\begin{cases} \dfrac{\zeta_i^k - \bar{\zeta}_i^{k-1}}{\Delta t} = \dfrac{1}{2}(\sigma_i^k)^2 \delta_x^2(\zeta)_i^k + I_i^k + e_i^k, \\ \zeta_i^0 = 0, \end{cases}$$

where

$$I_i^k = -\frac{u(t^{k-1}, \bar{\mathbf{x}}_i^k) - u(t^{k-1}, \bar{x}_i^k)}{\Delta t}$$
$$+ \frac{1}{2}[\sigma^2(t^k, x_i, u_i^{k-1}) - \sigma^2(t^k, x_i, w_i^{k-1})]\delta_x^2(u)_i^k + [\widehat{b}(u)_i^k - \widehat{b}_i^k].$$

It is clear that, by (1.6) and (2.11), for some constant $C > 0$ that is independent of k and i, it holds that

$$\begin{cases} \left| \dfrac{1}{2}[\sigma^2(t^k, x_i, u_i^{k-1}) - \sigma^2(t^k, x_i, w_i^{k-1})]\delta_x^2(u)_i^k + [\widehat{b}(u)_i^k - \widehat{b}_i^k] \right| \le C|\zeta_i^{k-1}|, \\ \left| \dfrac{u(t^{k-1}, \bar{\mathbf{x}}_i^k) - u(t^{k-1}, \bar{x}_i^k)}{\Delta t} \right| \le C\Delta t. \end{cases}$$

Consequently we have

(2.20)
$$|I_i^k| \le C(|\zeta_i^{k-1}| + \Delta t).$$

Now by (2.19) we have

$$\zeta_i^k = \bar{\zeta}_i^{k-1} + \left\{ \frac{1}{2}(\sigma_i^k)^2 \delta_x^2(\zeta)_i^k + I_i^k + e_i^k \right\} \Delta t.$$

Considering the "localized" solution of u (described in the previous subsection) if necessary, we assume without loss generality that the maximum absolute value of ζ_i^k occurs at an "interior" mesh point $x_{i(k)}^k$, where $-R < i(k)h < R$ for some large $R > 0$. Now, if we set $\|\zeta^k\| = \max_i |\zeta_i^k|$, then at $i(k)$ we have $\delta_x^2(\zeta)_{i(k)}^k \le 0$. Applying Lemma 2.1 and (2.20) we have

(2, 21)
$$\|\zeta^k\| \le \max_i |\bar{\zeta}_i^{k-1}| + \max_i \left\{ |I_i^k| + |e_i^k| \right\} \Delta t$$
$$\le \max_i |\bar{\zeta}_i^{k-1}| + C\|\zeta^{k-1}\|\Delta t + C(h + \Delta t)\Delta t,$$

where C is again a generic constant. Note that the constant C is independent of the localization, therefore by taking the limit we see that (2.21) should hold for the "global solution" as well.

In order to estimate $\max_i |\bar{\zeta}_i^{k-1}|$, we let $I_1(u)(t^k, \cdot)$ denote the linear interpolate of the grid values $\{u_i^k\}_{i=-\infty}^{\infty}$ and $w^k(\cdot)$ the linear interpolate of $\{w_i^k\}_{i=-\infty}^{\infty}$, then

$$(2.22) \quad \max_i |\bar{\zeta}_i^{k-1}| \le \max_i |\zeta_i^{k-1}| + \max_i |u(t^{k-1}, \bar{x}_i^k) - I_1(u)(t^{k-1}, \bar{x}_i^k)|.$$

Apply the *Peano Kernel* Theorem (cf. e.g., Davis [1]) to show that

$$\max_i |u(t^{k-1}, \bar{x}_i^k) - I_1(u)(t^{k-1}, \bar{x}_i^k)| \le Ch^*h,$$

where $h^* = \mathcal{O}(\Delta t)$ and $C > 0$ is independent of k and i. This, together with (3.27), amounts to saying that (2.21) can be rewritten as

$$(2.23) \quad \begin{aligned} \|\zeta^k\| &\le \|\zeta^{k-1}\| + C\|\zeta^{k-1}\|\Delta t + C(h + \Delta t)\Delta t, \\ &= \|\zeta^{k-1}\|(1 + C\Delta t) + C(h + \Delta t)\Delta t, \end{aligned}$$

where C is independent of k. It then follows from the Gronwall lemma and the bound on $\|\zeta^0\|$ that $\|\zeta^k\| \le C(h + \Delta t)$, proving the theorem. $\quad\square$

§2.1.3. The approximating solutions $\{u^{(n)}\}_{n=1}^{\infty}$

We now construct for each n an approximate solution $u^{(n)}$ as follows. for each $n \in \mathbb{N}$ let $\Delta t = T/n$, and $h = 2\|b\|_\infty \Delta t$. Since $h > C\Delta t$ implies that $|\bar{x}_i^k - x_i| \le \|b\|_\infty \Delta t < h$, $\bar{x}_i^k$ do not go beyond the interval (x_{i-1}^k, x_{i+1}^k) for each i. Now define

$$(2.24) \quad u^{(n)}(t, x) = w^{\frac{2\|b\|_\infty T}{n}, \frac{T}{n}}(t, x), \qquad (t, x) \in [0, T] \times \mathbb{R},$$

where $w^{h, \Delta t}$ is defined by (2.9). Our main theorem of this section is the following.

Theorem 2.3. *Suppose that (A1)—(A3) hold. Then, the sequence $\{u^{(n)}(\cdot, \cdot)\}$ enjoys the following properties:*

(1) for fixed $x \in \mathbb{R}$, $u^{(n)}(\cdot, x)$ is left continuous;

(2) for fixed $t \in [0, T]$, $u^{(n)}(t, \cdot)$ is Lipschitz, uniformly in t and n (i.e., the Lipschitz constant is independent of t and n);

(3) $\sup_{t,x} |u^{(n)}(t, x) - u(t, x)| = \mathcal{O}(\frac{1}{n})$.

Proof. The property (1) is obvious by definition (2.9). To see (3), we note that

$$u^{(n)}(t, x) - u(t, x) = [w^0(x) - u(0, x)]\mathbf{1}_{\{0\}}(t) + \sum_{k=1}^{N} [w^k(x) - u(t, x)]\mathbf{1}_{(t^{k-1}, t^k]}(t).$$

Since for each fixed $t \in (t^{k-1}, t^k]$, $k > 0$ or $t = 0$, we have $u^{(n)}(t, x) = w^k(x)$ for $k > 0$ or $k = 0$ if $t = 0$. Thus,

$$\sup_x |w^k(x) - u(t, x)|$$

$$\le \|\zeta^k\| + \sup_x |I_1(u)(t^k, x) - u(t^k, x)| + \sup_x |u(t^k, x) - u(t, x)|$$

$$\le \|\zeta^k\| + o(h + \Delta t) + \|u_t\|_\infty \Delta t = \mathcal{O}(h + \Delta t) = \mathcal{O}(\frac{1}{n}),$$

by virtue of Theorem 2.2 and the definitions of h and Δt. This proves (3).

To show (2), let n and t be fixed, and assume that $t \in (t^k, t^{k+1}]$. Then, $u^{(n)}(t, x) = w^k(x)$ is obviously Lipschitz in x. So it remains to determine the Lipschitz constant of every w^k. Let x^1 and x^2 be given. We may assume that $x^1 \in [x_i, x_{i+1})$ and $x^2 \in [x_j, x_{j+1})$, with $i < j$. For $i < \ell < j - 1$, Theorem 2.2 implies that

$$
\begin{aligned}
|w^k(x_\ell) - w^k(x_{\ell+1})| &\leq |w^k(x_\ell) - u(t^k, x_\ell)| \\
(2.25) \qquad &+ |u(t^k, x_\ell) - u(t^k, x_{\ell+1})| + |u(t^k, x_{\ell+1}) - w^k(x_{\ell+1})| \\
&\leq 2\|\zeta^k\| + \|u_x\|_\infty |x_\ell - x_{\ell+1}| \leq Kh = K(x_{\ell+1} - x_\ell),
\end{aligned}
$$

where K is a constant independent of k, ℓ and n. Further, for $x^1 \in [x_i, x_{i+1})$,

$$
w^k(x^1) = w^k(x_{i+1}) + \frac{w^k(x_{i+1}) - w^k(x_i)}{x_{i+1} - x_i}(x^1 - x_{i+1}).
$$

Hence,

$$
|w^k(x^1) - w^k(x_{i+1})| = \left| \frac{w^k(x_{i+1}) - w^k(x_i)}{x_{i+1} - x_i} \right| |x^1 - x_{i+1}| \leq K|x^1 - x_{i+1}|,
$$

where K is the same as that in (2.25). Similarly,

$$
|w^k(x^2) - w^k(x_j)| \leq K|x^2 - x_j|.
$$

Combining the above gives

$$
\begin{aligned}
&|w^k(x^1) - w^k(x^2)| \\
&\leq |w^k(x^1) - w^k(x_{i+1})| + \sum_{\ell=1}^{j-1} |w^k(x_\ell) - w^k(x_{\ell+1})| + |w^k(x_j) - w^k(x^2)| \\
&\leq K\left\{ (x_{i+1} - x^1) + \sum_{\ell=1}^{j-1} (x_{\ell+1} - x_\ell) + (x^2 - x_{j+1}) \right\} = K|x^2 - x^1|.
\end{aligned}
$$

Since the constant K is independent of t and n, the theorem is proved. $\qquad\square$

§2.2. General case

In order to approximate the adapted solution $\Theta = (X, Y, Z)$ to the general FBSDE (1.1), we need to approximate also component Z. In fact this component is particularly important in some application, for instance, it is the hedging strategy in an option pricing problem (see Chapter 1). The main difficulty is, in light of the Four Step Scheme, we need also to approximate the derivative of the solution θ of the PDE (2.4), which in general is more difficult. Our idea is to reduce the PDE (2.4) to a system of PDEs so that θ_x becomes a part of the solution but not the derivative of the solutions.

To be more precise let us assume that b and $\widehat{b}$ both depend on z, thus PDE (2.4) becomes

(2.25)
$$\begin{cases} 0 = \theta_t + \dfrac{1}{2}\sigma^2(t, x, \theta)\theta_{xx} + b(t, x, \theta, -\sigma(t, x, \theta)\theta_x)\theta_x \\ \qquad + \widehat{b}(t, x, \theta, -\sigma(t, x, \theta)\theta_x); \\ \theta(T, x) = g(x). \end{cases}$$

Define b_0 and $\widehat{b}_0$ by

(2.26)
$$\begin{aligned} b_0(t, x, y, z) &= b(t, x, y, -\sigma(t, x, y)z); \\ \widehat{b}_0(t, x, y, z) &= \widehat{b}(t, x, y, -\sigma(t, x, y)z). \end{aligned}$$

One can check that, if σ, b and $\widehat{b}$ satisfy (A1)–(A3), then so do the functions σ, b_0 and $\widehat{b}_0$. Further, if we again set $u(t, x) = \theta(T - t, x)$, $\forall(t, x)$, then (2.25) becomes

(2.27)
$$\begin{cases} u_t = \dfrac{1}{2}\bar{\sigma}^2(t, x, u)u_{xx} + \bar{b}_0(t, x, u, u_x)u_x + \widehat{\bar{b}}_0(t, x, u, u_x); \\ u(0, x) = g(x). \end{cases}$$

We will again drop the sign "–" in the sequel. Now define $v(t, x) = u_x(t, x)$. Using standard "difference quotient" argument (see, e.g., Gilbarg-Trudinger [1]) one can show that under (A1)–(A3) v is a solution to the "differentiated" equation of (2.27). In other words, (u, v) satisfies a parabolic system:

(2.28)
$$\begin{cases} u_t = \dfrac{1}{2}\bar{\sigma}^2(t, x, u)u_{xx} + b_0(t, x, u, v)u_x + \widehat{b}_0(t, x, u, v); \\ v_t = \dfrac{1}{2}\bar{\sigma}^2(t, x, u)v_{xx} + B_0(t, x, u, v)v_x + \widehat{B}_0(t, x, u, v); \\ u(0, x) = g(x), \quad v(0, x) = g'(x), \end{cases}$$

where

(2.29)
$$\begin{cases} B_0(t, x, y, z) = \sigma(t, x, y)[\sigma_x(t, x, y) + \sigma_y(t, x, y)z] + b(t, x, y, z) \\ \qquad + b_z(t, x, y, z) + \widehat{b}_z(t, x, y, z); \\ \widehat{B}_0(t, x, y, z) = [b_x(t, x, y, z) + b_y(t, x, y, z)z]z + \widehat{b}(t, x, y, z)z \\ \qquad + \widehat{b}_x(t, x, y, z). \end{cases}$$

We should point out that, unlike in the previous case, the functions B_0 and $\widehat{B}_0$ in (2.29) are neither uniformly bounded nor uniformly Lipschitz, thus more careful consideration should be given before we make arguments parallel to the previous special case. First let us modify (2.29) as follows. Let K be the constant in (1.6) and let $\varphi_K \in C^\infty(\mathbb{R})$ be a "truncation function" such that $\varphi_K(z) = z$ for $|z| \leq K$, $\varphi_K(z) = 0$ for $|z| > K + 1$, and $|\varphi_K'(z)| \leq C$ for some (generic) constant $C > 0$. Define

$$B_0^K(t, x, y, z) \overset{\Delta}{=} B_0(t, x, y, \varphi_K(z)); \quad \widehat{B}_0^K(t, x, y, z) \overset{\Delta}{=} \widehat{B}_0(t, x, y, \varphi_K(z)).$$

Then B_0^K and $\widehat{B}_0^K$ are uniformly bounded and uniform Lipschitz in all variables. Now consider the "truncated" version of (2.28), that is, we replace B_0 and $\widehat{B}_0$ by B_0^K and $\widehat{B}_0^K$ in (2.28). Applying Lemma 4.2.1 we know that this truncated version of (2.28) has a unique classical solution, say, (u^K, v^K), that is uniformly bounded. But since $\|v\|_\infty \leq K$ by (1.6), (u, v) is also a (classical) solution to the truncated version of (2.28), thus we must have $(u, v) \equiv (u^K, v^K)$ by uniqueness. Consequently, we need only approximate the solution to the truncated version of (2.28), which reduces the technical difficulty considerably. For notational simplicity, from now on we will not distinguish (2.28) and its truncated version unless specified. In fact, as we will see later, such a truncation will be used only once in the error analysis.

§2.2.1. Numerical scheme

Following the idea presented in §2.1, we first determine the characteristics of the first order system

$$\begin{cases} u_t - b_0(t, x, u, v)u_x = 0; \\ v_t - B_0(t, x, u, v)v_x = 0. \end{cases}$$

It is easy to check that the two characteristic curves $\mathcal{C}_i : (t, x_i(t))$, $i = 1, 2$, are determined by the ODEs

$$\begin{cases} dx_1(t) = -b_0(t, x_1(t), u(t, x_1(t)), v(t, x_1(t)))dt; \\ dx_2(t) = -B_0(t, x_2(t), u(t, x_2(t)), v(t, x_2(t)))dt. \end{cases}$$

Let τ_1 and τ_2 be the arc-lengths along $\mathcal{C}_1$ and $\mathcal{C}_2$, respectively. Then,

$$d\tau_1 = \psi_1(t, x_1(t))dt; \qquad d\tau_2 = \psi_2(t, x_2(t))dt,$$

where

$$\begin{cases} \psi_1(t, x) = [1 + b_0^2(t, x, u(t, x), v(t, x))]^{1/2}; \\ \psi_2(t, x) = [1 + B_0^2(t, x, u(t, x), v(t, x))]^{1/2}. \end{cases}$$

Thus, along $\mathcal{C}_1$ and $\mathcal{C}_2$, respectively,

$$\psi_1 \frac{\partial}{\partial \tau_1} = \left\{ \frac{\partial}{\partial t} - b_0 \frac{\partial}{\partial x} \right\}; \qquad \psi_2 \frac{\partial}{\partial \tau_2} = \left\{ \frac{\partial}{\partial t} - B_0 \frac{\partial}{\partial x} \right\},$$

and (2.28) can be simplified to

$$(2.30) \qquad \begin{cases} \psi_1 \dfrac{\partial u}{\partial \tau_1} = \dfrac{1}{2}\sigma^2(t, x, u)u_{xx} + \widehat{b}_0(t, x, u, v); \\[2mm] \psi_2 \dfrac{\partial v}{\partial \tau_2} = \dfrac{1}{2}\sigma^2(t, x, u)v_{xx} + \widehat{B}_0(t, x, u, v). \end{cases}$$

Numerical Scheme.

For any $n \in \mathbb{N}$, let $\Delta t = T/n$. Let $h > 0$ be given. Let $t^k = k\Delta t$, $k = 0, 1, 2, \cdots$, and $x_i = ih$, $i = \cdots, -1, 0, 1, \cdots$, as before.

Step 0: Set $U_i^0 = g(x_i)$, $V_i^0 = g'(x_i)$, $\forall i$, and extend U^0 and V^0 to all $x \in \mathbb{R}$ by linear interpolation.

Next, suppose that U^{k-1}, V^{k-1} are defined such that $U^{k-1}(x_i) = U_i^{k-1}$, $V^{k-1}(x_i) = V_i^{k-1}$, and let

(2.31)
$$
\begin{cases}
(\widehat{b}_0)_i^k = \widehat{b}_0(t^k, x_i, U_i^{k-1}, V_i^{k-1}); \\
(\widehat{B}_0)_i^k = \widehat{B}_0(t^k, x_i, U_i^{k-1}, V_i^{k-1}); \\
\sigma_i^k = \sigma(t^k, x_i, U_i^{k-1}); \\
\bar{x}_i^k = x_i + b_0(t^k, x_i, U_i^{k-1}, V_i^{k-1})\Delta t; \\
\bar{\mathbf{x}}_i^k = x_i + B_0(t^k, x_i, U_i^{k-1}, V_i^{k-1})\Delta t,
\end{cases}
$$

and $\bar{U}_i^{k-1} = U^{k-1}(\bar{x}_i^k)$, $\bar{V}_i^{k-1} = V^{k-1}(\bar{\mathbf{x}}_i^k)$.

Step k: Determine the k-th step grid values (U^k, V^k) by the system of difference equations

(2.32)
$$
\begin{cases}
\dfrac{U_i^k - \bar{U}_i^{k-1}}{\Delta t} = \dfrac{1}{2}(\sigma_i^k)^2 \delta_x^2(U)_i^k + (\widehat{b}_0)_i^k; \\
\dfrac{V_i^k - \bar{V}_i^{k-1}}{\Delta t} = \dfrac{1}{2}(\sigma_i^k)^2 \delta_x^2(V)_i^k + (\widehat{B}_0)_i^k.
\end{cases}
$$

We then extend the grid values $\{U_i^k\}$ and $\{V_i^k\}$ to the functions $U^k(x)$ and $V^k(x)$, $x \in \mathbb{R}$, by linear interpolation.

§2.2.2. Error analysis

We follow the arguments in §2.1. First, we evaluate the first equation in (2.30) along $\mathcal{C}_1$ and the second one along $\mathcal{C}_2$ to get an analogue of (2.12):

$$
\begin{cases}
\dfrac{u_i^k - \widehat{u}_i^{k-1}}{\Delta t} = \dfrac{1}{2}(\sigma(u)_i^k)^2 \delta_x^2(u)_i^k + \widehat{b}_0(u,v)_i^k + (e_1)_i^k; \\
\dfrac{v_i^k - \widehat{v}_i^{k-1}}{\Delta t} = \dfrac{1}{2}(\sigma(u)_i^k)^2 \delta_x^2(u)_i^k + \widehat{B}_0(u,v)_i^k + (e_2)_i^k,
\end{cases}
$$

where $u_i^k = u(t^k, x_i)$, $v_i^k = v(t^k, x_i)$ (recall that $(u, v) = (u, u_x)$ is the true solution of (2.29)), and $\widehat{u}_i^{k-1} = u(t^{k-1}, \widehat{x}_i)$, $\widehat{v}_i^k = v(t^{k-1}, \widehat{\mathbf{x}}_i)$, with

$$
\widehat{x}_i^k = x_i + b_0(t^k, x_i, u_i^{k-1}, v_i^{k-1})\Delta t; \qquad \widehat{\mathbf{x}}_i^k = x_i + B_0(t^k, x_i, u_i^{k-1}, v_i^{k-1})\Delta t.
$$

Also, $\sigma(u)_i^k$, $\widehat{b}_0(u,v)_i^k$ and $\widehat{B}_0(u,v)_i^k$ are analogous to σ_i^k, $(\widehat{b}_0)_i^k$ and $(\widehat{B}_0)_i^k$, except that U_i^{k-1} and V_i^{k-1} are replaced by u_i^{k-1} and v_i^{k-1}.

Estimating the error $\{(e_1)_i^k\}$ and $\{(e_2)_i^k\}$ in the same fashion as in Lemma 2.1 we obtain that

(2.33)
$$
\sup_{k,i}\{|(e_1)_i^k| + |(e_2)_i^k|\} \le \mathcal{O}(h + \Delta t).
$$

We now define as we did in (2.9) the approximate solutions $U^{(n)}$ and $V^{(n)}$ by

(2.34)

$$U^{(n)}(t,x) = \begin{cases} \displaystyle\sum_{k=1}^{n} U^k(x) 1_{(\frac{(k-1)T}{n}, \frac{kT}{n}]}(t), & t \in (0,T]; \\ U^0(x), & t = 0; \end{cases}$$

$$V^{(n)}(t,x) = \begin{cases} \displaystyle\sum_{k=1}^{n} V^k(x) 1_{(\frac{(k-1)T}{n}, \frac{kT}{n}]}(t), & t \in (0,T]; \\ V^0(x), & t = 0. \end{cases}$$

Let $\xi(t,x) = u(t,x) - U^n(t,x)$ and $\zeta(t,x) = v(t,x) - V^n(t,x)$. We can derive the analogue of (2.19):

$$\begin{cases} \dfrac{\xi_i^k - \widehat{\xi}_i^{k-1}}{\Delta t} = \dfrac{1}{2}(\sigma_i^k)^2 \delta_x^2(\xi)_i^k + (I_1)_i^k + (e_1)_i^k; \\[2mm] \dfrac{\zeta_i^k - \widehat{\zeta}_i^{k-1}}{\Delta t} = \dfrac{1}{2}(\sigma_i^k)^2 \delta_x^2(\zeta)_i^k + (I_2)_i^k + (e_2)_i^k, \end{cases}$$

where

$$(I_1)_i^k = -\frac{u(t^{k-1}, \widehat{x}_i^k) - u(t^{k-1}, \bar{x}_i^k)}{\Delta t} + [\widehat{b}_0(u,v)_i^k - (\widehat{b}_0)_i^k]$$
$$+ \frac{1}{2}[\sigma^2(t^k, x_i, u_i^{k-1}) - \sigma^2(t^k, x_i, U_i^{k-1})]\delta_x^2(u)_i^k;$$

$$(I_2)_i^k = -\frac{v(t^{k-1}, \widehat{x}_i^k) - v(t^{k-1}, \bar{x}_i^k)}{\Delta t} + [\widehat{B}_0(u,v)_i^k - (\widehat{B}_0)_i^k]$$
$$+ \frac{1}{2}[\sigma^2(t^k, x_i, v_i^{k-1}) - \sigma^2(t^k, x_i, V_i^{k-1})]\delta_x^2(v)_i^k;$$

Using the uniform Lipschitz property of $\widehat{b}_0$ in y and z, one shows that

(2.35) $|(I_1)_i^k| \le C_2\{|\xi_i^{k-1}| + |\zeta_i^{k-1}|\} + C_3(h + \Delta t), \qquad \forall k, i.$

To estimate $(I_2)_i^k$, we will assume that $(\{U_i^k\}, \{V_i^k\})$ is uniformly bounded, otherwise we consider the truncated version version of (2.28). Thus $\widehat{B}_0^K$ is uniform Lipschitz. Thus,

$$|\widehat{B}_0(u,v)_i^k - (\widehat{B}_0)_i^k| \le C_4(|\xi_i^{k-1}| + |\zeta_i^{k-1}|), \qquad \forall k, i,$$

where C_4 depends only on the bounds of u, v, $\{U_i^k\}$, $\{V_i^k\}$, and that of σ, b, $\widehat{b}$ and their partial derivatives. Consequently,

(2.36) $|(I_2)_i^k| \le C_2'\{|\xi_i^{k-1}| + |\zeta_i^{k-1}|\} + C_3'(h + \Delta t), \qquad \forall k, i.$

Use of the maximum principle and the estimates (2.33), (2.35) and (2.36) leads to

$$\|\xi^k\| \le \|\xi^{k-1}\| + C_2(\|\xi^{k-1}\| + \|\zeta^{k-1}\|)\Delta t + C_5(h + \Delta t)\Delta t;$$
$$\|\zeta^k\| \le \|\zeta^{k-1}\| + C_2'(\|\xi^{k-1}\| + \|\zeta^{k-1}\|)\Delta t + C_5'(h + \Delta t)\Delta t;$$

Add the two inequalities above and apply Gronwall's lemma; we see that

$$\sup_k(\|\xi^k\| + \|\zeta^k\|) = \mathcal{O}(h + \Delta t).$$

Applying the arguments similar to those in Theorem 2.3 we can derive the following theorem.

Theorem 2.4. *Suppose that (A1)–(A3) hold. Then,*

$$\sup_{(t,x)}\{|U^{(n)}(t,x) - u(t,x)| + |V^{(n)}(t,x) - u_x(t,x)|\} = \mathcal{O}(\frac{1}{n}).$$

Moreover, for each fixed $x \in \mathbb{R}$, $U^{(n)}(\cdot, x)$ and $V^{(n)}(\cdot, x)$ are left-continuous; for fixed $t \in [0, T]$, $U^{(n)}(t, \cdot)$ and $V^{(n)}(t, \cdot)$ are uniformly Lipschitz, with the same Lipschitz constant that is independent of n.

§3. Numerical Approximation of the Forward SDE

Having derive the numerical solution of the PDE (1.5), we are now ready to complete the final step: approximating the Forward SDE (1.4). Recall that the FSDE to be approximated has the following form:

$$(3.1) \qquad X_t = x + \int_0^t \tilde{b}(s, X_s)ds + \int_0^t \tilde{\sigma}(s, X_s)dW_s,$$

where

$$\tilde{b}(t, x) = b(t, x, \theta(t, x), -\sigma(t, x, \theta(t, x)\theta_x(t, x)) = b_0(t, x, \theta(t, x), \theta_x(t, x));$$
$$\tilde{\sigma}(t, x) = \sigma(t, x, \theta(t, x)).$$

for $(t, x) \in [0, T] \times \mathbb{R}$.

To define the approximate SDEs, we need some notations. For each $n \in \mathbb{N}$, set $\Delta t_n = T/n$, $t^{n,k} = k\Delta t_n$, $k = 0, 1, 2, \cdots, n$, and

$$(3.2) \qquad \begin{cases} \eta^n(t) = \sum_{k=0}^{n-1} t^{n,k} 1_{[t^{n,k}, t^{n,k+1})}(t), & t \in [0, T); \\ \eta^n(T) = T. \end{cases}$$

Next, for each n, let $(U^{(n)}, V^{(n)})$ be the approximate solution to the PDE (1.5), defined by (2.35) (in the special case we may consider only $u^{(n)}$ defined by (2.24)). Set

$$(3.3) \qquad \theta^n(t, x) = U^{(n)}(T - t, x), \qquad \theta_x^n(t, x) = V^{(n)}(T - t, x),$$

and

$$\tilde{b}^n(t, x) = b_0(t, x, \theta^n(t, x), \theta_x^n(t, x)); \qquad \tilde{\sigma}^n(t, x) = \sigma(t, x, \theta^n(t, x)).$$

By Theorem 2.4 we know that θ^n is right continuous in t and uniformly Lipschitz in x, with the Lipschitz constant being independent of t and n;

thus, so also are the functions $\tilde{b}^n$ and $\tilde{\sigma}^n$. We henceforth assume that there exists a constant K such that, for all t and n,

$$(3.4) \quad |\tilde{b}^n(t, x) - \tilde{b}^n(t, x')| + |\tilde{\sigma}^n(t, x) - \tilde{\sigma}^n(t, x')| \leq K|x - x'|, \quad x, x' \in \mathbb{R}.$$

Also, from Theorem 3.4,

$$(3.5) \quad \sup_{t,x} |\tilde{b}^n(t, x) - \tilde{b}(t, x)| + \sup_{t,x} |\tilde{\sigma}^n(t, x) - \tilde{\sigma}(t, x)| = \mathcal{O}(\frac{1}{n}).$$

We now introduce two SDEs: the first one is a *discretized SDE* given by

$$(3.6) \quad \bar{X}^n_t = x + \int_0^t \tilde{b}^n(\cdot, \bar{X}^n_{\cdot})_{\eta^n(s)} ds + \int_0^t \tilde{\sigma}^n(\cdot, \bar{X}^n_{\cdot})_{\eta^n(s)} dW_s,$$

where η^n is defined by (3.2). The other is an *intermediate approximate SDE* given by

$$(3.7) \quad X^n_t = x + \int_0^t \tilde{b}^n(s, X^n_s) ds + \int_0^t \tilde{\sigma}^n(s, X^n_s) dW_s.$$

It is clear from the properties of $\tilde{b}^n$ and $\tilde{\sigma}^n$ mentioned above that both SDEs (3.6) and (3.7) above possess unique strong solutions.

We shall estimate the differences $\bar{X}^n_t - X^n_t$ and $X^n - X$, separately.

Lemma 3.1. *Assume (A1)—(A3). Then,*

$$E\left\{ \sup_{0 \leq t \leq T} |\bar{X}^n_t - X^n_t|^2 \right\} = \mathcal{O}(\frac{1}{n}).$$

Proof. To simplify notation, we shall suppress the sign "~" for the coefficients in the sequel. We first rewrite (3.6) as follows:

$$\bar{X}^n_t = X_0 + u^n_t + \int_0^t b^n(s, \bar{X}^n_s) ds + \int_0^t \sigma^n(s, \bar{X}^n_s) dW_s,$$

where

$$u^n_t = \int_0^t [b^n(\cdot, \bar{X}^n_{\cdot})_{\eta_n(s)} - b^n(s, \bar{X}^n_s)] ds + \int_0^t [\sigma^n(\cdot, \bar{X}^n_{\cdot})_{\eta_n(s)} - \sigma^n(s, \bar{X}^n_s)] dW_s.$$

Applying Doob's inequality, Jensen's inequality, and using the Lipschitz property of the coefficients (3.4) we have

$$E\left\{ \sup_{s \leq t} |X^n_s - \bar{X}^n_s|^2 \right\}$$

$$(3.8) \quad \leq 3E\left\{ \sup_{s \leq t} |u^n_s|^2 \right\} + 3K^2 t \int_0^t E\{|X^n_s - \bar{X}^n_s|^2\} ds$$

$$+ 12K^2 \int_0^t E\{|X^n_s - \bar{X}^n_s|^2\} ds.$$

Now, set $\alpha_n(t) = E\{\sup_{s \leq t} |X_s^n - \bar{X}_s^n|^2\}$. Then, from (3.8),

$$\alpha_n(t) \leq 3E\{\sup_{s \leq t} |u_s^n|^2\} + 3K^2(T+4) \int_0^t \alpha_n(s)ds,$$

and Gronwall's inequality leads to

(3.9) $$E\{\sup_{s \leq t} |X_s^n - \bar{X}_s^n|^2\} \leq 3e^{3K^2(T+4)} E\{\sup_{s \leq t} |u_s^n|^2\}.$$

We now estimate $E\{\sup_{s \leq t} |u_s^n|^2\}$. Note that if $s \in [t^{n,k}, t^{n,k+1})$, for some $1 \leq k < n$, then $\eta^n(s) = k\Delta t_n$ (whence $T - \eta^n(s) = (n-k)\Delta t_n$, as $T = n\Delta t_n$) and $T - s \in ((n-k-1)\Delta t_n, (n-k)\Delta t_n]$. Thus, by definitions (2.9) and (3.2), for every $x \in \mathbb{R}$

$$\theta^n(\eta^n(s), x) = u^{(n)}(T - \eta^n(s), x) = u^{(n)}((n-k)\Delta t_n, x)$$
$$= u^{(n)}(T - s, x) = 0^n(s, x).$$

More generally, for all $(s, x) \in [0, T] \times \mathbb{R}$,

$$b^n(s, x) = b(s, x, \theta^n(s, x)) = b(s, x, \theta^n(\eta^n(s), x)).$$

Using this fact, it is easily seen that

$$\left| \int_0^t b^n(\cdot, \bar{X}^n)_{\eta_n(s)} - b^n(s, X_s^n)ds \right|$$

$$\leq \int_0^t \left| b(\eta^n(s), \bar{X}_{\eta^n(s)}^n, \theta^n(\eta^n(s), \bar{X}_{\eta^n(s)}^n)) - b(s, X_s^n, \theta^n(s, X_s^n)) \right| ds$$

$$\leq \int_0^t \left\{ \left| b(\eta^n(s), \bar{X}_{\eta^n(s)}^n, \theta^n(s, \bar{X}_{\eta^n(s)}^n)) - b(s, X_s^n, \theta^n(s, \bar{X}_{\eta^n(s)}^n)) \right| \right.$$

$$\left. + \left| b(s, X_s^n, \theta^n(s, \bar{X}_{\eta^n(s)}^n)) - b(s, X_s^n, \theta^n(s, X_s^n)) \right| \right\} ds$$

$$= I_1 + I_2.$$

Using the boundedness of the functions b_t, b_x and b_y, we see that

$$\begin{cases} I_1 \leq \int_0^t \left\{ \|b_t\|_\infty |\eta^n(s) - s| + \|b_x\|_\infty |\bar{X}_{\eta^n(s)}^n - X_s^n| \right\} ds, \\ I_2 \leq K\|b_y\|_\infty \cdot \int_0^t |\bar{X}_{\eta^n(s)}^n - X_s^n| ds. \end{cases}$$

Thus,

$$\left| \int_0^t b^n(\cdot, \bar{X}^n)_{\eta^n(s)} - b^n(s, X_s^n)ds \right| \leq \tilde{K} \int_0^t \left\{ |\eta^n(s) - s| + |\bar{X}_{\eta^n(s)}^n - X_s^n| \right\} ds,$$

where $\tilde{K}$ depends only on K, $\|b_t\|_\infty$, $\|b_x\|_\infty$ and $\|b_y\|_\infty$. Since

$$\int_0^t |\eta^n(s) - s| ds = \sum_{k=0}^{n-1} \int_{t^k \wedge t}^{t^{k+1} \wedge t} (s - t^k)ds \leq \frac{1}{2} \sum_{k=0}^{n-1} (\Delta t_n)^2 = \frac{T^2}{2n},$$

$$\text{(3.10)} \qquad E\Big\{ \sup_{u \le t} \Big| \int_0^t b^n(\cdot, \bar{X}^n_{\cdot})_{\eta^n(s)} - b^n(s, X^n_s) ds \Big|^2 \Big\}$$

$$\le 2\widetilde{K}^2 \Big\{ T \int_0^t E|\bar{X}^n_{\eta^n(s)} - X^n_s|^2 ds + \frac{T^4}{4n^2} \Big\}.$$

Using the same reasoning for σ with Doob's inequality, we can see that

$$\text{(3.11)} \qquad E\Big\{ \sup_{u \le t} \Big| \int_0^t \sigma^n(\cdot, \bar{X}^n_{\cdot})_{\eta^n(s)} - \sigma^n(s, X^n_s) dW_s \Big|^2 \Big\}$$

$$\le 8\widetilde{K}^2 \Big\{ \int_0^t E|\bar{X}^n_{\eta^n(s)} - X^n_s|^2 ds + \int_0^t (s - \eta^n(s))^2 ds \Big\}$$

$$\le 8\widetilde{K}^2 \Big\{ \int_0^t E|\bar{X}^n_{\eta^n(s)} - X^n_s|^2 ds + \frac{T}{3n^2} \Big\}$$

Combining (3.10) and (3.11), we get

$$E\{ \sup_{s \le t} |u^k_s|^2 \} \le \widetilde{K}^2 (4T + 16) \int_0^t E|\bar{X}^n_{\eta^n(s)} - X^n_s|^2 ds + \widetilde{K}^2 T (T + \frac{16}{3}) \frac{1}{n^2}.$$

Thus, by (3.9),

$$\text{(3.12)} \qquad E\Big\{ \sup_{s \le t} |X^n_s - \bar{X}^n_s|^2 \Big\}$$

$$\le 3e^{3K^2(T+4)} \Big\{ \widetilde{K}^2 (4T + 16) \int_0^t E|\bar{X}^n_{\eta^n(s)} - X^n_s|^2 ds$$

$$+ \widetilde{K}^2 T (T + \frac{16}{3}) \frac{1}{n^2} \Big\}.$$

Finally, noting that $|\bar{X}^n_{\eta^n(s)} - X^n_s| \le |\bar{X}^n_{\eta^n(s)} - \bar{X}^n_s| + |\bar{X}^n_s - X^n_s|$ and that

$$\bar{X}^n_{\eta_k(s)} - \bar{X}^n_s = b^n(\cdot, \bar{X}^n_{\cdot})_{\eta^n(s)}(s - \eta^n(s)) + \sigma(\cdot, \bar{X}^n_{\cdot})_{\eta^n(s)}(W_s - W_{\eta^n(s)}),$$

we see as before that

$$\int_0^t E|\bar{X}^n_{\eta^n(s)} - \bar{X}^n_s|^2 ds \le 2 \int_0^t \Big\{ \|b\|^2_\infty (s - \eta^n(s))^2 + \|\sigma\|^2_\infty |s - \eta^n(s)| \Big\} ds$$

$$\le \frac{2\|b\|^2_\infty T}{3} \cdot \frac{1}{n^2} + \|\sigma\|^2_\infty T \cdot \frac{1}{n}.$$

Therefore, (3.12) becomes

$$\text{(3.13)} \quad E\Big\{ \sup_{s \le t} |X^n_s - \bar{X}^n_s|^2 \Big\} \le C_1 \frac{1}{n} + C_2 \frac{1}{n^2} + C_3 \int_0^t E\Big\{ \sup_{r \le s} |\bar{X}^n_r - X^n_r|^2 \Big\} ds,$$

where C_1, C_2 and C_3 are constants depending only on the coefficients b, σ and K and can be calculated explicitly from (3.12). Now, we conclude from (3.13) and Gronwall's inequality that

$$\alpha_n(t) \le \beta_n e^{C_T}, \qquad \forall t \in [0, T],$$

where $\beta_n = C_1 n^{-1} + C_2 n^{-2}$ and $C_T = C_3 T$. In particular, by slightly changing the constants, we have

$$\alpha_n(T) = E\left\{ \sup_{0 \le t \le T} |\bar{X}_t^n - X_s^n|^2 \right\} \le \frac{C_1}{n} + \frac{C_2}{n^2} = \mathcal{O}\left(\frac{1}{n}\right),$$

proving the lemma.   □

The main result of this chapter is the following theorem.

Theorem 3.2. *Suppose that the standing assumptions (A1)—(A3) hold. Then, the adapted solution (X, Y, Z) to the FBSDE (1.1) can be approximated by a sequence of adapted processes $(\bar{X}^n, \bar{Y}^n, \bar{Z}^n)$, where $\bar{X}^n$ is the solution to the discretized SDE (3.6) and, for $t \in [0, T]$,*

$$\bar{Y}_t^n := \theta^n(t, \bar{X}_t^n); \qquad \bar{Z}_t^n := -\sigma(t, \bar{X}_t^n, \theta^n(t, \bar{X}_t^n)) \theta_x^n(t, \bar{X}_t^n),$$

with θ^n and θ_x^n being defined by (3.3) and $U^{(n)}$ and $V^{(n)}$ by (2.34). Furthermore,

$$(3.14) \quad E\left\{ \sup_{0 \le t \le T} |\bar{X}_t^n - X_t| + \sup_{0 \le t \le T} |\bar{Y}_t^n - Y_t| + \sup_{0 \le t \le T} |\bar{Z}_t^n - Z_t| \right\} = \mathcal{O}\left(\frac{1}{\sqrt{n}}\right).$$

Moreover, if f is C^2 and uniformly Lipschitz, then for n large enough,

$$(3.15) \qquad \left| E\{f(\bar{X}_T^n, \bar{Z}_T^n)\} - E\{f(X_T, Z_T)\} \right| \le \frac{K}{n},$$

for a constant K.

Proof. Recall that at the beginning of the proof of Lemma 3.1, we have suppressed the sign "~" for $\tilde{b}$ and $\tilde{\sigma}$ to simplify notation. Set

$$\varepsilon^n(t) = \left\{ \sup_x |b^n(t, x) - b(t, x)|^2 + \sup_x |\sigma^n(t, x) - \sigma(t, x)|^2 \right\},$$

where b, b^n, σ and σ^n are defined by (3.1) and (3.3). Then, from (3.5) we know that $\sup_t |\varepsilon^n(t)| = \mathcal{O}(\frac{1}{n^2})$. Now, applying Lemma 3.1, we have

$$E\left\{ \sup_{s \le t} |\bar{X}_s^n - X_s|^2 \right\} \le 2E\left\{ \sup_{s \le t} |\bar{X}_s^n - X_s^n|^2 \right\} + 2E\left\{ \sup_{s \le t} |X_s^n - X_s|^2 \right\}$$

$$= \mathcal{O}\left(\frac{1}{n}\right) + 2E\left\{ \sup_{s \le t} |X_s^n - X_s|^2 \right\}.$$

Further, observe that

$$E\left\{\sup_{s\leq t}|X_s^n - X_s|^2\right\}$$

$$\leq 2T\int_0^t E\left|b^n(s,X_s^n) - b(s,X_s)\right|^2 ds + 8\int_0^t E\left|\sigma^n(s,X_s^n) - \sigma(s,X_s)\right|^2 ds$$

$$\leq 4T\int_0^t E\left|b^n(s,X_s^n) - b^n(s,X_s)\right|^2 ds$$

$$+ 16\int_0^t E\left|\sigma^n(s,X_s^n) - \sigma^n(s,X_s)\right|^2 ds + 4(T+4)\int_0^t \varepsilon_n(s)ds$$

$$\leq 4(T+4)K^2\int_0^t E\left\{\sup_{r\leq s}|X_r^n - X_r|^2\right\}ds + 4(T+4)\int_0^t \varepsilon_n(s)ds.$$

Applying Gronwall's inequality, we get

$$(3.16)\qquad E\left\{\sup_{s\leq t}|X_s^n - X_s|^2\right\} \leq 4(T+4)\int_0^t \varepsilon_n(s)ds \cdot e^{4(T+4)K^2} \leq \frac{\widetilde{C}}{n^2},$$

where $\widetilde{C}$ is a constant depending only on K and T. Now, note that the functions θ and θ^n are both uniformly Lipschitz in x. So, if we denote their Lipschitz constants by the same L, then

$$E\left\{\sup_{0\leq t\leq T}|Y_t - \bar{Y}_t^n|^2\right\}$$

$$\leq 2E\left\{\sup_{0\leq t\leq T}|\theta(t,X_t) - \theta^n(t,\bar{X}_t^n)|^2\right\}$$

$$+ 2E\left\{\sup_{0\leq t\leq T}|\theta^n(t,\bar{X}_t^n) - \theta(t,\bar{Y}_t^n)|^2\right\}$$

$$\leq 2L^2 E\left\{\sup_{0\leq t\leq T}|X_t - \bar{X}_t^n|^2\right\} + 2\sup_{(t,x)}|\theta(t,x) - \theta^n(t,x)|^2 = O\left(\frac{1}{n}\right),$$

by Theorem 3.4 and (3.16). The estimate (3.14) then follows from an easy application of Cauchy-Schwartz inequality. To prove (3.15), note that Theorem 2.3 implies that, for n large enough, $\sup_{(t,x)}|\theta^n(t,x) - \theta(t,x)| \leq Cn^{-1}$, for some (generic) constant $C > 0$. We modify $\bar{X}_t^n$ as defined by (3.6) by fixing n and approximating the solution X^n of (3.7) by a standard Euler scheme indexed by k:

$$\bar{X}_t^{n,k} = x + \int_0^t b(\cdot,\bar{X}_{\cdot}^{n,k})_{\eta^k(s)}ds + \int_0^t \sigma(\cdot,\bar{X}_{\cdot}^{n,k})_{\eta^k(s)}dW_s.$$

It is then standard (see, for example, Kloeden-Platen [1, p.460]) that

$$(3.17)\qquad\qquad \left|E\{f(X_T^n)\} - E\{f(\bar{X}_T^{n,k})\}\right| \leq \frac{C_1}{k}.$$

On the other hand, we have

$$|E\{f(X_T)\} - E\{f(X_T^n)\}| \le KE\{|X_T - X_T^n|\}$$

(3.18)
$$\le E\left\{\sup_{0\le t\le T}|X_t - X_t^n|\right\} \le \frac{C_2}{n}$$

for Lipschitzian f, by (3.16). Therefore, noting that $\bar{X}_t^n$ as defined by (3.6) is just $\bar{X}_t^{n,n}$, the triangle inequality, (3.17) and (3.18) lead to (3.15). □

Comments and Remarks

The main body of this book is built on the works of the authors, with various collaboration with other researchers, on this subject since 1993. Some significant results of other researchers are also included to enhance the book. However, due to the limitation of our information, we inevitably might have overlooked some new development in this field while writing this book, for which we deeply regret.

In Chapter 1, the results on the pure BSDEs, especially the fundamental well-posedness result, are based on the method introduced in the seminal paper of Pardoux-Peng [1]. The results on nonsolvability of FBSDEs are inspired by the example of Antonelli [1]. The well-posedness results of FBSDEs over small duration is also based in the spirit of the work of Antonelli [1]. The whole Chapter 2 is based on the paper of Yong [4].

In Chapter 3 we begin to consider a general form of the FBSDE (1) with an arbitrarily given $T > 0$. The main references for this chapter are based on the works of Ma-Yong [1], virtually the first result regarding solvability of FBSDE in this generality; and Ma-Yong [4], in which the notion of *approximate solvability* is introduced. A direct consequence of the method of optimal control is the Four Step Scheme presented in Chapter 4. The finite horizon case is initiated by Ma-Protter-Yong [1]; and the infinite horizon case is the theoretical part of the work on "Black's Consol Rate Conjecture" presented later in Chapter 8, by Duffie-Ma-Yong [1].

Chapter 5 can be viewed either as a tool needed to extend the Four Step Scheme to the situation when the coefficients are allowed to be random, or as an independent subject in stochastic partial differential equations. The main results come from the papers of Ma-Yong [2] and [3]; and the applications in finance (e.g, the stochastic Black-Scholes formula) are collected in Chapter 8.

The method of continuation of Chapter 6 is based on the paper of Hu-Peng [2], and its generalization by Yong [1]. The method adopted a widely used idea in the theory of partial differential equations. Compared to the Four Step Scheme, this method allows the randomness of the coefficients and the degeneracy of the forward diffusion, but requires some analysis which readers might find difficult in a different way.

Chapter 7 is based on the work of Cvitanic-Ma [2]. The idea for the forward SDER using the solution mapping of Skorohod problem is due to Anderson-Orey [1], while the Lipschitz property of such solution mapping is adopted from Dupuis-Ishii [1]. The proof of the backward SDER is a modification of the arguments of Pardoux-Rascanu [1], [2], as well as some arguments from Buckdahn-Hu [1]. The proof of the existence and uniqueness of FBSDER adopted the idea of Pardoux-Tang [1], a generalized method of contraction mapping theorem, which can be viewed as an independent method for solving FBSDE as well.

Chapter 8 collects some successful applications of the FBSDEs developed so far. The integral representation theorem is due to Ma-Protter-Yong [1]; the Nonlinear Feynman-Kac formula is in the spirit of Peng [4], but the argument of the proof follows more closely those of Cvitanic-Ma [2]. The Black's consol rate conjecture is due to Duffie-Ma-Yong [1]; while hedging contingent claims for large investors comes from Cvitanic-Ma [1] for unconstraint case, and from Buckdahn-Hu [1] for constraint case. The section on stochastic Black-Scholes formula is based on the results of Ma-Yong [2] and [3], and the American game option is from Cvitanic-Ma [2].

Finally, the numerical method presented in Chapter 9 is essentially the paper of Douglas-Ma-Protter [1], with slight modifications. We should point out that, to our best knowledge, the scheme presented here is the only numerical method for (strongly coupled) FBSDEs discovered so far, and even when reduced to the pure BSDE case, it is still one of the very few existing numerical methods that can be found in the literature.

In summary, FBSDE is a new type of stochastic differential equations that has its own mathematical flavor and many applications. Like a usual two-point boundary value problem, there is no generic theory for its solvability, and many interesting insights of the equations has yet to be discovered. In the meantime, although the theory exists only for such a short period of time (recall that the first paper on FBSDE was published in 1993!), many topics in theoretical and applied mathematics have already been found closely related to it, and its applicability is quite impressive. It is our hope that by presenting a lecture notes in the series of LNM, more attention would be drawn from the mathematics community, and the beauty of the problem would be further exposed.

References

Ahn, H., Muni, A., and Swindle, G.,
[1] Misspecified asset pricing models and robust hedging strategies, preprint.

Anderson, R. F. and Orey, S.,
[1] Small random perturbation of dynamical systems with reflecting boundary, *Nagoya Math. J.*, 60 (1976), 189–216.

Antonelli, F.,
[1] Backward-forward stochastic differential equations, *Ann. Appl. Prob.*, 3 (1993), 777–793.

Bailey, P. B., Shampine, L. F., and Waltman, P. E.,
[1] *Nonlinear Two Point Boundary Value Problems*, Academic Press, New York, 1968.

Barbu, V.,
[1] *Nonlinear Semigroups and Differential Equations in Banach Spaces*, Noordhoff Internation Publishing, 1976.

Barles, G., Buckdahn, R., and Pardoux, E.,
[1] Backward stochastic differential equations and integral-partial differential equations, *Stochastics and Stochastics Reports*, 60 (1997), 57–83.

Bellman, R., and Wing, G. M.,
[1] *An Introduction to Invariant Imbedding*, John Wiley & Sons, New York, 1975.

Bensoussan, A.,
[1] Stochastic maximum principle for distributed parameter systems, *J. Franklin Inst.*, 315 (1983), 387–406.
[2] Maximum principle and dynamic programming approaches of the optimal control of partially observed diffusions, *Stochastics*, 9 (1983), 169–222.
[3] On the theory of option pricing, *Acta Appl. Math.*, 2 (1984), 139–158.
[4] *Perturbation Methods in Optimal Control*, John-Wiley & Sons, New York, 1988.
[5] *Stochastic Control of Partially Observable Systems*, Cambridge University Press, 1992.

Bergman, Y.Z., Grundy, B.D., and Wiener, Z.,
[1] General Properties of Option Pricing, Preprint, 1996.

Bismut, J. M.,
[1] Théorie Probabiliste du Contrôle des Diffusions, *Mem. Amer. Math. Soc.* 176, Providence, Rhode Island, 1973.

[2] An introductory approach to duality in optimal stochastic control, *SIAM Rev.*, 20 (1978), 62–78.

Black, F., and Scholes, M.,

[1] The pricing of options and corporate liability, *J. Polit. Economy*, 81 (1973), 637–659.

Borkar, V. S.,

[1] *Optimal Control of Diffusion Processes*, Longman Scientific & Technical, 1989.

Brennan, M., and Schwartz, E.,

[1] A continuous time approach to the pricing of bonds, *J. Banking & Finance*, 3 (1979), 133-155.

Buckdahn, R.,

[1] Backward stochastic differential equations driven by a martingale, preprint.

Buckdahn, R., and Hu, Y.,

[1] Hedging contingent claims for a large investor in an incomplete market, *Advances in Applied Probability*, 30 (1998), 239–255.

Buckdahn, R., and Pardoux, E.,

[1] BSDE's with jumps and associated integral-stochastic differential equations, preprint.

Caffarelli, L. A., and Friedman, A.,

[1] Partial regularity of the zero-set of semilinear and superlinear elliptic equations, *J. Diff. Eqs.*, 60 (1985), 420–433.

Cannarsa, P., and Soner, H. M.,

[1] On the singularities of the viscosity solutions to Hamilton-Jacobi-Bellman equations, *Indiana Univ. Math. J.*, 36 (1987), 501–524.

Casti, J., and Kalaba, R.,

[1] *Imbedding Methods in Applied Mathematics*, Addison-Wesley Pub., London, 1973.

Chen, S., and Yong, J.,

[1] Stochastic linear quadratic optimal control problems, preprint.

Courant, R., and Hilbert, D.,

[1] *Methods of Mathematical Physics, Vol.1*, Interscience, New York, 1953.

Crandall, M.G., Ishii, H., and Lions, P.L.,

[1] User's guide to viscosity solutions of second order partial differential equations, *Bull. Amer. Soc.*, 27 (1992), 1–67.

Cvitanic, J., and Karatzas, I.,

[1] Hedging contingent claims with constrained portfolios, *Ann. of Appl. Probab.* 3 (1993), 652–681.

[2] Backward stochastic differential equations with reflection and Dynkin games, *Ann. of Proba.*, 24 (1996), 2024–2056.

Cvitanic, J., and Ma, J.,

[1] Hedging options for a large investor and forward-backward SDEs, *Ann. Appl. Probab.*, 6 (1996), 370–398.
[2] Forward-backward stochastic differential equations with reflecting boundary conditions, preprint.

Darling, R. W. R.,

[1] Martingales on non–compact manifolds: maximal inequalities and prescribed limits, preprint.
[2] Constructing Gamma–martingales with prescribed limit using backward SDE, preprint.

Darling, R. W. R., and Peng, S.,

[1] An useful nonlinear version of conditional expectation, preprint.

Davis, P.,

[1] *Interpolation and Approximation*, Dover, New York, 1975.

Deimling, K.,

[1] *Nonlinear Functional Analysis*, Springer-Verlag, Berlin, 1988.

Dellacherie, C., and Meyer, P. A.,

[1] *Probabilities and Potentials B*, North Holland, 1982.

Detemple. J.,

[1] Intertemporal asset pricing with incomplete markets and nontraded assets, preprint.

Donnelly, H., and Fefferman, C.,

[1] Nodal sets of eigenfunctions on Riemannian manifolds, *Invent. Math.*, 93 (1988), 161–183.

Douglas, J.,Jr., Ma, J., and Protter, P.,

[1] Numerical methods for forward-backward stochastic differential equations, *Ann. Appl. Probab.*, 6 (1996), 940–968.

Duffie, D.,

[1] *Security Markets: Stochastic Models*, Boston, Academic Press, 1988.
[2] *Dynamic Asset Pricing Theory*, Princeton Univ. Press, 1992.

Duffie, D., Epstein, .L,

[1] Stochastic differential utility, *Econometrica*, 60 (1992), 353-394.
[2] Asset pricing with stochastic differential utilities, *Review of Financial Studies*, 5 (1992), 411–436.

Duffie, D., Geoffard, P. Y., and Skiadas, C.,

[1] Efficient and equilibrium allocations with stochastic differential utility, *J. Math. Economics*, 23 (1994), 133–146.

Duffie, D., and Lions, P. L.,

[1] PDE solutions of stochastic differential utility, *J. of Mathematical Economics*, 21 (1993), 577–606.

Duffie, D., Ma, J., and Yong, J.,

[1] Black's consol rate conjecture, *IMA preprint #1164*, July, 1993; also, *Ann. Appl. Probab.*, 5 (1995), 356–382.

Dupuis, P. and Ishii, H.,

[1] On Lipschitz continuity of the solution mapping to the Skorokhod problem, with applications, *Stoch. & Stoch. Rep.*, 35 (1991), 31–62.

El Karoui, N., Huu Nguyen, D., and Jeanblanc-Picqué, M.,

[1] Compactification methods in the control of degenerate diffusions: existence of an optimal control, *Stochastics*, 20 (1987), 169–219.

[2] Existence of an optimal Markovian filter for the control under partial observations, *SIAM J. Control & Optim.*, 26, (1988), 1025–1061.

El Karoui, N., and Jeanblanc-Picqué, M.,

[1] Optimization of consumption with labor income, preprint.

El Karoui, N., Jeanblanc-Picqué, M., and Shreve, S. E.,

[1] Robustness of the Black and Scholes formula, preprint.

El Karoui, N., Kapoudjian, C., Pardoux, E., Peng, S., and Quenez, M. C.,

[1] Reflected solutions of backward SDE's and related obstacle problems for PDE's, *Ann. of Proba.*, 25 (1997), 702–737.

El Karoui, N., Peng, S., and Quenez, M.-C.,

[1] Backward stochastic differential equation in finance, *Mathematical Finance*, 7 (1997), 1–71.

El Karoui, N., and Quenez, M.-C.,

[1] Dynamic programming and pricing of contingent claims in an incomplete market, preprint.

Epstein, L.,

[1] The global stability of efficient intertemporal allocations, *Econometrica*, 55 (1987), 329-355.

Fleming, W. H.,

[1] The Cauchy problem for a nonlinear first order differential equation, *J. Diff. Eqs.*, 5 (1969), 515–530.

[2] *Generalized Solutions in Optimal Stochastic Control*, Differential games and control theory II, Lect. Notes in Pure and Appl. Math., 30, Dekker, 1977.

Fleming, W. F., and Soner, M.,

[1] *Controlled Markov Processes and Viscosity Solutions*, Springer-Verlag, New York, 1993.

Friedman, A.,

[1] *Partial Differential Equations of Parabolic Type*, Prentice-Hall, Inc., Englewood Cliffs, N.J., 1964.

[2] *Stochastic Differential Equations and Applications, Vol. 1*, Academic Press, 1975.

Geoffard, P. Y.,

[1] Discounting and optimizing, preprint.

Gilbarg, D., and Trudinger, N. S.,

[1] *Elliptic Partial Differential Equations of Second Order*, Springer-Verlag, Berlin, 1977.

Hamadene, S.

[1] Backward-forward stochastic differential equations and stochastic games, preprint.

Hamadene, S., and Lepeltier, J. P.,

[1] Zero-sum stochastic differential games and backward equations, *Sys. & Control Lett.*, 24 (1995), 259–263.

Hardt, R., and Simon, L.,

[1] Nodal sets for solutions of elliptic equations, *J. Diff. Geom.*, 30 (1989), 505–522.

Harrison, M., and Kreps, D.

[1] Martingales and arbitrage in multiperiod securities markets, *J. Economic Theory*, 20 (1979), 381–408.

Haussmann, U. G.,

[1] On the Integral Representation of Functionals of Itô's Processes, *Stochastics*, 3 (1979), 17–27.

Hiriart-Urruty, J-B., and Lemaréchal, C.,

[1] *Convex Analysis and Minimization Algorithms, I.*, Springer-Verlag, 1993.

Hogan, M.

[1] Problems in certain two factor term structure models, *Ann. Appl. Probab.*, 3 (1993), 576–581.

Hörmander, L.,

[1] *The Analysis of Linear Partial Differential Operators, I.*, Springer-Verlag, Berlin, 1983.

Hu, Y.,

[1] N-person differential games governed by semilinear stochastic evolution systems, *Appl. Math. Optim.*, 24 (1991), 257–271.

[2] Probabilistic interpretation of a system of quasilinear elliptic partial differential equations under Neumann boundary conditions, *Stoch. Proc. Appl.*, 48 (1993), 107–121.

Hu, Y., and Peng, S.,

[1] Adapted solution of backward stochastic evolution equation, *Stoch. Anal. & Appl.*, 9 (1991), 445–459.
[2] Solution of forward-backward stochastic differential equations, *Probab. Theory & Rel. Fields*, 103 (1995), 273–283.

Ikeda, N., and Watanabe, S.,

[1] *Stochastic Differential Equations and Diffusion Processes*, North Holland, Amsterdam, 1981.

John, F.,

[1] *Partial Differential Equations*, Springer-Verlag, 1975.

Karatzas, I., and Shreve, S. E.,

[1] *Brownian Motion and Stochastic Calculus*, Springer-Verlag, Berlin, 1988.

Kloeden, P.E. and Platen, E.,

[1] *Numerical Solution of Stochastic Differential Equations*, Springer-Verlag, 1992.

Koopmans, T.

[1] Stationary ordinary utility and impatience, *Econometrics*, 28 (1960), 287–309.

Krylov, N. V.,

[1] *Controlled Diffusion Processes*, Springer-Verlag, New York, 1980.
[2] *Nonlinear Elliptic and Parabolic Equations of the Second Order*, Reidel, Dordrecht, Holland, 1987.

Krylov, N. N., and Rozovskii, B. L.,

[1] Stochastic partial differential equations and diffusion processes, *Uspekhi Mat. Nauk*, 37 (1982), No.6, 75–95.

Kunita, H.,

[1] *Stochastic Flows and Stochastic Differential Equations*, Cambridge University Press, 1990.

Ladyzenskaja, O. A., Solonnikov, V. A., and Ural'ceva, N. N.,

[1] *Linear and Quasilinear Equations of Parabolic Type*, AMS, Providence, RI., 1968.

Lepeltier, M., and San Martin, J.

[1] Backward stochastic differential equations with non-Lipschitz coefficients, *Stat. & Prob. Letters*, **32** (1997), 425–430.

Li, X., and S. Tang,

[1] General necessary conditions for partially observed optimal stochastic control, *J. Appl. Prob.*, 32 (1995), 1118–1137.

Lin, F. H.,

[1] Nodal sets of solutions of elliptic and parabolic equations, *Comm. Pure Appl. Math.*, 44 (1991), 287–308.

Lions, J. L.,

[1] *Équations Différantielles Opérationnelles et Problèmes aux Limites*, Springer-Verlag, Berlin, 1961, 2nd Edition.

Ma, C.,

[1] Market equilibrium with heterogeneous recursive utility maximizing agents, *Economic Theory*, 3 (1993), 243–266.

[2] Intertemporal recursive utility in the presence of mixed Poisson-Brownian uncertainty, preprint.

[3] Valuation of derivative securities with mixed Poisson–Brownian information and recursive utility, preprint.

Ma, J., Protter, P., and Yong, J.,

[1] Solving forward-backward stochastic differential equations explicitly — a four step scheme, *IMA preprint #1146*, June, 1993; also, *Prob.& Related Fields*, 98 (1994), 339–359.

Ma, J., and Yong, J.,

[1] Solvability of forward backward SDEs and the nodal set of Hamilton-Jaccobi-Bellman Equations, *IMA preprint #1117*, March, 1993; also, *Chin. Ann. Math.*, 16B (1995), 279–298.

[2] Adapted solution of a degenerate backward SPDE, with applications, *Stochastic Processes and their Applications*, 70 (1997), 59–84.

[3] On linear, degenerate backward stochastic partial differential equations, *Probab. Theory Relat. Fields*, 113 (1999), 135–170.

[4] Approximate solvability of forward-backward stochastic differential equations, preprint.

Meyer, G. H.,

[1] *Initial Value Methods for Boundary Value Problems — Theory and Application of Invariant Imbedding*, Academic Press, New York, 1973.

Nualart, D.,

[1] Noncausal Stochastic Integrals and Calculus, *Stochastic Analysis and Related Topics*, Korezlioglu, H. and Ustunel, A. S. eds. Lect. Notes Math., vol.1316, Springer-Verlag, Berlin, 1986, 80–129.

[2] *Malliavin Calculus and Related Topics*, Springer-Verlag, 1995.

Nualart, D., and Pardoux, E.,

[1] Stochastic calculus with anticipating integrands, *Prob. Th. Rel. Fields*, 78 (1988), 535–581.

Ocone, D.,

[1] Malliavin's Calculus and Stochastic Integral Representations of Functionals of Diffusion Processes, *Stochastics*, 12 (1984), 161–185.

Ocone, D., and Pardoux, E.,

[1] Linear stochastic differential equations with boundary conditions, *Probab. Th. Rel. Fields*, 82 (1989), 489–526.

[2] A stochastic Feynman-Kac formula for anticipating SPDE's, and application to nonlinear smoothing, *Stoch. & Stoch. Rep.*, 45 (1993), 79–126.

Pardoux, E.,

[1] Stochastic partial differential equations and filtering of diffusion processes, *Stochastics*, 3 (1979), 127–167.

[2] Backward stochastic differential equations and applications, *Proc. ICM*, 1994.

Pardoux, E., and Peng, S.,

[1] Adapted solution of a backward stochastic differential equation, *Systems & Control Lett.*, 14 (1990), 55–61.

[2] Backward stochastic differential equations and quasilinear parabolic partial differential equations, *Lecture Notes in CIS* **176**, 200–217, Springer 1992.

[3] Backward doubly stochastic differential equations and systems of quasilinear parabolic SPDEs, *Probab. Theory & Related Fields*, 98 (1994), 209–227.

[4] Some Backward stochastic differential equations with non-Lipschitz coefficients, preprint.

Pardoux, E., and Protter, P.,

[1] A two-sided stochastic integral and its calculus, *Probab. Th. Rel. Fields*, 76 (1987), 15–49.

Pardoux, E., and Rascanu, A.,

[1] Backward stochastic differential equations with subdifferential operator and related variational inequalities, *Stochastic Process. Appl.*, **76** (1998), no. 2, 191–215.

[2] Backward stochastic differential equations with maximal monotone operators, preprint.

Pardoux, P., and Tang, S.,

[1] The study of forward-backward stochastic differential equation and its application in quasilinear PDEs, preprint.

Peng, S.,

[1] A general stochastic maximum principle for optimal control problems, *SIAM J. Control & Optim.*, 28 (1990), 966–979.

[2] A generalized Hamilton-Jacobi-Bellman equation, *Lecture Notes in CIS*, *184*, *Li & Yong eds*, 126–134, Springer-Verlag, Berlin, 1991.

[3] Probabilistic interpretation for systems of quasilinear parabolic partial differential equations, *Stoch. & Stoch. Rep.*, 37 (1991), 61–74.

[4] A nonlinear Feynman-Kac formula and applications, *Proc. Symposium of System Sciences and Control Theory, Chen & Yong eds.* 173–184, World Scientific, Singapore, 1992.

[5] Stochastic Hamilton-Jacobi-Bellman equations, *SIAM J. Control & Optim.*, 30 (1992), 284–304.

[6] A generalized dynamic programming principle and Hamilton-Jacobi-Bellmen equation, *Stoch. & Stoch. Rep.*, 38 (1992), 119–134.

[7] New development in stochastic maximum principle and related backward stochastic differential equations, *Proc. 31st CDC*, Tucson 1992.

[8] Backward stochastic differential equation and its application in optimal control, *Appl. Math. & Optim.*, 27 (1993), 125–144.

[9] Backward SDE and related g-expectation, preprint.

[10] Adapted solution of backward stochastic differential equations and related partial differential equations, preprint.

Peng, S., and Wu, Z.,

[1] Fully coupled forward-backward stochastic differential equations, preprint.

Picard, J.,

[1] Martingales on Riemannian manifolds with prescribed limits, *J. Funct. Anal.*, 99 (1991), 223–261.

[2] Baricentres et martingales sur une variété, *Ann. de l'IHP*, to appear.

Protter, P.,

[1] *Stochastic Integration and Differential Equations, A New Approach*, Springer-Verlag, Berlin, 1990.

Rozovskii, B. L.,

[1] *Stochastic Evolution Systems: Linear Theory and Applications to Nonlinear Filtering*, Kluwer Academic Publishers, Boston, 1990.

Schroder, M. and Skiadas, C.,

[1] Optimal consumption and portfolio selection with stochastic differential utility, preprint.

Scott, M. R.,

[1] *Invariant Imbedding and Its Applications to Ordinary Differential Equations — An Introduction*, Addison-Wesley Pub., London, 1973.

Shreve, S. E., and Soner, H. M.,

[1] Optimal investment and consumption with transaction costs, to appear.

Sowers, R.,

[1] Short-time geometry of random heat kernels, preprint.

Tang, S.,

[1] Optimal control of stochastic systems in Hilbert space with random jumps, Ph. D Thesis, Fudan Univ., 1992.

Tang, S., and X. Li,

[1] Necessary conditions for optimal control of stochastic systems with random jumps, *SIAM J. Control & Optim.*, 32 (1994), 1447–1475.

Thalmaier, A.,

[1] Martingales on Riemannian manifolds and the nonlinear heat equation, *Stochastic Analysis and Applications, Davies et al eds.*, World Scientific Press, Singapore, 1996, 429–440.

[2] Brownian motion and the formation of singularities in the heat flow for harmonic maps, *Probab. Theory Relat. Fields*, 105 (1996), 335–367.

Wiegner, M.,

[1] Global solutions to a class of strongly coupled parabolic systems, *Math. Ann.*, 292 (1992), 711–727.

Wonham, W. H.,

[1] *Linear Multivariable Control: A Geometric Approach*, Springer-Verlag, Berlin, 1980.

Yong, J.,

[1] Finding adapted solutions of forward-backward stochastic differential equations — method of continuation, *Prob. Theory & Rel. Fields*, 107 (1997), 537–572.

[2] Relations among ODEs, PDEs, FSDEs, BSDEs, and FBSDEs, *Proc. 36th CDC*, 2779–2784.

[3] Some results on the reachable sets of linear stochastic systems, *Proc. 37th CDC*, 1335–1340.

[4] Linear forward-backward stochastic differential equations, *Appl. Math. Optim.*, 39 (1999), 93–119.

[5] European type contingent claims in an incomplete market with constrained wealth and portfolio, *Math. Finance*, 9 (1999), 387–412.

Yong, J., and Zhou, X. Y.,

[1] *Stochastic Controls: Hamiltonian Systems and HJB Equations*, Springer-Verlag, New York, 1999.

Zhou, X. Y.,

[1] A unified treatment of maximum principle and dynamic programming in stochastic controls, *Stoch. & Stoch. Rep.*, 36 (1991), 137–161.

[2] A duality analysis on stochastic partial differential equations, *J. Funct. Anal.*, 103 (1992), 275–293.

[3] On the necessary conditions of optimal controls for stochastic partial differential equations, *SIAM J. Control Optim.*, 31 (1993), 1462–1478.

Index

Lecture Notes in Mathematics

For information about earlier volumes
please contact your bookseller or Springer
LNM Online archive: springerlink.com

Vol. 1757: R. R. Phelps, Lectures on Choquet's Theorem (2001)

Vol. 1758: N. Monod, Continuous Bounded Cohomology of Locally Compact Groups (2001)

Vol. 1759: Y. Abe, K. Kopfermann, Toroidal Groups (2001)

Vol. 1760: D. Filipović, Consistency Problems for Heath-Jarrow-Morton Interest Rate Models (2001)

Vol. 1761: C. Adelmann, The Decomposition of Primes in Torsion Point Fields (2001)

Vol. 1762: S. Cerrai, Second Order PDE's in Finite and Infinite Dimension (2001)

Vol. 1763: J.-L. Loday, A. Frabetti, F. Chapoton, F. Goichot, Dialgebras and Related Operads (2001)

Vol. 1764: A. Cannas da Silva, Lectures on Symplectic Geometry (2001)

Vol. 1765: T. Kerler, V. V. Lyubashenko, Non-Semisimple Topological Quantum Field Theories for 3-Manifolds with Corners (2001)

Vol. 1766: H. Hennion, L. Hervé, Limit Theorems for Markov Chains and Stochastic Properties of Dynamical Systems by Quasi-Compactness (2001)

Vol. 1767: J. Xiao, Holomorphic Q Classes (2001)

Vol. 1768: M. J. Pflaum, Analytic and Geometric Study of Stratified Spaces (2001)

Vol. 1769: M. Alberich-Carramiñana, Geometry of the Plane Cremona Maps (2002)

Vol. 1770: H. Gluesing-Luerssen, Linear Delay-Differential Systems with Commensurate Delays: An Algebraic Approach (2002)

Vol. 1771: M. Émery, M. Yor (Eds.), Séminaire de Probabilités 1967-1980. A Selection in Martingale Theory (2002)

Vol. 1772: F. Burstall, D. Ferus, K. Leschke, F. Pedit, U. Pinkall, Conformal Geometry of Surfaces in S^4 (2002)

Vol. 1773: Z. Arad, M. Muzychuk, Standard Integral Table Algebras Generated by a Non-real Element of Small Degree (2002)

Vol. 1774: V. Runde, Lectures on Amenability (2002)

Vol. 1775: W. H. Meeks, A. Ros, H. Rosenberg, The Global Theory of Minimal Surfaces in Flat Spaces. Martina Franca 1999. Editor: G. P. Pirola (2002)

Vol. 1776: K. Behrend, C. Gomez, V. Tarasov, G. Tian, Quantum Comohology. Cetraro 1997. Editors: P. de Bartolomeis, B. Dubrovin, C. Reina (2002)

Vol. 1777: E. García-Río, D. N. Kupeli, R. Vázquez-Lorenzo, Osserman Manifolds in Semi-Riemannian Geometry (2002)

Vol. 1778: H. Kiechle, Theory of K-Loops (2002)

Vol. 1779: I. Chueshov, Monotone Random Systems (2002)

Vol. 1780: J. H. Bruinier, Borcherds Products on O(2,1) and Chern Classes of Heegner Divisors (2002)

Vol. 1781: E. Bolthausen, E. Perkins, A. van der Vaart, Lectures on Probability Theory and Statistics. Ecole d' Eté de Probabilités de Saint-Flour XXIX-1999. Editor: P. Bernard (2002)

Vol. 1782: C.-H. Chu, A. T.-M. Lau, Harmonic Functions on Groups and Fourier Algebras (2002)

Vol. 1783: L. Grüne, Asymptotic Behavior of Dynamical and Control Systems under Perturbation and Discretization (2002)

Vol. 1784: L. H. Eliasson, S. B. Kuksin, S. Marmi, J.-C. Yoccoz, Dynamical Systems and Small Divisors. Cetraro, Italy 1998. Editors: S. Marmi, J.-C. Yoccoz (2002)

Vol. 1785: J. Arias de Reyna, Pointwise Convergence of Fourier Series (2002)

Vol. 1786: S. D. Cutkosky, Monomialization of Morphisms from 3-Folds to Surfaces (2002)

Vol. 1787: S. Caenepeel, G. Militaru, S. Zhu, Frobenius and Separable Functors for Generalized Module Categories and Nonlinear Equations (2002)

Vol. 1788: A. Vasil'ev, Moduli of Families of Curves for Conformal and Quasiconformal Mappings (2002)

Vol. 1789: Y. Sommerhäuser, Yetter-Drinfel'd Hopf algebras over groups of prime order (2002)

Vol. 1790: X. Zhan, Matrix Inequalities (2002)

Vol. 1791: M. Knebusch, D. Zhang, Manis Valuations and Prüfer Extensions I: A new Chapter in Commutative Algebra (2002)

Vol. 1792: D. D. Ang, R. Gorenflo, V. K. Le, D. D. Trong, Moment Theory and Some Inverse Problems in Potential Theory and Heat Conduction (2002)

Vol. 1793: J. Cortés Monforte, Geometric, Control and Numerical Aspects of Nonholonomic Systems (2002)

Vol. 1794: N. Pytheas Fogg, Substitution in Dynamics, Arithmetics and Combinatorics. Editors: V. Berthé, S. Ferenczi, C. Mauduit, A. Siegel (2002)

Vol. 1795: H. Li, Filtered-Graded Transfer in Using Non-commutative Gröbner Bases (2002)

Vol. 1796: J.M. Melenk, hp-Finite Element Methods for Singular Perturbations (2002)

Vol. 1797: B. Schmidt, Characters and Cyclotomic Fields in Finite Geometry (2002)

Vol. 1798: W.M. Oliva, Geometric Mechanics (2002)

Vol. 1799: H. Pajot, Analytic Capacity, Rectifiability, Menger Curvature and the Cauchy Integral (2002)

Vol. 1800: O. Gabber, L. Ramero, Almost Ring Theory (2003)

Vol. 1801: J. Azéma, M. Émery, M. Ledoux, M. Yor (Eds.), Séminaire de Probabilités XXXVI (2003)

Vol. 1802: V. Capasso, E. Merzbach, B. G. Ivanoff, M. Dozzi, R. Dalang, T. Mountford, Topics in Spatial Stochastic Processes. Martina Franca, Italy 2001. Editor: E. Merzbach (2003)

Vol. 1803: G. Dolzmann, Variational Methods for Crystalline Microstructure – Analysis and Computation (2003)

Vol. 1804: I. Cherednik, Ya. Markov, R. Howe, G. Lusztig, Iwahori-Hecke Algebras and their Representation Theory. Martina Franca, Italy 1999. Editors: V. Baldoni, D. Barbasch (2003)

Vol. 1805: F. Cao, Geometric Curve Evolution and Image Processing (2003)

Vol. 1806: H. Broer, I. Hoveijn. G. Lunther, G. Vegter, Bifurcations in Hamiltonian Systems. Computing Singularities by Gröbner Bases (2003)

Vol. 1807: V. D. Milman, G. Schechtman (Eds.), Geometric Aspects of Functional Analysis. Israel Seminar 2000-2002 (2003)

Vol. 1808: W. Schindler, Measures with Symmetry Properties (2003)

Vol. 1809: O. Steinbach, Stability Estimates for Hybrid Coupled Domain Decomposition Methods (2003)

Vol. 1810: J. Wengenroth, Derived Functors in Functional Analysis (2003)

Vol. 1811: J. Stevens, Deformations of Singularities (2003)

Vol. 1812: L. Ambrosio, K. Deckelnick, G. Dziuk, M. Mimura, V. A. Solonnikov, H. M. Soner, Mathematical Aspects of Evolving Interfaces. Madeira, Funchal, Portugal 2000. Editors: P. Colli, J. F. Rodrigues (2003)

Vol. 1813: L. Ambrosio, L. A. Caffarelli, Y. Brenier, G. Buttazzo, C. Villani, Optimal Transportation and its

Applications. Martina Franca, Italy 2001. Editors: L. A. Caffarelli, S. Salsa (2003)

Vol. 1814: P. Bank, F. Baudoin, H. Föllmer, L.C.G. Rogers, M. Soner, N. Touzi, Paris-Princeton Lectures on Mathematical Finance 2002 (2003)

Vol. 1815: A. M. Vershik (Ed.), Asymptotic Combinatorics with Applications to Mathematical Physics. St. Petersburg, Russia 2001 (2003)

Vol. 1816: S. Albeverio, W. Schachermayer, M. Talagrand, Lectures on Probability Theory and Statistics. Ecole d'Eté de Probabilités de Saint-Flour XXX-2000. Editor: P. Bernard (2003)

Vol. 1817: E. Koelink, W. Van Assche (Eds.), Orthogonal Polynomials and Special Functions. Leuven 2002 (2003)

Vol. 1818: M. Bildhauer, Convex Variational Problems with Linear, nearly Linear and/or Anisotropic Growth Conditions (2003)

Vol. 1819: D. Masser, Yu. V. Nesterenko, H. P. Schlickewei, W. M. Schmidt, M. Waldschmidt, Diophantine Approximation. Cetraro, Italy 2000. Editors: F. Amoroso, U. Zannier (2003)

Vol. 1820: F. Hiai, H. Kosaki, Means of Hilbert Space Operators (2003)

Vol. 1821: S. Teufel, Adiabatic Perturbation Theory in Quantum Dynamics (2003)

Vol. 1822: S.-N. Chow, R. Conti, R. Johnson, J. Mallet-Paret, R. Nussbaum, Dynamical Systems. Cetraro, Italy 2000. Editors: J. W. Macki, P. Zecca (2003)

Vol. 1823: A. M. Anile, W. Allegretto, C. Ringhofer, Mathematical Problems in Semiconductor Physics. Cetraro, Italy 1998. Editor: A. M. Anile (2003)

Vol. 1824: J. A. Navarro González, J. B. Sancho de Salas, $\mathscr{C}^{\infty}$ – Differentiable Spaces (2003)

Vol. 1825: J. H. Bramble, A. Cohen, W. Dahmen, Multiscale Problems and Methods in Numerical Simulations, Martina Franca, Italy 2001. Editor: C. Canuto (2003)

Vol. 1826: K. Dohmen, Improved Bonferroni Inequalities via Abstract Tubes. Inequalities and Identities of Inclusion-Exclusion Type. VIII, 113 p, 2003.

Vol. 1827: K. M. Pilgrim, Combinations of Complex Dynamical Systems. IX, 118 p, 2003.

Vol. 1828: D. J. Green, Gröbner Bases and the Computation of Group Cohomology. XII, 138 p, 2003.

Vol. 1829: E. Altman, B. Gaujal, A. Hordijk, Discrete-Event Control of Stochastic Networks: Multimodularity and Regularity. XIV, 313 p, 2003.

Vol. 1830: M. I. Gil', Operator Functions and Localization of Spectra. XIV, 256 p, 2003.

Vol. 1831: A. Connes, J. Cuntz, E. Guentner, N. Higson, J. E. Kaminker, Noncommutative Geometry, Martina Franca, Italy 2002. Editors: S. Doplicher, L. Longo (2004)

Vol. 1832: J. Azéma, M. Émery, M. Ledoux, M. Yor (Eds.), Séminaire de Probabilités XXXVII (2003)

Vol. 1833: D.-Q. Jiang, M. Qian, M.-P. Qian, Mathematical Theory of Nonequilibrium Steady States. On the Frontier of Probability and Dynamical Systems. IX, 280 p, 2004.

Vol. 1834: Yo. Yomdin, G. Comte, Tame Geometry with Application in Smooth Analysis. VIII, 186 p, 2004.

Vol. 1835: O.T. Izhboldin, B. Kahn, N.A. Karpenko, A. Vishik, Geometric Methods in the Algebraic Theory of Quadratic Forms. Summer School, Lens, 2000. Editor: J.-P. Tignol (2004)

Vol. 1836: C. Năstăsescu, F. Van Oystaeyen, Methods of Graded Rings. XIII, 304 p, 2004.

Vol. 1837: S. Tavaré, O. Zeitouni, Lectures on Probability Theory and Statistics. Ecole d'Eté de Probabilités de Saint-Flour XXXI-2001. Editor: J. Picard (2004)

Vol. 1838: A.J. Ganesh, N.W. O'Connell, D.J. Wischik, Big Queues. XII, 254 p, 2004.

Vol. 1839: R. Gohm, Noncommutative Stationary Processes. VIII, 170 p, 2004.

Vol. 1840: B. Tsirelson, W. Werner, Lectures on Probability Theory and Statistics. Ecole d'Eté de Probabilités de Saint-Flour XXXII-2002. Editor: J. Picard (2004)

Vol. 1841: W. Reichel, Uniqueness Theorems for Variational Problems by the Method of Transformation Groups (2004)

Vol. 1842: T. Johnsen, A. L. Knutsen, K_3 Projective Models in Scrolls (2004)

Vol. 1843: B. Jefferies, Spectral Properties of Noncommuting Operators (2004)

Vol. 1844: K.F. Siburg, The Principle of Least Action in Geometry and Dynamics (2004)

Vol. 1845: Min Ho Lee, Mixed Automorphic Forms, Torus Bundles, and Jacobi Forms (2004)

Vol. 1846: H. Ammari, H. Kang, Reconstruction of Small Inhomogeneities from Boundary Measurements (2004)

Vol. 1847: T.R. Bielecki, T. Björk, M. Jeanblanc, M. Rutkowski, J.A. Scheinkman, W. Xiong, Paris-Princeton Lectures on Mathematical Finance 2003 (2004)

Vol. 1848: M. Abate, J. E. Fornaess, X. Huang, J. P. Rosay, A. Tumanov, Real Methods in Complex and CR Geometry, Martina Franca, Italy 2002. Editors: D. Zaitsev, G. Zampieri (2004)

Vol. 1849: Martin L. Brown, Heegner Modules and Elliptic Curves (2004)

Vol. 1850: V. D. Milman, G. Schechtman (Eds.), Geometric Aspects of Functional Analysis. Israel Seminar 2002-2003 (2004)

Vol. 1851: O. Catoni, Statistical Learning Theory and Stochastic Optimization (2004)

Vol. 1852: A.S. Kechris, B.D. Miller, Topics in Orbit Equivalence (2004)

Vol. 1853: Ch. Favre, M. Jonsson, The Valuative Tree (2004)

Vol. 1854: O. Saeki, Topology of Singular Fibers of Differential Maps (2004)

Vol. 1855: G. Da Prato, P.C. Kunstmann, I. Lasiecka, A. Lunardi, R. Schnaubelt, L. Weis, Functional Analytic Methods for Evolution Equations. Editors: M. Iannelli, R. Nagel, S. Piazzera (2004)

Vol. 1856: K. Back, T.R. Bielecki, C. Hipp, S. Peng, W. Schachermayer, Stochastic Methods in Finance, Bressanone/Brixen, Italy, 2003. Editors: M. Fritelli, W. Runggaldier (2004)

Vol. 1857: M. Émery, M. Ledoux, M. Yor (Eds.), Séminaire de Probabilités XXXVIII (2005)

Vol. 1858: A.S. Cherny, H.-J. Engelbert, Singular Stochastic Differential Equations (2005)

Vol. 1859: E. Letellier, Fourier Transforms of Invariant Functions on Finite Reductive Lie Algebras (2005)

Vol. 1860: A. Borisyuk, G.B. Ermentrout, A. Friedman, D. Terman, Tutorials in Mathematical Biosciences I. Mathematical Neurosciences (2005)

Vol. 1861: G. Benettin, J. Henrard, S. Kuksin, Hamiltonian Dynamics – Theory and Applications, Cetraro, Italy, 1999. Editor: A. Giorgilli (2005)

Vol. 1862: B. Helffer, F. Nier, Hypoelliptic Estimates and Spectral Theory for Fokker-Planck Operators and Witten Laplacians (2005)

Vol. 1863: H. Führ, Abstract Harmonic Analysis of Continuous Wavelet Transforms (2005)

Vol. 1864: K. Efstathiou, Metamorphoses of Hamiltonian Systems with Symmetries (2005)

Vol. 1865: D. Applebaum, B.V. R. Bhat, J. Kustermans, J. M. Lindsay, Quantum Independent Increment Processes I. From Classical Probability to Quantum Stochastic Calculus. Editors: M. Schürmann, U. Franz (2005)

Vol. 1866: O.E. Barndorff-Nielsen, U. Franz, R. Gohm, B. Kümmerer, S. Thorbjønsen, Quantum Independent Increment Processes II. Structure of Quantum Lévy Processes, Classical Probability, and Physics. Editors: M. Schürmann, U. Franz, (2005)

Vol. 1867: J. Sneyd (Ed.), Tutorials in Mathematical Biosciences II. Mathematical Modeling of Calcium Dynamics and Signal Transduction. (2005)

Vol. 1868: J. Jorgenson, S. Lang, $\text{Pos}_n(\text{R})$ and Eisenstein Series. (2005)

Vol. 1869: A. Dembo, T. Funaki, Lectures on Probability Theory and Statistics. Ecole d'Eté de Probabilités de Saint-Flour XXXIII-2003. Editor: J. Picard (2005)

Vol. 1870: V.I. Gurariy, W. Lusky, Geometry of Müntz Spaces and Related Questions. (2005)

Vol. 1871: P. Constantin, G. Gallavotti, A.V. Kazhikhov, Y. Meyer, S. Ukai, Mathematical Foundation of Turbulent Viscous Flows, Martina Franca, Italy, 2003. Editors: M. Cannone, T. Miyakawa (2006)

Vol. 1872: A. Friedman (Ed.), Tutorials in Mathematical Biosciences III. Cell Cycle, Proliferation, and Cancer (2006)

Vol. 1873: R. Mansuy, M. Yor, Random Times and Enlargements of Filtrations in a Brownian Setting (2006)

Vol. 1874: M. Yor, M. Émery (Eds.), In Memoriam Paul-André Meyer - Séminaire de probabilités XXXIX (2006)

Vol. 1875: J. Pitman, Combinatorial Stochastic Processes. Ecole d'Eté de Probabilités de Saint-Flour XXXII-2002. Editor: J. Picard (2006)

Vol. 1876: H. Herrlich, Axiom of Choice (2006)

Vol. 1877: J. Steuding, Value Distributions of L-Functions (2007)

Vol. 1878: R. Cerf, The Wulff Crystal in Ising and Percolation Models, Ecole d'Eté de Probabilités de Saint-Flour XXXIV-2004. Editor: Jean Picard (2006)

Vol. 1879: G. Slade, The Lace Expansion and its Applications, Ecole d'Eté de Probabilités de Saint-Flour XXXIV-2004. Editor: Jean Picard (2006)

Vol. 1880: S. Attal, A. Joye, C.-A. Pillet, Open Quantum Systems I, The Hamiltonian Approach (2006)

Vol. 1881: S. Attal, A. Joye, C.-A. Pillet, Open Quantum Systems II, The Markovian Approach (2006)

Vol. 1882: S. Attal, A. Joye, C.-A. Pillet, Open Quantum Systems III, Recent Developments (2006)

Vol. 1883: W. Van Assche, F. Marcellàn (Eds.), Orthogonal Polynomials and Special Functions, Computation and Application (2006)

Vol. 1884: N. Hayashi, E.I. Kaikina, P.I. Naumkin, I.A. Shishmarev, Asymptotics for Dissipative Nonlinear Equations (2006)

Vol. 1885: A. Telcs, The Art of Random Walks (2006)

Vol. 1886: S. Takamura, Splitting Deformations of Degenerations of Complex Curves (2006)

Vol. 1887: K. Habermann, L. Habermann, Introduction to Symplectic Dirac Operators (2006)

Vol. 1888: J. van der Hoeven, Transseries and Real Differential Algebra (2006)

Vol. 1889: G. Osipenko, Dynamical Systems, Graphs, and Algorithms (2006)

Vol. 1890: M. Bunge, J. Funk, Singular Coverings of Toposes (2006)

Vol. 1891: J.B. Friedlander, D.R. Heath-Brown, H. Iwaniec, J. Kaczorowski, Analytic Number Theory, Cetraro, Italy, 2002. Editors: A. Perelli, C. Viola (2006)

Vol. 1892: A. Baddeley, I. Bárány, R. Schneider, W. Weil, Stochastic Geometry, Martina Franca, Italy, 2004. Editor: W. Weil (2007)

Vol. 1893: H. Hanßmann, Local and Semi-Local Bifurcations in Hamiltonian Dynamical Systems, Results and Examples (2007)

Vol. 1894: C.W. Groetsch, Stable Approximate Evaluation of Unbounded Operators (2007)

Vol. 1895: L. Molnár, Selected Preserver Problems on Algebraic Structures of Linear Operators and on Function Spaces (2007)

Vol. 1896: P. Massart, Concentration Inequalities and Model Selection, Ecole d'Eté de Probabilités de Saint-Flour XXXIII-2003. Editor: J. Picard (2007)

Vol. 1897: R.A. Doney, Fluctuation Theory for Lévy Processes, Ecole d'Eté de Probabilités de Saint-Flour XXXV-2005. Editor: J. Picard (2007)

Vol. 1898: H.R. Beyer, Beyond Partial Differential Equations, On linear and Quasi-Linear Abstract Hyperbolic Evolution Equations (2007)

Vol. 1899: Séminaire de Probabilités XL. Editors: C. Donati-Martin, M. Émery, A. Rouault, C. Stricker (2007)

Vol. 1900: E. Bolthausen, A. Bovier (Eds.), Spin Glasses (2007)

Vol. 1901: O. Wittenberg, Intersections de deux quadriques et pinceaux de courbes de genre 1, Intersections of Two Quadrics and Pencils of Curves of Genus 1 (2007)

Vol. 1902: A. Isaev, Lectures on the Automorphism Groups of Kobayashi-Hyperbolic Manifolds (2007)

Vol. 1903: G. Kresin, V. Maz'ya, Sharp Real-Part Theorems (2007)

Recent Reprints and New Editions

Vol. 1618: G. Pisier, Similarity Problems and Completely Bounded Maps. 1995 – 2nd exp. edition (2001)

Vol. 1629: J.D. Moore, Lectures on Seiberg-Witten Invariants. 1997 – 2nd edition (2001)

Vol. 1638: P. Vanhaecke, Integrable Systems in the realm of Algebraic Geometry. 1996 – 2nd edition (2001)

Vol. 1702: J. Ma, J. Yong, Forward-Backward Stochastic Differential Equations and their Applications. 1999 – Corr. 3rd printing (2007)

Vol. 830: J.A. Green, Polynomial Representations of GL_n, with an Appendix on Schensted Correspondence and Littelmann Paths by K. Erdmann, J.A. Green and M. Schocker 1980 – 2nd corr. and augmented edition (2007)